Specialization of Quadratic and Symmetric Bilinear Forms

Algebra and Applications

Volume 11

Managing Editor:

Alain Verschoren
University of Antwerp, Belgium

Series Editors:

Alice Fialowski
Eötvös Loránd University, Hungary

Eric Friedlander
Northwestern University, USA

John Greenlees
Sheffield University, UK

Gerhard Hiss
Aachen University, Germany

Ieke Moerdijk
Utrecht University, The Netherlands

Idun Reiten
Norwegian University of Science and Technology, Norway

Christoph Schweigert
Hamburg University, Germany

Mina Teicher
Bar-llan University, Israel

Algebra and Applications aims to publish well written and carefully refereed monographs with up-to-date information about progress in all fields of algebra, its classical impact on commutative and noncommutative algebraic and differential geometry, K-theory and algebraic topology, as well as applications in related domains, such as number theory, homotopy and (co)homology theory, physics and discrete mathematics.

Particular emphasis will be put on state-of-the-art topics such as rings of differential operators, Lie algebras and super-algebras, group rings and algebras, C^*-algebras, Kac-Moody theory, arithmetic algebraic geometry, Hopf algebras and quantum groups, as well as their applications. In addition, Algebra and Applications will also publish monographs dedicated to computational aspects of these topics as well as algebraic and geometric methods in computer science.

Manfred Knebusch

Specialization of Quadratic and Symmetric Bilinear Forms

Translated by Thomas Unger

 Springer

Prof. Dr. Manfred Knebusch
Department of Mathematics
University of Regensburg
Universitätsstr. 31
93040 Regensburg
Germany
manfred.knebusch@mathematik.uni-regensburg.de

Dr. Thomas Unger
School of Mathematical Sciences
University College Dublin
Belfield
Dublin 4
Ireland
thomas.unger@ucd.ie

ISBN 978-1-84882-241-2   e-ISBN 978-1-84882-242-9
DOI 10.1007/978-1-84882-242-9
Springer London Dordrecht Heidelberg New York

British Library Cataloguing in Publication Data
A catalogue record for this book is available from the British Library

Library of Congress Control Number: 2010931863

Mathematics Subject Classification (2010): 11E04, 11E08, 11E81, 11E39

Cover design: deblik, Berlin

Printed on acid-free paper

Springer is part of Springer Science+Business Media (www.springer.com)

Dedicated to the memory of my teachers

Emil Artin 1898–1962

Hel Braun 1914–1986

Ernst Witt 1911–1991

Preface

This pretty limerick first came to my ears in May 1998 during a talk by T.Y. Lam on field invariants from the theory of quadratic forms.[1] It is—poetic exaggeration allowed—a suitable motto for this monograph.

What is it about? At the beginning of the seventies I drew up a specialization theory of quadratic and symmetric bilinear forms over fields [32]. Let $\lambda : K \to L \cup \infty$ be a place. Then one can assign a form $\lambda_*(\varphi)$ to a form φ over K in a meaningful way if φ has "good reduction" with respect to λ (see §1.1). The basic idea is to simply apply the place λ to the coefficients of φ, which must therefore be in the valuation ring of λ.

The specialization theory of that time was satisfactory as long as the field L, and therefore also K, had characteristic $\neq 2$. It served me in the first place as the foundation for a theory of generic splitting of quadratic forms [33], [34]. After a very modest beginning, this theory is now in full bloom. It became important for the understanding of quadratic forms over fields, as can be seen from the book [26] of Izhboldin–Kahn–Karpenko–Vishik for instance. One should note that there exists a theory of (partial) generic splitting of central simple algebras and reductive algebraic groups, parallel to the theory of generic splitting of quadratic forms (see [29] and the literature cited there).

In this book I would like to present a specialization theory of quadratic and symmetric bilinear forms with respect to a place $\lambda : K \to L \cup \infty$, without the assumption that char $L \neq 2$. This is where complications arise. We have to make a distinction

---

[1] "Some reflections on quadratic invariants of fields", 3 May 1998 in Notre Dame (Indiana) on the occasion of O.T. O'Meara's 70th birthday.

between bilinear and quadratic forms and study them both over fields and valuation rings. From the viewpoint of reductive algebraic groups, the so-called regular quadratic forms (see below) are the natural objects. But, even if we are interested only in such forms, we have to know a bit about specialization of nondegenerate symmetric bilinear forms, since they occur as "multipliers" of quadratic forms: if φ is such a bilinear form and ψ is a regular quadratic form, then we can form a tensor product $\varphi \otimes \psi$, see §1.5. This is a quadratic form, which is again regular when ψ has even dimension (dim ψ = number of variables occurring in ψ). However—and here already we run into trouble—when dim ψ is odd, $\varphi \otimes \psi$ is not necessarily regular.

Even if we only want to understand quadratic forms over a field K of characteristic zero, it might be necessary to look at specializations with respect to places from K to fields of characteristic 2, especially in arithmetic investigations. When K itself has characteristic 2, an often more complicated situation may occur, for which we are not prepared by the available literature. Certainly fields of characteristic 2 were already allowed in my work on specializations in 1973 [32], but from today's point of view satisfactory results were only obtained for symmetric bilinear forms. For quadratic forms there are gaping holes. We have to study quadratic forms over a valuation ring in which 2 is not a unit. Even the beautiful and extensive book of Ricardo Baeza [6] doesn't give us enough for the theory of specializations, although Baeza even allows semilocal rings instead of valuation rings. He studies only quadratic forms whose associated bilinear forms are nondegenerate. This forces those forms to have even dimension.

Let me now discuss the contents of this book. After an introduction to the problem in §1.1, which can be understood without any previous knowledge of quadratic and bilinear forms, the specialization theory of symmetric bilinear forms is presented in §1.2–§1.3. There are good, generally accessible sources available for the foundations of the algebraic theory of symmetric bilinear forms. Therefore many results are presented without a proof, but with a reference to the literature instead. As an important application, the outlines of the theory of generic splitting in characteristic $\neq 2$ are sketched in §1.4, nearly without proofs.

From §1.5 onwards we address the theory of quadratic forms. In characteristic 2 fewer results can be found in the literature for such forms than for bilinear forms, even at the basic level. Therefore we present most of the proofs. We also concern ourselves with the so-called "weak specialization" (see §1.1) and get into areas which may seem strange even to specialists in the theory of quadratic forms. In particular we have to require a quadratic form over K to be "obedient" in order to weakly specialize it with respect to a place $\lambda : K \to L \cup \infty$ (see §1.7). I have never encountered such a thing anywhere in the literature.

At the end of Chapter 1 we reach a level in the specialization theory of quadratic forms that facilitates a generic splitting theory, useful for many applications. In the first two sections (§2.1, §2.2) of Chapter 2 we produce such a generic splitting theory in two versions, both of which deserve interest in their own right.

We call a quadratic form φ over a field k *nondegenerate* when its quasilinear part (cf. Arf [3]), which we denote by $QL(\varphi)$, is anisotropic. We further call—deviating from Arf [3]—φ *regular* when $QL(\varphi)$ is at most one-dimensional and *strictly regular* when $QL(\varphi) = 0$ (cf. §1.6, Definition 1.59). When k has characteristic $\neq 2$, every nondegenerate form is strictly regular, but in characteristic 2 the quasilinear part causes complications. For in this case φ can become degenerate under a field extension $L \supset k$. Only in the regular case is this impossible.

In §2.1 we study the splitting behaviour of a regular quadratic form φ over k under field extensions, while in §2.2 any nondegenerate form φ, but only separable extensions of k are allowed. The theory of §2.1 incorporates the theory of §1.4, so the missing proofs of §1.4 are subsequently filled in.

Until the end of §2.2 our specialization theory is based on an obvious "canonical" concept of *good reduction* of a form φ over a field K (quadratic or symmetric bilinear) to a valuation ring $\mathfrak{o}$ of K, similar to what is known under this name in other areas of mathematics (e.g. abelian varieties). There is nothing wrong with this theory; however, for many applications it is too limited.

This is particularly clear when studying specializations with respect to a place $\lambda : K \to L \cup \infty$ with char $K = 0$, char $L = 2$. If φ is a nondegenerate quadratic form over K with good reduction with respect to λ, then the specialization $\lambda_*(\varphi)$ is automatically strictly regular. However, we would like to have a more general specialization concept, in which forms with quasilinear part $\neq 0$ can arise over L. Conversely, if the place λ is surjective, i.e. $\lambda(K) = L \cup \infty$, we would like to "lift" *every* nondegenerate quadratic form ψ over L with respect to λ to a form φ over K, i.e. to find a form φ over K which specializes to ψ with respect to λ. Then we could use the theory of forms over K to make statements about ψ.

We present such a general specialization theory in §2.3. It is based on the concept of "*fair reduction*", which is less orthodox than good reduction, but which nevertheless possesses quite satisfying properties.

Next, in §2.4, we present a theory of generic splitting, which unites the theories of §1.4, §2.1 and §2.2 under one roof and which incorporates fair reduction. This theory is deepened in §2.5 and §2.6 through the study of generic splitting towers, and thus we reach the end of Chapter 2.

Chapter 3 (§3.1–§3.13) is a long chapter in which we present a panorama of results about quadratic forms over fields for which specialization and generic splitting of forms play an important role. This only scratches the surface of applications of the specialization theory of Chapters 1 and 2. Certainly many more results can be unearthed.

We return to the foundations of specialization theory in the final short Chapter 4 (§4.1–§4.5). Quadratic and bilinear forms over a field can be specialized with respect to a more general "quadratic place" $\Lambda : K \to L \cup \infty$ (defined in §4.1) instead of a usual place $\lambda : K \to L \cup \infty$. This represents a considerable broadening of the specialization theory of Chapters 1 and 2. Of course we require again "obedience" from a quadratic form q over K in order for its specialization $\Lambda_*(q)$ to reasonably exist. It then turns out that the generic splitting behaviour of $\Lambda_*(q)$ is governed by the splitting behaviour of q and Λ, in so far as good or fair reduction is present in a weak sense, as elucidated for ordinary places in Chapter 2.

Why are quadratic places of interest, compared with ordinary places? To answer this question we observe the following. If a form q over K has bad reduction with respect to a place $\lambda : K \to L \cup \infty$, it often happens that λ can be "enlarged" to a quadratic place $\Lambda : K \to L \cup \infty$ such that q has good or fair reduction with respect to Λ in a weak sense, and the splitting properties of q are handed down to $\Lambda_*(q)$ while there is no form $\lambda_*(q)$ available for which this would be the case. The details of such a notion of reduction are much more tricky compared with what happens in Chapters 1 and 2. The central term which renders possible a unified theory of generic splitting of quadratic forms is called "stably conservative reduction", see §4.4.

One must get used to the fact that for bilinear forms there is in general no Witt cancellation rule, in contrast to quadratic forms. Nevertheless the specialization theory is in many respects easier for bilinear forms than for quadratic forms.

On the other hand we do not have any theory of generic splitting for symmetric bilinear forms over fields of characteristic 2. Such a theory might not even be possible in a meaningful way. This may well be connected to the fact that the automorphism groups of such forms can be very far from being reductive groups (which may also account for the absence of a good cancellation rule).

This book is intended for audiences with different interests. For a mathematician with perhaps only a little knowledge of quadratic or symmetric bilinear forms, who just wants to get an impression of specialization theory, it suffices to read §1.1–§1.4. The theory of generic splitting in characteristic $\neq 2$ will acquaint the reader with an important application area.

From §1.5 onwards the book is intended for scholars working in the algebraic theory of quadratic forms, and also for specialists in the area of algebraic groups. They have always been given something to look at by the theory of quadratic forms.

On reaching §2.2 of the book, readers can lean back in their chair and take a well-deserved break. They will have learned about the specialization theory, which is based on the concept of good reduction, and will have gained a certain perspective on specific phenomena in characteristic 2. Furthermore, they will have been introduced to the foundations of generic splitting and so will have seen the specialization theory in action. Admittedly, readers will not yet have seen independent applications of the weak specialization theory (§1.3, §1.7), for this theory has only appeared up to then as an auxiliary one.

The remaining sections §2.3–§2.6 of Chapter 2 develop the specialization theory sufficiently far to allow an understanding of the classical algebraic theory of quadratic forms (as presented in the books of Lam [43], [44] and Scharlau [55]) without the usual restriction that the characteristic should be different from 2. Precisely this happens in Chapter 3 where readers will also obtain sufficient illustrations to enable them to relieve other classical theorems from the characteristic $\neq 2$ restriction, although this is often a nontrivial task.

The final Chapter 4 is ultimately intended for mathematicians who want to embark on a more daring expedition in the realm of quadratic forms over fields. It cannot be mere coincidence that the specialization theory for quadratic places works just as well as the specialization theory for ordinary places. It is therefore a safe prediction that quadratic places will turn out to be generally useful and important in a future theory of quadratic forms over fields.

Regensburg,   *Manfred Knebusch*
June 2007

Postscript (October 2009)

I had the very good luck to find a translator of the German text into English, who, besides having two languages, could also understand the mathematical content of this book in depth. I met Professor Thomas Unger within the framework of the European network "Linear Algebraic Groups, Algebraic K-theory, and Related Topics", and most of the translation and our collaboration has been done under the auspices of this network, which we acknowledge gratefully.

I further owe deep thanks to my former secretary Rosi Bonn, who typed the whole German text in various versions and large parts of the English text.

The German text, *Spezialisierung von quadratischen und symmetrischen bilinearen Formen*, can be found on my homepage.[2]

---

[2] http://www-nw.uni-regensburg.de/~.knm22087.mathematik.uni-regensburg.de

Contents

Chapter 1
Fundamentals of Specialization Theory

1.1 Introduction: on the Problem of Specialization of Quadratic and Bilinear Forms

Let φ be a nondegenerate symmetric bilinear form over a field K, in other words

$$\varphi(x,y) = \sum_{i,j=1}^{n} a_{ij} x_i y_j,$$

where $x = (x_1, \ldots, x_n) \in K^n$ and $y = (y_1, \ldots, y_n) \in K^n$ are vectors, (a_{ij}) is a symmetric $(n \times n)$-matrix with coefficients $a_{ij} = a_{ji} \in K$ and $\det(a_{ij}) \neq 0$. We would like to write $\varphi = (a_{ij})$. The number of variables n is called the *dimension of* φ, $n = \dim \varphi$.

Let also $\lambda : K \to L \cup \infty$ be a place, $\mathfrak{o} = \mathfrak{o}_\lambda$ the valuation ring associated to K and $\mathfrak{m}$ the maximal ideal of $\mathfrak{o}$. We denote the group of units of $\mathfrak{o}$ by $\mathfrak{o}^*$, $\mathfrak{o}^* = \mathfrak{o} \setminus \mathfrak{m}$.

We would like λ to "specialize" φ to a bilinear form $\lambda_*(\varphi)$ over L. When is this possible in a reasonable way? If all $a_{ij} \in \mathfrak{o}$ and if $\det(a_{ij}) \in \mathfrak{o}^*$, then one can associate the nondegenerate form $(\lambda(a_{ij}))$ over L to φ. This naive idea leads us to the following:

Definition 1.1. We say that φ has *good reduction with respect to* λ when φ is isometric to a form (c_{ij}) over K with $c_{ij} \in \mathfrak{o}$, $\det(c_{ij}) \in \mathfrak{o}^*$. We then call the form $(\lambda(c_{ij}))$ "the" *specialization of* φ *with respect to* λ. We denote this specialization by $\lambda_*(\varphi)$.

Note. $\varphi = (a_{ij})$ is isometric to (c_{ij}) if and only if there exists a matrix $S \in \mathrm{GL}(n, K)$ with $(c_{ij}) = {}^t S (a_{ij}) S$. In this case we write $\varphi \cong (c_{ij})$.

We also allow the case $\dim \varphi = 0$, standing for the unique bilinear form on the zero vector space, the form $\varphi = 0$. We agree that the form $\varphi = 0$ has good reduction and set $\lambda_*(\varphi) = 0$.

Problem 1.2. Is this definition meaningful? Up to isometry $\lambda_*(\varphi)$ should be independent of the choice of the matrix (c_{ij}).

M. Knebusch, *Specialization of Quadratic and Symmetric Bilinear Forms*,
Algebra and Applications 11, DOI 10.1007/978-1-84882-242-9_1,
© Springer-Verlag London Limited 2010

We shall later see that this is indeed the case, provided $2 \notin \mathfrak{m}$, so that L has characteristic $\neq 2$. If L has characteristic 2, then $\lambda_*(\varphi)$ is well-defined up to "stable isometry" (see §1.3).

Problem 1.3. Is there a meaningful way in which one can associate a symmetric bilinear form over L to φ, when φ has bad reduction?

With regard to this problem we would like to recall a classical result of T.A. Springer, which leads us to suspect that finding a solution to the problem is not completely beyond hope. Let $v : K \to \mathbb{Z} \cup \infty$ be a discrete valuation of a field K with associated valuation ring $\mathfrak{o}$. Let π be a generator of the maximal ideal $\mathfrak{m}$ of $\mathfrak{o}$, so that $\mathfrak{m} = \pi\mathfrak{o}$. Finally, let $k = \mathfrak{o}/\mathfrak{m}$ be the residue class field of $\mathfrak{o}$ and $\lambda : K \to k \cup \infty$ the canonical place with valuation ring $\mathfrak{o}$. We suppose that $2 \notin \mathfrak{m}$, so that char $k \neq 2$ is.

Let φ be a nondegenerate symmetric bilinear form over K. Then there exists a decomposition $\varphi \cong \varphi_0 \perp \pi\varphi_1$, where φ_0 and φ_1 have good reduction with respect to λ. Indeed, we can choose a diagonalization $\varphi \cong \langle a_1, \ldots, a_n \rangle$. {As usual $\langle a_1, \ldots, a_n \rangle$ denotes the diagonal matrix $\begin{pmatrix} a_1 & & 0 \\ & \ddots & \\ 0 & & a_n \end{pmatrix}$.} Then we can arrange that $v(a_i) = 0$ or 1 for each i, by multiplying the a_i by squares and renumbering indices to get $a_i \in \mathfrak{o}^*$ for $1 \leq i \leq t$ and $a_i = \pi\varepsilon_i$, where $\varepsilon_i \in \mathfrak{o}^*$ for $t < i \leq n$. {Possibly $t = 0$, so that $\varphi_0 = 0$, or $t = n$, so that $\varphi_1 = 0$.}

Theorem 1.4 (Springer 1955 [56]). *Let K be complete with respect to the discrete valuation v. If φ is anisotropic (i.e. there is no vector $x \neq 0$ in K^n with $\varphi(x, x) = 0$), then the forms $\lambda_*(\varphi_0)$ and $\lambda_*(\varphi_1)$ are anisotropic and up to isometry independent of the choice of decomposition $\varphi \cong \varphi_0 \perp \pi\varphi_1$.*

Conversely, if ψ_0 and ψ_1 are anisotropic forms over k, then there exists up to isometry a unique anisotropic form φ over K with $\lambda_*(\varphi_0) \cong \psi_0$ and $\lambda_*(\varphi_1) \cong \psi_1$.

Given any place $\lambda : K \to L \cup \infty$ and any form φ over K, Springer's theorem suggests to look for a "weak specialization" $\lambda_W(\varphi)$ by orthogonally decomposing φ in a form φ_0 with good reduction and a form φ_1 with "extremely bad" reduction, subsequently forgetting φ_1 and setting $\lambda_W(\varphi) = \lambda_*(\varphi_0)$.

Given an arbitrary valuation ring $\mathfrak{o}$, this sounds like a daring idea. Nonetheless we shall see in §1.3 that a weak specialization can be defined in a meaningful way. Admittedly $\lambda_W(\varphi)$ is not uniquely determined by φ and λ up to isometry, but up to so-called Witt equivalence. In the situation of Springer's theorem, $\lambda_W(\varphi)$ is then the Witt class of φ_0 and $\lambda_W(\pi\varphi)$ the Witt class of φ_1.

A *quadratic form* q of dimension n over K is a function $q : K^n \to K$, defined by a homogeneous polynomial of degree 2,

$$q(x) = \sum_{1 \leq i \leq j \leq n} a_{ij} x_i x_j$$

($x = (x_1, \ldots, x_n) \in K^n$). We can associate (a possibly degenerate) symmetric bilinear form

$$B_q(x, y) = q(x + y) - q(x) - q(y) = \sum_{i=1}^{n} 2a_{ii}x_iy_i + \sum_{i<j} a_{ij}(x_iy_j + x_jy_i)$$

to q. It is clear that $B_q(x, x) = 2q(x)$ for all $x \in K^n$.

If char $K \neq 2$, then any symmetric bilinear form φ over K corresponds to just one quadratic form q over K with $B_q = \varphi$, namely $q(x) = \frac{1}{2}\varphi(x, x)$. In this way we can interpret a quadratic form as a symmetric bilinear form and vice versa. In characteristic 2, however, quadratic forms and symmetric bilinear forms are very different objects.

Problem 1.5. Let $\lambda : K \to L \cup \infty$ be a place.

(a) To which quadratic forms q over K can we associate "specialized" quadratic forms $\lambda_*(q)$ over L in a meaningful way?

(b) Let char $L = 2$ and char $K \neq 2$, hence char $K = 0$. Should one specialize a quadratic form q over K with respect to λ as a quadratic form, or rather as a symmetric bilinear form?

In what follows we will present a specialization theory for arbitrary nondegenerate symmetric bilinear forms (§1.3), but only for a rather small class of quadratic forms, the so-called "obedient" quadratic forms (§1.7). Problem 1.5(b) will be answered unequivocally. If q is obedient, B_q will determine a really boring bilinear form $\lambda_*(B_q)$ (namely a hyperbolic form) which gives almost no information about q. However, $\lambda_*(q)$ can give important information about q. If possible, a specialization in the quadratic sense is thus to be preferred over a specialization in the bilinear sense.

1.2 An Elementary Treatise on Symmetric Bilinear Forms

In this section a "form" will always be understood to be a *nondegenerate* symmetric bilinear form over a field. So let K be a field.

Theorem 1.6 ("Witt decomposition").

(a) *Any form φ over K has a decomposition*

$$\varphi \cong \varphi_0 \perp \begin{pmatrix} a_1 & 1 \\ 1 & 0 \end{pmatrix} \perp \cdots \perp \begin{pmatrix} a_r & 1 \\ 1 & 0 \end{pmatrix}$$

with φ_0 anisotropic and $a_1, \dots, a_r \in K$ ($r \geq 0$).

(b) *The isometry class of φ_0 is uniquely determined by φ. (Therefore $\dim \varphi_0$ and the number r are uniquely determined.)*

To clarify these statements, let us recall the following:

(1) A form φ_0 over K is called *anisotropic* if $\varphi_0(x, x) \neq 0$ for all vectors $x \neq 0$.

(2) If char $K \neq 2$, then we have for every $a \in K^*$ that

$$\begin{pmatrix} a & 1 \\ 1 & 0 \end{pmatrix} \cong \begin{pmatrix} 0 & 1 \\ 1 & 0 \end{pmatrix} \cong \langle 1, -1 \rangle \cong \langle a, -a \rangle.$$

If char $K = 2$, however, and $a \neq 0$, then $\begin{pmatrix} a & 1 \\ 1 & 0 \end{pmatrix} \ncong \begin{pmatrix} 0 & 1 \\ 1 & 0 \end{pmatrix}$. Indeed if $\varphi = \begin{pmatrix} 0 & 1 \\ 1 & 0 \end{pmatrix}$ we have $\varphi(x, x) = 0$ for every vector $x \in K^2$, while this is not the case for $\varphi = \begin{pmatrix} a & 1 \\ 1 & 0 \end{pmatrix}$. In characteristic 2 we still have $\begin{pmatrix} a & 1 \\ 1 & 0 \end{pmatrix} \cong \langle a, -a \rangle$ $(a \in K^*)$, but $\begin{pmatrix} a & 1 \\ 1 & 0 \end{pmatrix}$ *need not* be isometric to $\begin{pmatrix} 1 & 1 \\ 1 & 0 \end{pmatrix} \cong \langle 1, -1 \rangle$.

(3) The form $\begin{pmatrix} 0 & 1 \\ 1 & 0 \end{pmatrix}$ is given the name "hyperbolic plane" (even in characteristic 2), and every form φ, isometric to an orthogonal sum $r \times \begin{pmatrix} 0 & 1 \\ 1 & 0 \end{pmatrix}$ of r copies of $\begin{pmatrix} 0 & 1 \\ 1 & 0 \end{pmatrix}$, is called "*hyperbolic*" $(r \geq 0)$.

(4) Forms which are isometric to an orthogonal sum $\begin{pmatrix} a_1 & 1 \\ 1 & 0 \end{pmatrix} \perp \cdots \perp \begin{pmatrix} a_r & 1 \\ 1 & 0 \end{pmatrix}$ are called *metabolic* $(r \geq 0)$. If char $K \neq 2$, then every metabolic form is hyperbolic. This is not the case if char $K = 2$.

(5) If char $K = 2$, then φ is hyperbolic exactly when *every* vector x of the underlying vector space K^n is isotropic, i.e. $\varphi(x, x) = 0$. If φ is not hyperbolic, we can always find an orthogonal basis such that $\varphi \cong \langle a_1, \ldots, a_n \rangle$ for suitable $a_i \in K^*$.

One can find a proof of Theorem 1.6 in any book about quadratic forms when char $K \neq 2$ (see in particular [10], [43], [55]). Part (b) of the theorem is then an immediate consequence of Witt's Cancellation Theorem. There is no general cancellation theorem in characteristic 2, as the following example shows:

$$\begin{pmatrix} a & 1 \\ 1 & 0 \end{pmatrix} \perp \langle -a \rangle \cong \begin{pmatrix} 0 & 1 \\ 1 & 0 \end{pmatrix} \perp \langle -a \rangle \tag{1.1}$$

for all $a \in K^*$. If e, f, g is a basis of K^3 which has the left-hand side of (1.1) as value matrix, then $e + g, f, g$ will be a basis which has the right-hand side of (1.1) as value matrix. For characteristic 2 one can find proofs of Theorem 1.6 and the other statements we made in [50, Chap. I and Chap. III, §1], [31, §8], [49, §4]. The following is clear from formula (1.1):

Lemma 1.7. *If a form φ with $\dim \varphi = 2r$ is metabolic, then there exists a form ψ such that $\varphi \perp \psi \cong r \times \begin{pmatrix} 0 & 1 \\ 1 & 0 \end{pmatrix} \perp \psi$.*

Definition 1.8.

(a) In the situation of Theorem 1.6, we call the form φ_0 the *kernel form* of φ and r the *(Witt) index* of φ. We write $\varphi_0 = \ker(\varphi)$, $r = \text{ind}(\varphi)$. {In the literature one frequently sees the notation $\varphi_0 = \varphi_{an}$ ("anisotropic part" of φ).}

(b) Two forms φ, ψ over K are called *Witt equivalent*, denoted by $\varphi \sim \psi$, if $\ker \varphi \cong \ker \psi$. We write $\varphi \approx \psi$ when $\ker \varphi \cong \ker \psi$ and $\dim \varphi = \dim \psi$. On the basis of the next theorem, we then call φ and ψ *stably isometric*.

Theorem 1.9. *$\varphi \approx \psi$ exactly when there exists a form χ such that $\varphi \perp \chi \cong \psi \perp \chi$.*

We omit the proof. It is easy when one uses Theorem 1.6, Lemma 1.7 and the following lemma.

Lemma 1.10. *The form $\chi \perp (-\chi)$ is metabolic for every form χ.*

Proof. From Theorem 1.6(a) we may suppose that χ is anisotropic. If χ is different from the zero form, then $\chi \cong \langle a_1, \ldots, a_n \rangle$ with elements $a_i \in K^*$ $(n \geq 1)$. Finally, $\langle a_i \rangle \perp \langle -a_i \rangle \cong \left(\begin{smallmatrix} a_i & 1 \\ 1 & 0 \end{smallmatrix} \right)$. $\qquad\qquad\qquad\square$

As is well-known, Witt's Cancellation Theorem (already mentioned above) is valid if char $K \neq 2$. It says that two stably isometric forms are already isometric: $\varphi \approx \psi \Rightarrow \varphi \cong \psi$.

Let φ be a form over K. We call the equivalence class of φ with respect to the relation $\sim$, introduced above, the *Witt class* of φ and denote it by $\{\varphi\}$. We can add Witt classes together as follows:

$$\{\varphi\} + \{\psi\} := \{\varphi \perp \psi\}.$$

The class $\{0\}$ of the zero form, whose members are exactly the metabolic forms, is the neutral element of this addition. From Lemma 1.10 it follows that $\{\varphi\} + \{-\varphi\} = 0$. In this way, the Witt classes of forms over K form an abelian group, which we denote by $W(K)$. We can also multiply Witt classes together:

$$\{\varphi\} \cdot \{\psi\} := \{\varphi \otimes \psi\}.$$

Remark. The definition of the tensor product $\varphi \otimes \psi$ of two forms φ, ψ belongs to the domain of linear algebra [10, §1, No. 9]. For diagonalizable forms we have

$$\langle a_1, \ldots, a_n \rangle \otimes \langle b_1, \ldots, b_m \rangle \cong \langle a_1 b_1, \ldots, a_1 b_m, a_2 b_1, \ldots, a_n b_m \rangle.$$

We also have $\langle a_1, \ldots, a_n \rangle \otimes \left(\begin{smallmatrix} 0 & 1 \\ 1 & 0 \end{smallmatrix} \right) \cong n \times \left(\begin{smallmatrix} 0 & 1 \\ 1 & 0 \end{smallmatrix} \right)$. Finally, for a form $\left(\begin{smallmatrix} b & 1 \\ 1 & 0 \end{smallmatrix} \right)$ with $b \neq 0$ we have

$$\langle a \rangle \otimes \begin{pmatrix} b & 1 \\ 1 & 0 \end{pmatrix} \cong \langle a \rangle \otimes \langle b, -b \rangle \cong \langle ab, -ab \rangle \cong \begin{pmatrix} ab & 1 \\ 1 & 0 \end{pmatrix}.$$

Now it is clear that the tensor product of any given form and a metabolic form is again metabolic. {For a conceptual proof of this see [31, §3], [50, Chap. I].} Therefore the Witt class $\{\varphi \otimes \psi\}$ is completely determined by the classes $\{\varphi\}$, $\{\psi\}$, independent of the choice of representatives φ, ψ.

With this multiplication, $W(K)$ becomes a commutative ring. The identity element is $\{\langle 1 \rangle\}$. We call $W(K)$ the *Witt ring* of K. For char $K \neq 2$ this ring was already introduced by Ernst Witt in 1937 [58].

We would like to describe the ring $W(K)$ by generators and relations. In characteristic $\neq 2$ this was already known by Witt [oral communication] and is implicitly contained in his work [58, Satz 7].

First we must recall the notion of determinant of a form. For $a \in K^*$, the isometry class of a one-dimensional form $\langle a \rangle$ will again be denoted by $\langle a \rangle$. The tensor product

$\langle a \rangle \otimes \langle b \rangle$ will be abbreviated by $\langle a \rangle \langle b \rangle$. We have $\langle a \rangle \langle b \rangle = \langle ab \rangle$ and $\langle a \rangle \langle a \rangle = \langle 1 \rangle$. In this way the isometry classes form an abelian group of exponent 2, which we denote by $Q(K)$. Given $a, b \in K^*$, it is clear that $\langle a \rangle = \langle b \rangle$ exactly when $b = ac^2$ for a $c \in K^*$. So $Q(K)$ is just the *group of square classes* K^*/K^{*2} in disguise. We identify $Q(K) = K^*/K^{*2}$.

It is well-known that for a given form $\varphi = (a_{ij})$ the square class of the determinant of the symmetric matrix (a_{ij}) depends only on the isometry class of φ. We denote this square class by $\det(\varphi)$, so $\det(\varphi) = \langle \det(a_{ij}) \rangle$, and call it the *determinant of* φ. A slight complication arises from the fact that the determinant is not compatible with Witt equivalence. To remedy this, we introduce the *signed determinant*

$$d(\varphi) := \langle -1 \rangle^{\frac{n(n-1)}{2}} \cdot \det(\varphi)$$

($n := \dim \varphi$). One can easily check that $d(\varphi \perp \left(\begin{smallmatrix} a & 1 \\ 1 & 0 \end{smallmatrix} \right)) = d(\varphi)$, for any $a \in K$. Hence $d(\varphi)$ depends only on the Witt class $\{\varphi\}$. The signed determinant $d(\varphi)$ also has a disadvantage though. In contrast with $\det(\varphi)$, $d(\varphi)$ does not behave completely well with respect to the orthogonal sum. Let $\nu(\varphi)$ denote the *dimension index* of φ, $\nu(\varphi) = \dim \varphi + 2\mathbb{Z} \in \mathbb{Z}/2\mathbb{Z}$. Then we have (cf. [55, I §2])

$$d(\varphi \perp \psi) = \langle -1 \rangle^{\nu(\varphi)\nu(\psi)} d(\varphi) d(\psi).$$

Let us now describe $W(K)$ by means of generators and relations. Every one-dimensional form $\langle a \rangle$ satisfies $d(\langle a \rangle) = \langle a \rangle$. This innocent remark shows that the map from $Q(K)$ to $W(K)$, which sends every isometry class $\langle a \rangle$ to its Witt class $\{\langle a \rangle\}$, is injective. We can thus interpret $Q(K)$ as a subgroup of the group of units of the ring $W(K)$, $Q(K) \subset W(K)^*$.

$W(K)$ is additively generated by the subset $Q(K)$, since every non-hyperbolic form can be written as $\langle a_1, \ldots, a_n \rangle = \langle a_1 \rangle \perp \cdots \perp \langle a_n \rangle$. Hence $Q(K)$ is a system of generators of $W(K)$. There is an obviously surjective ring homomorphism

$$\Phi : \mathbb{Z}[Q(K)] \twoheadrightarrow W(K)$$

from the group ring $\mathbb{Z}[Q(K)]$ to $W(K)$. Recall that $\mathbb{Z}[Q(K)]$ is the ring of formal sums $\sum_g n_g g$ with $g \in Q(K)$, $n_g \in \mathbb{Z}$, and almost all $n_g = 0$. Φ associates to such a sum the in $W(K)$ constructed sum $\sum_g n_g g$.

The elements of the kernel of Φ are the relations on $Q(K)$ we are looking for. We can write down some of those relations immediately: for every $a \in K^*$, $\langle a \rangle + \langle -a \rangle$ is clearly a relation. For $a, b \in K^*$ and given $\lambda, \mu \in K^*$, the form $\langle a, b \rangle$ represents the element $c := \lambda^2 a + \mu^2 b$. If $c \neq 0$, then we can find another element $d \in K^*$ with $\langle a, b \rangle \cong \langle c, d \rangle$. Comparing determinants shows that $\langle d \rangle = \langle abc \rangle$. Hence

$$\langle a \rangle + \langle b \rangle - \langle c \rangle - \langle abc \rangle = (\langle a \rangle + \langle b \rangle)(\langle 1 \rangle - \langle c \rangle)$$

is also a relation. We have the technically important:

Theorem 1.11. *The ideal* Ker Φ *of the ring* $\mathbb{Z}[Q(K)]$ *is additively generated (i.e. as abelian group) by the elements* $\langle a \rangle + \langle -a \rangle$, $a \in K^*$ *and the elements* $\langle a \rangle + \langle b \rangle - \langle c \rangle - \langle abc \rangle$ *with* $a, b \in K^*$, $\langle b \rangle \neq \langle -a \rangle$, $c = \lambda^2 a + \mu^2 b$ *with* $\lambda, \mu \in K^*$.

Remark. Ker Φ is therefore generated as an ideal by the element $\langle 1 \rangle + \langle -1 \rangle$ and the elements $(\langle 1 \rangle + \langle a \rangle)(1 - \langle c \rangle)$ with $\langle a \rangle \neq \langle -1 \rangle$, $c = 1 + \lambda^2 a$ with $\lambda \in K^*$. For application in the next section, the additive description of Ker Φ above is more favourable though.

A proof of Theorem 1.11, which also works in characteristic 2, can be found in [31, §5], [38, §1], [35, II, §4] (even over semi-local rings instead of over fields[1]), [50, p. 85]. For characteristic $\neq 2$ the proof is a bit simpler, since every form has an orthogonal basis in this case, see [55, I § 9].

1.3 Specialization of Symmetric Bilinear Forms

In this section, a "form" will again be understood to be a nondegenerate symmetric bilinear form. Let $\lambda : K \to L \cup \infty$ be a place from the field K to a field L. Let $\mathfrak{o} = \mathfrak{o}_\lambda$ be the valuation ring associated to λ and $\mathfrak{m}$ its maximal ideal. As usual for rings, $\mathfrak{o}^*$ stands for the group of units of $\mathfrak{o}$, so that $\mathfrak{o}^* = \mathfrak{o} \setminus \mathfrak{m}$. This is the set of all $x \in K$ with $\lambda(x) \neq 0, \infty$.

We will now denote the Witt class of a one-dimensional form $\langle a \rangle$ over K (or L) by $\{a\}$. The group of square classes $Q(\mathfrak{o}) = \mathfrak{o}^*/\mathfrak{o}^{*2}$ can be embedded in $Q(K) = K^*/K^{*2}$ in a natural way via $a\mathfrak{o}^{*2} \mapsto aK^{*2}$. We interpret $Q(\mathfrak{o})$ as a subgroup of $Q(K)$, so $Q(\mathfrak{o}) = \{\langle a \rangle \mid a \in \mathfrak{o}^*\} \subset Q(K)$. Our specialization theory is based on the following:

Theorem 1.12. *There exists a well-defined additive map* $\lambda_W : W(K) \to W(L)$, *given by* $\lambda_W(\{a\}) = \{\lambda(a)\}$ *if* $a \in \mathfrak{o}^*$, *and* $\lambda_W(\{a\}) = 0$ *if* $\langle a \rangle \notin Q(\mathfrak{o})$ *(i.e.* $(aK^{*2}) \cap \mathfrak{o}^* = \emptyset$).[2]

Proof. (Copied from [32, §3].) Our place λ is a combination of the canonical place $K \to (\mathfrak{o}/\mathfrak{m}) \cup \infty$ with respect to $\mathfrak{o}$, and a field extension $\bar{\lambda} : \mathfrak{o}/\mathfrak{m} \hookrightarrow L$. Thus it suffices to prove the theorem for the canonical place. So let $L = \mathfrak{o}/\mathfrak{m}$ and $\lambda(a) = \bar{a} := a + \mathfrak{m}$ for $a \in \mathfrak{o}$.

We have a well-defined additive map $\Lambda : \mathbb{Z}[Q(K)] \to W(L)$ such that $\Lambda(\langle a \rangle) = \{\bar{a}\}$ if $a \in \mathfrak{o}^*$, and $\Lambda(\langle a \rangle) = 0$ if $\langle a \rangle \notin Q(\mathfrak{o})$. Clearly Λ vanishes on all elements $\langle a \rangle + \langle -a \rangle$ with $a \in K^*$. According to Theorem 1.11 we will be finished if we can show that Λ also disappears on every element

$$Z = \langle a_1 \rangle + \langle a_2 \rangle - \langle a_3 \rangle - \langle a_4 \rangle$$

with $a_i \in K^*$ and $\langle a_1, a_2 \rangle \cong \langle a_3, a_4 \rangle$.

---

[1] The case where K has only two elements, $K = \mathbb{F}_2$, is not covered by the more general theorems there. The statement of Theorem 1.11 for $K = \mathbb{F}_2$ is trivial however, since K has only one square class $\langle 1 \rangle$ and $\langle 1, 1 \rangle \sim 0$.

[2] The letter W in the notation λ_W refers to "Witt" or "weak", see §1.1 and §1.7.

This will be the case when the four square classes $\langle a_i \rangle$ are not all in $Q(\mathfrak{o})$. Suppose from now on, without loss of generality, that $a_1 \in \mathfrak{o}^*$. Then we have $Z = \langle a_1 \rangle y$, where

$$y = 1 + \langle c \rangle - \langle b \rangle - \langle bc \rangle$$

is an element such that $\langle 1, c \rangle \cong \langle b, bc \rangle$. So $b = u^2 + w^2 c$ for elements $u, w \in K$. Clearly the equation $\Lambda(\langle a \rangle x) = \{\overline{a}\}\Lambda(x)$ is satisfied for any $a \in \mathfrak{o}^*$, $x \in \mathbb{Z}[Q(K)]$. Therefore it is enough to verify that $\Lambda(y) = 0$. We suppose without loss of generality that u and w are not both zero, otherwise we already have that $y = 0$.

Let us first treat the case $\langle c \rangle \in Q(\mathfrak{o})$, so without loss of generality $c \in \mathfrak{o}^*$. Then we have

$$\Lambda(y) = (1 + \{\overline{c}\})\Lambda(1 - \langle b \rangle).$$

If $\{\overline{c}\} = \{-1\}$, we are done. So suppose from now on that $\{\overline{c}\} \neq \{-1\}$. Then the form $\langle 1, \overline{c} \rangle$ is anisotropic over L. Since we are allowed to replace u and v by gu and gv for some $g \in K^*$, we may additionally assume that u and v are both in $\mathfrak{o}$, but not both in $\mathfrak{m}$. Since $\langle 1, \overline{c} \rangle$ is anisotropic, we have $\overline{b} = \overline{u}^2 + \overline{c}\overline{w}^2 \neq 0$ and

$$\Lambda(y) = (1 + \{\overline{c}\})(1 - \{\overline{u}^2 + \overline{c}\overline{w}^2\}) = 0.$$

The case which remains to be tackled is when the square class cK^{*2} doesn't contain a unit from $\mathfrak{o}$. Then $u^{-2}w^2 c$ is definitely not a unit and either $b = u^2(1 + d)$ or $b = w^2 c(1 + d)$ with $d \in \mathfrak{m}$. Hence $\Lambda(1 - \langle \overline{b} \rangle)$ is 0 or $1 - \{\overline{c}\}$, and both times $\Lambda(y) = 0$. $\square$

Scholium 1.13. *The map $\lambda_W : W(K) \to W(L)$ can be described very conveniently as follows: Let φ be a form over K. If φ is hyperbolic (or, more generally metabolic), then $\lambda_W(\{\varphi\}) = 0$. If φ is not hyperbolic, then consider a diagonalization $\varphi \cong \langle a_1, a_2, \ldots, a_n \rangle$. Multiply each coefficient a_i for which it is possible by a square so that it becomes a unit in $\mathfrak{o}$, and leave the other coefficients as they are. Let for example $a_i \in \mathfrak{o}^*$ for $1 \leq i \leq r$ and $\langle a_i \rangle \notin Q(\mathfrak{o})$ for $r < i \leq n$ (possibly $r = 0$ or $r = n$). Then $\lambda_W(\{\varphi\}) = \{\langle \lambda(a_1), \ldots, \lambda(a_r) \rangle\}$.*

Let us now recall a definition from the Introduction §1.1.

Definition 1.14. We say that a form φ over K has *good reduction with respect to λ,* or that φ is *λ-unimodular* if φ is isometric to a form (a_{ij}) with $a_{ij} \in \mathfrak{o}$ and $\det(a_{ij}) \in \mathfrak{o}^*$. We call such a representation $\varphi \cong (a_{ij})$ a *λ-unimodular represention of φ* (or a unimodular representation with respect to the valuation ring $\mathfrak{o}$).

This definition can be interpreted geometrically as follows. We associate to φ a couple (E, B), consisting of an n-dimensional K-vector space E ($n = \dim \varphi$) and a symmetric bilinear form $B : E \times E \to K$ such that B represents the form φ after a choice of basis of E. We denote this by $\varphi \hat{=} (E, B)$. Since φ has good reduction with respect to λ, E contains a free $\mathfrak{o}$-submodule M of rank n with $E = KM$, i.e. $E = K \otimes_{\mathfrak{o}} M$, and with $B(M \times M) \subset \mathfrak{o}$, such that the restriction $B|M \times M : M \times M \to \mathfrak{o}$ is a *nondegenerate bilinear form over* $\mathfrak{o}$, i.e. gives rise to an isomorphism $x \mapsto B(x, -)$ from the $\mathfrak{o}$-module M to the dual $\mathfrak{o}$-module $\check{M} = \mathrm{Hom}_{\mathfrak{o}}(M, \mathfrak{o})$.

By means of Theorem 1.12 we can now quite easily find a solution of the first problem posed in §1.1.

Theorem 1.15. *Suppose that the form φ over K has good reduction with respect to λ. Let $\varphi \cong (a_{ij})$ be a unimodular representation of φ. Then the Witt class $\lambda_W(\{\varphi\})$ is represented by the (nondegenerate!) form $(\lambda(a_{ij}))$ over L. Consequently the form $(\lambda(a_{ij}))$ is up to stable isometry independent of the choice of unimodular representation. (Recall that, if two forms ψ and ψ' are Witt equivalent and $\dim \psi = \dim \psi'$, then $\psi \approx \psi'$.)*

To prove this theorem, we need the following easy lemma about lifting orthogonal bases.

Lemma 1.16. *Let M be a finitely generated free $\mathfrak{o}$-module, equipped with a nondegenerate symmetric bilinear form $B : M \times M \to \mathfrak{o}$. Let $k := \mathfrak{o}/\mathfrak{m}$ and let $\pi : M \to M/\mathfrak{m}M$ be the natural epimorphism from M to the k-vector space $M/\mathfrak{m}M$. Further, let $\overline{B}$ be the (again nondegenerate) bilinear form induced by B on $M/\mathfrak{m}M$, $\overline{B}(\pi(x), \pi(y)) := B(x, y) + \mathfrak{m}$. Suppose that the vector space $M/\mathfrak{m}M$ has a basis $\overline{e}_1, \ldots, \overline{e}_n$, orthogonal with respect to $\overline{B}$. Then M has a basis $e_1, \ldots, e_n$, orthogonal with respect to B, with $\pi(e_i) = \overline{e}_i$ $(1 \le i \le n)$.*

Proof. By induction on n, which obviously is the rank of the free $\mathfrak{o}$-module M. For $n = 1$ nothing has to be shown. So suppose that $n > 1$. We choose an element $e_1 \in M$ with $\pi(e_1) = \overline{e}_1$. Then $B(e_1, e_1) \in \mathfrak{o}^*$ since $\overline{B}(\overline{e}_1, \overline{e}_1) \ne 0$. Hence the restriction of B to the module $\mathfrak{o}e_1$ is a nondegenerate bilinear form on $\mathfrak{o}e_1$. Invoking a very simple theorem (e.g. [50, p. 5, Th. 3.2], Lemma 1.53 below) yields $M = (\mathfrak{o}e_1) \perp N$ with $N = (\mathfrak{o}e_1)^{\perp} = \{x \in M \mid B(x, e_1) = 0\}$. The restriction $\pi|N : N \to M/\mathfrak{m}M$ is then a homomorphism from N to $(k\overline{e}_1)^{\perp} = \bigoplus_{i=2}^{n} k\overline{e}_i$ with kernel $\mathfrak{m}N$. By our induction hypothesis, N contains an orthogonal basis $e_2, \ldots, e_n$ with $\pi(e_i) = \overline{e}_i$ $(2 \le i \le n)$ which can be completed by e_1 to form an orthogonal basis of M which has the required property. $\qquad\square$

Remark. Clearly the lemma and its proof remain valid when $\mathfrak{o}$ is an arbitrary local ring with maximal ideal $\mathfrak{m}$, instead of a valuation ring.

We also need the following:

Definition 1.17. A *bilinear $\mathfrak{o}$-module* is a couple (M, B) consisting of an $\mathfrak{o}$-module M and a symmetric bilinear form $B : M \times M \to \mathfrak{o}$. A bilinear module is called *free* when the $\mathfrak{o}$-module M is free of finite rank. If $e_1, \ldots, e_n$ is a basis of M, we write $(M, B) \cong (a_{ij})$ with $a_{ij} := B(e_i, e_j)$. If $e_1, \ldots, e_n$ is an orthogonal basis $(B(e_i, e_j) = 0$ for $i \ne j)$, then we also write $(M, B) \cong \langle a_1, \ldots, a_n \rangle$ with $a_i := B(e_i, e_i)$.

Note. The form B is nondegenerate exactly when $\det(a_{ij})$ is a unit in $\mathfrak{o}$, respectively when all a_i are units in $\mathfrak{o}$.

All this makes sense and remains correct when $\mathfrak{o}$ is an arbitrary commutative ring (with 1), instead of a valuation ring. As before, "$\cong$" stands for "isometric", also for bilinear modules.

Proof of Theorem 1.15. For $a \in \mathfrak{o}$, let $\bar{a}$ denote the image of a in $\mathfrak{o}/\mathfrak{m}$. We suppose for the moment that the bilinear space $(\bar{a}_{ij})$ over $\mathfrak{o}/\mathfrak{m}$ is not hyperbolic. Then it has an orthogonal basis. By the lemma, the bilinear module (a_{ij}) over $\mathfrak{o}$ also has an orthogonal basis. Hence over $\mathfrak{o}$,

$$(a_{ij}) \cong \langle a_1, \ldots, a_n \rangle \tag{1.2}$$

for certain $a_i \in \mathfrak{o}^*$. The isometry (1.2) is then also valid over K, so we have in $W(L)$

$$\lambda_W(\{\varphi\}) = \{\langle \lambda(a_1), \ldots, \lambda(a_n) \rangle\}.$$

On the other hand (1.2) implies that

$$(\bar{a}_{ij}) \cong \langle \bar{a}_1, \ldots, \bar{a}_n \rangle \quad \text{over } \mathfrak{o}/\mathfrak{m}.$$

If we now apply the (injective) homomorphism $\bar{\lambda} : \mathfrak{o}/\mathfrak{m} \to L$ induced by λ (thus we tensor with the field extension given by $\bar{\lambda}$), we obtain

$$(\lambda(a_{ij})) \cong \langle \lambda(a_1), \ldots, \lambda(a_n) \rangle$$

over L. Consequently the Witt class $\lambda_W(\{\varphi\})$ is represented by the form $(\lambda(a_{ij}))$.

Let us now tackle the remaining case, where the form $(\bar{a}_{ij})$ over $\mathfrak{o}/\mathfrak{m}$ is hyperbolic. We can apply what we just have proved to the form $\psi := \varphi \perp \langle 1 \rangle$. This gives us

$$\lambda_W(\{\psi\}) = \{(\lambda(a_{ij})) \perp \langle 1 \rangle\}$$
$$= \{(\lambda(a_{ij}))\} + \{\langle 1 \rangle\}$$

in $W(L)$. On the other hand we have $\lambda_W(\{\psi\}) = \lambda_W(\{\varphi\}) + \{\langle 1 \rangle\}$, and we find again that $\lambda_W(\{\varphi\}) = \{(\lambda(a_{ij}))\}$. □

Remark. If char $L \neq 2$, a hyperbolic form over $\mathfrak{o}/\mathfrak{m}$ also has an orthogonal basis, so that the distinction between the two cases above is unnecessary.

Definition 1.18. If φ has good reduction with respect to λ, $\varphi \cong (a_{ij})$ with $a_{ij} \in \mathfrak{o}$, $\det(a_{ij}) \in \mathfrak{o}^*$, we denote the form $(\lambda(a_{ij}))$ over L by $\lambda_*(\varphi)$ and call it "the" *specialization of φ with respect to λ.*

If char $L = 2$ we run into trouble with this definition, since $\lambda_*(\varphi)$ is only up to *stable isometry* uniquely determined by φ. We nevertheless use it, since it is so convenient. If char $L \neq 2$, $\lambda_*(\varphi)$ is up to isometry uniquely determined by φ.

Example 1.19. Every metabolic form φ over K has good reduction with respect to λ. Of course is $\lambda_*(\varphi) \sim 0$.

Proof. It suffices to prove this in the case dim $\varphi = 2$, so $\varphi = \left(\begin{smallmatrix} a & 1 \\ 1 & 0 \end{smallmatrix}\right)$ with $a \in K$. Let $\varphi \hat{=} (E, B)$ and let e, f be a basis of E with value matrix $\left(\begin{smallmatrix} a & 1 \\ 1 & 0 \end{smallmatrix}\right)$. Choose an element $c \in K^*$ with $ac^2 \in \mathfrak{o}$. Then $ce, c^{-1}f$ is a basis of E with value matrix $\left(\begin{smallmatrix} ac^2 & 1 \\ 1 & 0 \end{smallmatrix}\right)$. □

Theorem 1.20. *Let φ and ψ be forms over K, having good reduction with respect to λ. Then $\varphi \perp \psi$ also has good reduction with respect to λ and*

$$\lambda_*(\varphi \perp \psi) \approx \lambda_*(\varphi) \perp \lambda_*(\psi).$$

Proof. This is clear. □

Until now we got on with our specialization theory almost without any knowledge of bilinear forms over $\mathfrak{o}$. Except for the lemma above about the existence of orthogonal bases, we needed hardly anything from this area. We could even have avoided using this little bit of information if we had considered only diagonalized forms over fields.

We are still missing one important theorem of specialization theory (especially for applications later on): Theorem 1.26 below. For a proof of this theorem we need the basics of the theory of forms over valuation rings, which we will present next using a "geometric" point of view. In other words, we interpret a form φ over a field as an "inner product" on a vector space and use more generally "inner products" on modules over rings, while until now a form was usually interpreted as a polynomial in two sets of variables $x_1, \ldots, x_n, y_1, \ldots, y_n$.

For the moment we allow local rings instead of the valuation ring $\mathfrak{o}$, since this will not cost us anything extra. So let A be a local ring.

Definition 1.21. A *bilinear space* M over A is a free A-module M of finite rank, equipped with a symmetric bilinear form $B : M \times M \to A$ which is *nondegenerate*, i.e. which determines an isomorphism $x \mapsto B(x, -)$ from M on the dual module $\check{M} = \operatorname{Hom}_A(M, A)$.

Remark. We usually denote a bilinear space by the letter M. If confusion is possible, we write (M, B) or even (M, B_M).

In what follows, M denotes a bilinear space over A, with associated bilinear form B.

Definition 1.22. A *subspace* V of M is a submodule V of M which is a direct summand of M, i.e. for which there exists another submodule W of M with $M = V \oplus W$.

To a subspace V we can associate the orthogonal submodule

$$V^\perp = \{x \in M \mid B(x, V) = 0\},$$

and we have an exact sequence

$$0 \longrightarrow V^\perp \longrightarrow M \overset{\varphi}{\longrightarrow} \check{V} \longrightarrow 0.$$

Here $\check{V} = \mathrm{Hom}_A(V, A)$ and φ maps $x \in M$ to the linear form $y \mapsto B(x, y)$ on V. The sequence splits since $\check{V}$ is free. Thus $V^\perp$ is again a subspace of M.

Definition 1.23. A subspace V of M is called *totally isotropic* when $B(V, V) = \{0\}$, i.e. when $V \subset V^\perp$. V is called a *Lagrangian subspace* of M when $V = V^\perp$. If M contains a Lagrangian subspace, M is called *metabolic*. M is called *anisotropic* if it *does not* contain any totally isotropic subspace $V \neq \{0\}$.

Lemma 1.24.

(a) *Every bilinear space M over A has a decomposition*

$$M \cong M_0 \perp M_1$$

with M_0 anisotropic and M_1 metabolic.

(b) *Every metabolic space N over A is the orthogonal sum of spaces of the form* $\begin{pmatrix} a & 1 \\ 1 & 0 \end{pmatrix}$,

$$N \cong \begin{pmatrix} a_1 & 1 \\ 1 & 0 \end{pmatrix} \perp \cdots \perp \begin{pmatrix} a_r & 1 \\ 1 & 0 \end{pmatrix}$$

with $a_1, \ldots, a_r \in A$.

These statements can be inferred from more general theorems, which can be found in e.g. [6, §1], [31, §3], [35, I §3], [38, §1].

Remark. If 2 is a unit in A, Witt's Cancellation Theorem ([30], [54]) holds for bilinear spaces over A and every metabolic space over A is even hyperbolic, i.e. is an orthogonal sum $r \times \begin{pmatrix} 0 & 1 \\ 1 & 0 \end{pmatrix}$ of copies of the "hyperbolic plane" $\begin{pmatrix} 0 & 1 \\ 1 & 0 \end{pmatrix}$ over A. Now the anisotropic space M_0 in Lemma 1.24(a) is up to isometry uniquely determined by M. If $2 \notin A^*$ this is false in general.

If $\alpha : A \to C$ is a homomorphism from A to another local ring C, we can associate to a bilinear space $(M, B) = M$ over A a bilinear space $(C \otimes_A M, B') = C \otimes_A M$ over C as follows: the underlying free C-module is the tensor product $C \otimes_A M$ determined by α, and the C-bilinear form B' on this module is obtained from B by means of a basis extension, so

$$B'(c \otimes x, d \otimes y) = cd\,\alpha(B(x, y))$$

$(x, y \in M; c, d \in C)$. The form B' is again nondegenerate. If (a_{ij}) is the value matrix of B with respect to a basis $e_1, \ldots, e_n$ of M, then $(\alpha(a_{ij}))$ is the value matrix of B' with respect to the basis $1 \otimes e_1, \ldots, 1 \otimes e_n$ of $C \otimes_A M$.

If A doesn't contain any zero divisors and if K is the quotient field of A, we can in particular use the inclusion $A \hookrightarrow K$ to associate a bilinear space $K \otimes_A M$ to the bilinear space M over A. Now we can interpret M as an A-submodule of the K-vector space $K \otimes_A M$ ($x = 1 \otimes x$ for $x \in M$) and reconstruct B from B' by restriction, $B = B'|M \times M : M \times M \to A$.

Let us return to our place $\lambda : K \to L \cup \infty$ and the valuation ring $\mathfrak{o}$.

Lemma 1.25. *Let M be a bilinear space over $\mathfrak{o}$.*
(a) If $K \otimes_\mathfrak{o} M$ is isotropic, then M is isotropic.
(b) If $K \otimes_\mathfrak{o} M$ is metabolic, then M is metabolic.

Proof. Let $E := K \otimes_\mathfrak{o} M$. We interpret M as an $\mathfrak{o}$-submodule of E and have $E = KM$.

(a) If E is isotropic, there exists a subspace $W \neq \{0\}$ in E with $W \subset W^\perp$. The $\mathfrak{o}$-submodule $V := W \cap M$ of M satisfies $KV = W$ and so $V \neq \{0\}$. Furthermore $V \subset V^\perp$. The $\mathfrak{o}$-module M/V is torsion free and finitely generated, hence free. This is because every finitely generated ideal in $\mathfrak{o}$ is principal, cf. [13, VII, §4]. Therefore V is a totally isotropic subspace of E.

(b) If $W = W^\perp$, then $V = V^\perp$. Hence M is metabolic. $\qquad\square$

Now we are fully equipped to prove the following important theorem [32, Prop. 2.2].

Theorem 1.26. *Let φ and ψ be forms over K. If φ and $\varphi \perp \psi$ have good reduction with respect to λ, then ψ also has good reduction with respect to λ.*

Proof. Adopting geometric language, the statement says: Let F and G be bilinear spaces over K and $E := F \perp G$. If F and E have good reduction, i.e. $F \cong K \otimes_\mathfrak{o} N$, $E \cong K \otimes_\mathfrak{o} M$ for bilinear spaces N and M over $\mathfrak{o}$, then G has good reduction as well.

By Theorem 1.6 there is a decomposition $G = G_0 \perp G_1$ with G_0 anisotropic and G_1 metabolic. From above (cf. Example 1.19), G_1 has good reduction. Hence it suffices to show that G_0 has good reduction.

Now $E \perp (-F) \cong F \perp (-F) \perp G_0 \perp G_1$.[3] Since $F \perp (-F) \perp G_0$ is metabolic, but G_1 anisotropic, G_1 is the kernel space of $E \perp (-F)$. ("Kernel space" is the pendant of the word "kernel form" (= anisotropic part) in geometric language.) We decompose $M \perp (-N)$ following Lemma 1.24(a), $M \perp (-N) \cong R \perp S$ where R is anisotropic and S metabolic. Tensoring with K gives $E \perp (-F) \cong K \otimes_\mathfrak{o} R \perp K \otimes_\mathfrak{o} S$. Now $K \otimes_\mathfrak{o} S$ is metabolic and, according to Lemma 1.25, $K \otimes_\mathfrak{o} R$ is anisotropic. Hence $K \otimes_\mathfrak{o} R$ is also a kernel space of $E \perp (-F)$. Applying Theorem 1.6 gives $K \otimes_\mathfrak{o} R \cong G_0$, and we are finished. $\qquad\square$

Corollary 1.27. *Let φ and ψ be forms over K with $\varphi \sim \psi$. If φ has good reduction with respect to λ, ψ also has good reduction with respect to λ and $\lambda_*(\varphi) \sim \lambda_*(\psi)$. If furthermore $\varphi \approx \psi$, then $\lambda_*(\varphi) \approx \lambda_*(\psi)$.*

Proof. There are Witt decompositions $\varphi \cong \varphi_0 \perp \mu$, $\psi \cong \psi_0 \perp \nu$ with φ_0, ψ_0 anisotropic and μ, ν metabolic. As established above, μ and ν have good reduction with respect to λ and $\lambda_*(\mu), \lambda_*(\nu)$ are metabolic. By assumption φ has good reduction with respect to λ and φ_0 is isometric to ψ_0. Theorem 1.26 implies that φ_0 has good reduction with respect to λ. Therefore ψ_0, and hence ψ, has good reduction with respect to λ, and (according to Theorem 1.20)

$$\lambda_*(\varphi) \approx \lambda_*(\varphi_0) \perp \lambda_*(\mu) \sim \lambda_*(\varphi_0),$$
$$\lambda_*(\psi) \approx \lambda_*(\psi_0) \perp \lambda_*(\nu) \sim \lambda_*(\psi_0).$$

---

[3] If $E = (E, B)$ is a bilinear space, then $-E$ denotes the space $(E, -B)$.

Naturally $\lambda_*(\varphi_0) \approx \lambda_*(\psi_0)$, so $\lambda_*(\varphi) \sim \lambda_*(\psi)$. If $\varphi \approx \psi$, then φ and ψ have the same dimension and so $\lambda_*(\varphi) \approx \lambda_*(\psi)$. $\square$

Let us give a small illustration of Theorem 1.26.

Definition 1.28. Let φ and ψ be forms over a field K. If there exists a form χ over K with $\varphi \cong \psi \perp \chi$, we say that φ *represents* the form ψ and write $\psi < \varphi$.

For example, the one-dimensional forms represented by φ are exactly the square classes $\langle \varphi(x, x) \rangle$, where x runs through the anisotropic vectors of the space belonging to φ.

Theorem 1.29 (Substitution Principle). *Let k be a field and $K = k(t)$, where $t = (t_1, \ldots, t_r)$ is a set of indeterminates. Let $(f_{ij}(t))$ be a symmetric $(n \times n)$-matrix and $(g_{kl}(t))$ a symmetric $(m \times m)$-matrix, for polynomials $f_{ij}(t) \in k[t]$, and $g_{kl}(t) \in k[t]$. Let further be given a field extension $k \subset L$ and a point $c \in L^r$ with $\det(f_{ij}(c)) \neq 0$ and $\det(g_{kl}(c)) \neq 0$. If $\operatorname{char} k = 2$, also suppose that the form $(f_{ij}(c))$ is anisotropic over L.*

Claim: if $(g_{kl}(t)) < (f_{ij}(t))$ (as forms over K), then $(g_{kl}(c)) < (f_{ij}(c))$ (as forms over L).

Proof. Going from $k[t]$ to $L[t]$, we suppose without loss of generality that $L = k$. For every $s \in \{1, \ldots, r\}$ there is exactly one corresponding place $\lambda_s : k(t_1, \ldots, t_s) \to k(t_1, \ldots, t_{s-1}) \cup \infty$ with $\lambda_s(u) = u$ for all $u \in k(t_1, \ldots, t_{s-1})$ and $\lambda_s(t_s) = c_s$. {Read $k(t_1, \ldots, t_{s-1}) = k$ when $s = 1$.} The composition $\lambda_1 \circ \lambda_2 \circ \cdots \circ \lambda_s$ of these places is a place $\lambda : K \to k \cup \infty$ with $\lambda(a) = a$ for all $a \in k$ and $\lambda(t_i) = c_i$ for $i = 1, \ldots, r$. Let φ denote the form $(f_{ij}(t))$ over K and ψ the form $(g_{kl}(t))$ over K. {Note that obviously $\det(f_{ij}(t)) \neq 0$, $\det(g_{kl}(t)) \neq 0$.} The forms φ and ψ both have good reduction with respect to λ and $\lambda_*(\varphi) \approx (f_{ij}(c))$, $\lambda_*(\psi) \approx (g_{kl}(c))$.

Now let $\psi < \varphi$. Then there exists a form χ over K with $\psi \perp \chi \cong \varphi$. According to Theorem 1.26, χ has good reduction with respect to λ and according to Theorem 1.20, $\lambda_*(\psi) \perp \lambda_*(\chi) \approx \lambda_*(\varphi)$. Hence $\lambda_*(\psi) \perp \lambda_*(\chi) \cong \lambda_*(\varphi)$ if $\operatorname{char} k \neq 2$. If $\operatorname{char} k = 2$ and $\lambda_*(\varphi)$ is anisotropic, this remains true, since $\lambda_*(\varphi)$ is up to isometry the unique anisotropic form in the Witt class $\lambda_W(\varphi)$, and $\lambda_* (\psi) \perp \lambda_* (\chi)$ has the same dimension as $\lambda_*(\varphi)$. $\square$

Let us now return to our arbitrary place $\lambda : K \to L \cup \infty$ and to the conventions made at the beginning of the paragraph. The Lemmas 1.24 and 1.25 allow us to give an easier proof of Theorem 1.15, which is interesting in its own right.

Second proof of Theorem 1.15. We adopt the geometric language. Let E be a bilinear space over K, having good reduction with respect to λ, and let M and N be bilinear spaces over $\mathfrak{o}$ with $E \cong K \otimes_\mathfrak{o} M \cong K \otimes_\mathfrak{o} N$. We have to show that $L \otimes_\lambda M \approx L \otimes_\lambda N$, where the tensor product is taken over $\mathfrak{o}$, and L is regarded as an $\mathfrak{o}$-algebra via the homomorphism $\lambda|\mathfrak{o} : \mathfrak{o} \to L$. The space $K \otimes_\mathfrak{o} (M \perp (-N))$ is metabolic. According to Lemma 1.25, $M \perp (-N)$ is metabolic. Hence $L \otimes_\lambda (M \perp (-N)) = L \otimes_\lambda M \perp (-L \otimes_\lambda N)$ is also metabolic. Therefore $L \otimes_\lambda M \approx L \otimes_\lambda N$. $\square$

We can now describe the property "good reduction" and the specialization of a form by means of diagonal forms as follows.

Theorem 1.30. *Let φ be a form over K, $\dim \varphi = n$.*

(a) *The form φ has good reduction with respect to λ if and only if φ is Witt equivalent to a diagonal form $\langle a_1, \ldots, a_r \rangle$ with units $a_i \in \mathfrak{o}^*$. In this case $\dim \lambda_*(\varphi) = n$ and $\lambda_*(\varphi) \sim \langle \lambda(a_1), \ldots, \lambda(a_r) \rangle$. Furthermore one can choose $r = n + 2$.*

(b) *Let $2 \in \mathfrak{o}^*$, i.e. char $L \neq 2$. The form φ has good reduction with respect to λ if and only if φ is isometric to a diagonal form $\langle a_1, \ldots, a_n \rangle$ with $a_i \in \mathfrak{o}^*$. In this case $\lambda_*(\varphi) \cong \langle \lambda(a_1), \ldots, \lambda(a_n) \rangle$.*

Proof of part (a). If $\varphi \sim \langle a_1, \ldots, a_r \rangle$ with units $a_i \in \mathfrak{o}^*$, then by the Corollary φ has good reduction with respect to λ and $\lambda_*(\varphi) \sim \langle \lambda(a_1), \ldots, \lambda(a_r) \rangle$. Suppose now that φ has good reduction with respect to λ. The form φ corresponds to a bilinear space $K \otimes_{\mathfrak{o}} M$ over K, which comes from a bilinear space M over $\mathfrak{o}$. If $M/\mathfrak{m}M$ is not hyperbolic, then M has an orthogonal basis by Lemma 1.16. Therefore $\varphi \cong \langle a_1, \ldots, a_n \rangle$ with units $a_i \in \mathfrak{o}^*$. In general we consider the space $M' := M \perp \langle 1, -1 \rangle$ over $\mathfrak{o}$. Then $M'/\mathfrak{m}M'$ is definitely *not* hyperbolic. Hence $\varphi \perp \langle 1, -1 \rangle \cong \langle b_1, \ldots, b_{n+2} \rangle$ with units b_i. $\qquad \square$

The proof of part (b) is similar, but simpler since now $M/\mathfrak{m}M$ always has an orthogonal basis. We don't need the Corollary here.

Finally, we consider the specialization of tensor products of forms.

Theorem 1.31. *Let φ and ψ be two forms over K, which have good reduction with respect to λ. Then $\varphi \otimes \psi$ also has good reduction with respect to λ, and $\lambda_*(\varphi \otimes \psi) \approx \lambda_*(\varphi) \otimes \lambda_*(\psi)$.*

Proof. According to Theorem 1.30 we have the following Witt equivalences, $\varphi \sim \langle a_1, \ldots, a_m \rangle$, $\psi \sim \langle b_1, \ldots, b_n \rangle$, with units $a_i, b_j \in \mathfrak{o}^*$. Then

$$\varphi \otimes \psi \sim \langle a_1 b_1, a_1 b_2, \ldots, a_1 b_n, \ldots, a_m b_n \rangle.$$

Again according to Theorem 1.30, $\varphi \otimes \psi$ has good reduction and

$$\begin{aligned} \lambda_*(\varphi \otimes \psi) &\sim \langle \lambda(a_1)\lambda(b_1), \ldots, \lambda(a_m)\lambda(b_n) \rangle \\ &\cong \langle \lambda(a_1), \ldots, \lambda(a_m) \rangle \otimes \langle \lambda(b_1), \ldots, \lambda(b_n) \rangle \\ &\sim \lambda_*(\varphi) \otimes \lambda_*(\psi). \end{aligned}$$

Now the forms $\lambda_*(\varphi \otimes \psi)$ and $\lambda_*(\varphi) \otimes \lambda_*(\psi)$ both have the same dimension as $\varphi \otimes \psi$. Therefore $\lambda_*(\varphi \otimes \psi) \approx \lambda_*(\varphi) \otimes \lambda_*(\psi)$. $\qquad \square$

If we use a little more multilinear algebra, we can come up with the following conceptually more pleasing proof of Theorem 1.31.

Second proof of Theorem 1.31. Adopting geometric language, φ corresponds to a bilinear space $K \otimes_o M$ and ψ corresponds to a space $K \otimes_o N$ with bilinear spaces M and N over o. Hence $\varphi \otimes \psi$ corresponds to the space

$$(K \otimes_o M) \otimes_K (K \otimes_o N) \cong K \otimes_o (M \otimes_o N)$$

over K. Now the free bilinear module $M \otimes_o N$ is again nondegenerate (cf. e.g. [50, I §5]). Consequently $\varphi \otimes \psi$ has good reduction with respect to λ, and $\lambda_*(\varphi \otimes \psi)$ can be represented by the space $L \otimes_\lambda (M \otimes_o N)$, obtained from the space $M \otimes_o N$ by base extension to L using the homomorphism $\lambda|o : o \to L$. Now $L \otimes_\lambda (M \otimes_o N) \cong (L \otimes_\lambda M) \otimes_L (L \otimes_\lambda N)$. In other words, $\lambda_*(\varphi \otimes \psi) \cong \lambda_*(\varphi) \otimes \lambda_*(\psi)$, since $\lambda_*(\varphi) \cong L \otimes_\lambda M$, $\lambda_*(\psi) \cong L \otimes_\lambda N$. $\qquad\qquad\square$

1.4 Generic Splitting in Characteristic $\neq 2$

In this section we outline an important application area of the specialization theory developed in §1.3, namely the theory of generic splitting of bilinear forms. Many proofs will only be given in §1.7, after also having developed a specialization theory of quadratic forms.

Let k be a field and let φ be a *form* over k, which is just as before understood to be a nondegenerate symmetric bilinear form over k. Our starting point is the following simple

Observation. Let K and L be fields, containing k, and let $\lambda : K \to L \cup \infty$ be a place over k, i.e. with $\lambda(c) = c$ for all $c \in k$.

(a) Then $\varphi \otimes K$ has good reduction with respect to λ and $\lambda_*(\varphi \otimes K) \approx \varphi \otimes L$.[4]
    Indeed, if $\varphi \cong (a_{ij})$ with $a_{ij} \in k$, $\det(a_{ij}) \neq 0$, then also $\varphi \otimes K \cong (a_{ij})$, and this is a unimodular representation of φ with respect to λ, since k is contained in the valuation ring o of λ. So $\lambda_*(\varphi \otimes K) \approx (\lambda(a_{ij})) = (a_{ij})$ and since this symmetric matrix is now considered over L, we conclude that $\lambda_*(\varphi \otimes K) \approx \varphi \otimes L$.

(b) Suppose that $\varphi \otimes K$ has kernel form φ_1 and Witt index r_1, i.e.

$$\varphi \otimes K \approx \varphi_1 \perp r_1 \times H,$$

where H denotes from now on the hyperbolic "plane"[5] $\left(\begin{smallmatrix} 0 & 1 \\ 1 & 0 \end{smallmatrix}\right)$. According to Corollary 1.27, the form φ_1 has good reduction with respect to λ. Therefore, applying λ_* yields

$$\varphi \otimes L \approx \lambda_*(\varphi_1) \perp r_1 \times H.$$

Hence we conclude that $\mathrm{ind}(\varphi \otimes L) \geq \mathrm{ind}(\varphi \otimes K)$ and that

---

[4] $\varphi \otimes K$ is the form φ, considered over K instead of over k. If E is a bilinear space over k corresponding to φ, then $K \otimes_k E$—as described in §1.3—is a bilinear space corresponding to $\varphi \otimes K$.
[5] We do not make a notational distinction between the forms $\left(\begin{smallmatrix} 0 & 1 \\ 1 & 0 \end{smallmatrix}\right)$ over the different fields occurring here.

$$\lambda_*(\ker(\varphi \otimes K)) \sim \ker(\varphi \otimes L).$$

(Recall the terminology of Definition 1.8.)

Definition 1.32. We call two fields $K \supset k$, $L \supset k$ over k *specialization equivalent*, or just *equivalent*, if there exist places $\lambda : K \to L \cup \infty$ and $\mu : L \to K \cup \infty$ over k. We then write $K \sim_k L$.

The following conclusions can be drawn immediately from our observation:

Remark 1.33. If $K \sim_k L$, then every form φ over k satisfies:
(1) $\mathrm{ind}(\varphi \otimes K) = \mathrm{ind}(\varphi \otimes L)$.
(2) $\ker(\varphi \otimes K)$ has good reduction with respect to every place λ from K to L and $\lambda_*(\ker(\varphi \otimes K)) \cong \ker(\varphi \otimes L)$. (Note that "$\cong$" holds, not just "$\approx$"!)

From a technical point of view, it is a good idea to treat the following special case of Definition 1.32 separately:

Definition 1.34. We call a field extension $K \supset k$ *inessential* if there exists a place $\lambda : K \to k \cup \infty$ over k.

Obviously this just means that K is equivalent to k over k. In this case, Remark 1.33 becomes:

Remark 1.35. If K is an inessential extension of k, then every form φ over k satisfies:
(1) $\mathrm{ind}(\varphi \otimes K) = \mathrm{ind}(\varphi)$,
(2) $\ker(\varphi \otimes K) = \ker(\varphi) \otimes K$.

We will see that the forms φ and $\varphi \otimes K$ exhibit the "same" splitting behaviour with respect to an inessential extension K/k in an even broader sense (see Scholium 1.47 below).

Let us return to the general observation above. It gives rise to the following:

Problem 1.36. Let φ be a form over k, $\dim \varphi = n \geq 2$, and let t be an integer with $1 \leq t \leq \frac{n}{2}$. Does there exist a field extension $K \supset k$ which is "generic with respect to splitting off t hyperbolic planes", i.e. with the following properties?
(a) $\mathrm{ind}(\varphi \otimes K) \geq t$.
(b) If L is a field extension of k with $\mathrm{ind}(\varphi \otimes L) \geq t$, then there exists a place from K to L over k.

We first address this problem for the case $t = 1$. As before, let φ be a form over a field k.

Definition 1.37. An extension field $K \supset k$ is called a *generic zero field of* φ if the following conditions hold:
(a) $\varphi \otimes K$ is isotropic.
(b) For every field extension $k \hookrightarrow L$ with $\varphi \otimes L$ isotropic, there exists a place $\lambda : K \to L \cup \infty$ over k.

Note that if φ is isotropic, then the field k is of course itself a generic zero field of φ.

In the rest of this section, we assume that char $k \neq 2$. Now φ can also be viewed as a quadratic form,[6] $\varphi(x) := \varphi(x, x)$. We define a field extension $k(\varphi)$ of k, which is *a priori* suspected to be a generic zero field of φ.

Definition 1.38. Let dim $\varphi \geq 3$ or let dim $\varphi = 2$ and $\varphi \not\cong H$. Let $k(\varphi)$ denote the function field of the affine quadric $\varphi(X_1, \ldots, X_n) = 0$ (where $n = \dim \varphi$), i.e. the quotient field of the ring $A := k[X_1, \ldots, X_n]/(\varphi(X_1, \ldots, X_n))$ with indeterminates $X_1, \ldots, X_n$.

Observe that the polynomial $\varphi(X_1, \ldots, X_n)$ is irreducible. To prove this we may suppose that φ is diagonalized, $\varphi = \langle a_1, a_2, \ldots, a_n \rangle$. The polynomial $a_1 X_1^2 + a_2 X_2^2 + \cdots + a_n X_n^2$ is clearly not a product of two linear forms (since char $k \neq 2$).

If $\varphi = H$, then $\varphi(X_1, X_2) = X_1 X_2$. On formal grounds we then set $k(\varphi) = k(t)$ for an indeterminate t.

Let $x_1, \ldots, x_n$ be the images of the indeterminates $X_1, \ldots, X_n$ in A. Then $\varphi(x_1, \ldots, x_n) = 0$. Hence $\varphi \otimes k(\varphi)$ is isotropic. (This is also true for $\varphi \cong H$.) On top of that we have the following important

Theorem 1.39. *Let* dim $\varphi \geq 2$. *Then* $k(\varphi)$ *is a generic zero field of* φ.

Note that the assumption dim $\varphi \geq 2$ is necessary for the existence of a generic zero field, since forms of dimension ≤ 1 are never isotropic.

Theorem 1.39 can already be found in [33]. We will wait until §2.1 to prove it in the framework of a generic splitting theory of quadratic forms.

It is clear from above that every other generic zero field K of φ over k is equivalent to $k(\varphi)$. It is unknown which bilinear forms possess generic zero fields in characteristic 2.

Given an arbitrary form φ over k, we now construct a tower of fields $(K_r \mid 0 \leq r \leq h)$ together with anisotropic forms φ_r over K_r and numbers $i_r \in \mathbb{N}_0$ as follows: choose K_0 to be any inessential extension of the field k.[7] Let φ_0 be the kernel form of $\varphi \otimes K_0$ and i_0 the Witt index of φ. Then

$$\varphi \otimes K_0 \cong \varphi_0 \perp i_0 \times H.$$

If dim $\varphi_0 \leq 1$, then Stop! Otherwise choose a generic zero field $K_1 \supset K_0$ of φ_0. Let φ_1 be the kernel form of $\varphi_0 \otimes K_1$ and i_1 the Witt index of $\varphi_0 \otimes K_1$. Then

$$\varphi_0 \otimes K_1 \cong \varphi_1 \perp i_1 \times H.$$

If dim $\varphi_1 \leq 1$, then Stop! Otherwise choose a generic zero field $K_2 \supset K_1$ of φ_1. Let φ_2 be the kernel form of $\varphi_1 \otimes K_2$ and i_2 the Witt index of $\varphi_1 \otimes K_2$. Then

---

[6] In keeping with our earlier conventions, it would perhaps be better to write $\varphi(x) = \frac{1}{2}\varphi(x, x)$. However, the factor $\frac{1}{2}$ is not important for now.

[7] In earlier works (especially [33]) K_0 was always chosen to be k. From a technical point of view it is favourable to allow K_0 to be an inessential extension of k, just as in [37].

$$\varphi_1 \otimes K_2 \cong \varphi_2 \perp i_2 \times H,$$

etc.

The construction halts after $h \leq \left[\frac{\dim \varphi}{2}\right]$ steps with a Witt decomposition

$$\varphi_{h-1} \otimes K_h \cong \varphi_h \perp i_h \times H,$$

$\dim \varphi_h \leq 1$.

Definition 1.40. $(K_r \mid 0 \leq r \leq h)$ is called a *generic splitting tower of* φ. The number h is called the *height* of φ, denoted $h = h(\varphi)$. The number i_r is called the rth *higher index* of φ and the form φ_r the rth *higher kernel form* of φ.

Remark. Obviously φ_r is the kernel form of $\varphi \otimes K_r$ and $\mathrm{ind}(\varphi \otimes K_r) = i_0 + \cdots + i_r$. Note that $h = 0$ iff the form φ "splits", i.e. iff its kernel form is zero or one-dimensional.

We will see that the number h and the sequence $(i_0, \ldots, i_h)$ are independent of the choice of generic splitting tower and also that the forms φ_r are determined uniquely by φ in a precise way. For this the following theorem is useful.

Theorem 1.41. *Let ψ be a form over a field K. Let $\lambda : K \to L \cup \infty$ be a place, such that ψ has good reduction with respect to λ. Suppose that L (and therefore K) has characteristic $\neq 2$. Then $\lambda_*(\psi)$ is isotropic if and only if λ can be extended to a place $\mu : K(\psi) \to L \cup \infty$.*

If λ is a trivial place, i.e. a field extension, then this theorem says once more that $K(\psi)$ is a generic zero field of ψ (Theorem 1.39).

One direction of the assertion is trivial, just as for Theorem 1.39: if λ can be extended to a place $\mu : K(\psi) \to L \cup \infty$, then $\psi \otimes K(\psi)$ has good reduction with respect to μ and $\mu_*(\psi \otimes K(\psi)) \cong \lambda_*(\psi)$. Since $\psi \otimes K(\psi)$ is isotropic, $\lambda_*(\psi)$ is also isotropic.

The other direction will be established in §2.1. For a shorter and simpler proof, see [33] and the books [55], [39].

Theorem 1.42 (Corollary of Theorem 1.41). *Let φ be a form over a field k. Let $(K_r \mid 0 \leq r \leq h)$ be a generic splitting tower of φ with associated higher kernel forms φ_r and indices i_r. Suppose that φ has good reduction with respect to some place $\gamma : k \to L \cup \infty$. Suppose that L (and hence k) has characteristic $\neq 2$. Finally let $\lambda : K_m \to L \cup \infty$ be a place for some m, $0 \leq m \leq h$ which extends γ and which cannot be extended to K_{m+1} if $m < h$. Then φ_m has good reduction with respect to λ. The form $\gamma_*(\varphi)$ has kernel form $\lambda_*(\varphi_m)$ and Witt index $i_0 + \cdots + i_m$.*

Proof. There is an isometry

$$\varphi \otimes K_m \cong \varphi_m \perp (i_0 + \cdots + i_m) \times H. \tag{1.3}$$

The form $\varphi \otimes K_m$ has good reduction with respect to λ and $\lambda_*(\varphi \otimes K_m) = \gamma_*(\varphi)$. Using Theorem 1.26 and Theorem 1.20, (1.3) implies that φ_m has good reduction with respect to λ and

$$\gamma_*(\varphi) \cong \lambda_*(\varphi_m) \perp (i_0 + \cdots + i_m) \times H \qquad (1.4)$$

(cf. the observation at the beginning of this section). If $\lambda_*(\varphi_m)$ were isotropic, then we would have $m < h$, since $\dim \varphi_h \leq 1$. It would then follow from Theorem 1.41 that λ can be extended to a place from K_{m+1} to L, contradicting our assumption. Therefore $\lambda_*(\varphi_m)$ is anisotropic, so (1.4) is the Witt decomposition of $\gamma_*(\varphi)$. $\qquad \square$

Note that this theorem implies, in particular, that any attempt to successively extend the given place $\lambda : k \to L \cup \infty$ to a "storey" K_m of the generic splitting tower, which is as high as possible, always ends at the same number m.

Theorem 1.42 gives rise to a number of interesting statements about the splitting behaviour of forms and the extensibility of places.

Scholium 1.43. *Let φ be a form over k and $(K_r \mid 0 \leq r \leq h)$ a generic splitting tower of φ with kernel forms φ_r and indices i_r. Furthermore, let $k \subset L$ be a field extension. If we apply Theorem 1.42 to the trivial place $\gamma : k \hookrightarrow L$, we get:*

(1) *Let $\lambda : K_m \to L \cup \infty$ be a place over k $(0 \leq m \leq h)$, which* cannot *be extended to K_{m+1} in case $m < h$. Then φ_m has good reduction with respect to λ and $\lambda_*(\varphi_m)$ is the kernel form of $\varphi \otimes L$. The index of $\varphi \otimes L$ is $i_0 + \cdots + i_m$.*

(2) *If $\lambda' : K_r \to L \cup \infty$ is a place over k, then $r \leq m$ and λ' can be extended to a place $\mu : K_m \to L \cup \infty$.*

(3) *Given a number t with $1 \leq t \leq \left[\frac{\dim \varphi}{2}\right] = i_0 + \cdots + i_h$, let $m \in \mathbb{N}_0$ be minimal with $t \leq i_0 + \cdots + i_m$. Then K_m is a generic field extension of k with respect to splitting off t hyperbolic planes of φ (in the context of our problem above).*

Definition 1.44. *Let φ be a form over k. We call the set of indices $\mathrm{ind}(\varphi \otimes L)$, where L traverses all field extensions of k, the* splitting pattern *of φ, denoted by $\mathrm{SP}(\varphi)$.*

This definition also makes sense in characteristic 2, and it is *a priori* clear that $\mathrm{SP}(\varphi)$ consists of at most $\left[\frac{\dim \varphi}{2}\right] + 1$ elements. Usually the elements of $\mathrm{SP}(\varphi)$ are listed in ascending order. If $\mathrm{char}\, k \neq 2$ and $(i_r \mid 0 \leq r \leq h)$ is the sequence of higher indices of a generic splitting tower $(K_r \mid 0 \leq r \leq h)$ of φ, then Scholium 1.43 shows:

$$\mathrm{SP}(\varphi) = (i_0, i_0 + i_1, i_0 + i_1 + i_2, \ldots, i_0 + i_1 + \cdots + i_h).$$

Since the numbers i_r with $r > 0$ are all positive, it is evident that the height h and the higher indices i_r $(0 \leq r \leq h)$ are independent of the choice of generic splitting tower of φ. Obviously is $i_0 + \cdots + i_h = \left[\frac{\dim \varphi}{2}\right]$.

Scholium 1.45. *Let $(K_r \mid 0 \leq r \leq h)$ and $(K_r' \mid 0 \leq r \leq h)$ be two generic splitting towers of the form φ over k. Applying Scholium 1.43 to the field extensions $k \subset K_s$ and $k \subset K_r'$, yields: if $\lambda : K_r \to K_s' \cup \infty$ is a place over k, then $r \leq s$ and λ can be extended to a place $\mu : K_s \to K_s' \cup \infty$. The fields K_s and K_s' are equivalent over k. For every place $\mu : K_s \to K_s' \cup \infty$, the kernel form φ_s of $\varphi \otimes K_s$ has good reduction with respect to μ and $\mu_*(\varphi_s)$ is the kernel form φ_s' of $\varphi \otimes K_s'$.*

Scholium 1.46. *Let φ be a form over k and $\gamma : k \to L \cup \infty$ a place such that φ has good reduction with respect to γ. Applying Theorem 1.42 to the places $j \circ \gamma : k \to L \cup \infty$, being the composites of γ and trivial places $j : L \hookrightarrow L'$, yields:*

(1) $\mathrm{SP}(\gamma_*(\varphi)) \subset \mathrm{SP}(\varphi)$.

(2) *The higher kernel forms of $\gamma_*(\varphi)$ arise from certain higher kernel forms of φ by means of specialization. More precisely: if $(K_r \mid 0 \leq r \leq h)$ is a generic splitting tower of φ and $(K'_s \mid 0 \leq s \leq h')$ is a generic splitting tower of $\gamma_*(\varphi)$, then $h' \leq h$ and for every s with $0 \leq s \leq h'$ we have*

$$\mathrm{ind}(\gamma_*(\varphi) \otimes K'_s) = i_0 + \cdots + i_m$$

for some $m \in \{0, \ldots, h\}$. The number m is the biggest integer such that γ can be extended to $\lambda : K_m \to K'_s \cup \infty$. The kernel form φ_m of $\varphi \otimes K_m$ has good reduction with respect to every extension λ of this kind, and $\lambda_*(\varphi_m)$ is the kernel form of $\gamma_*(\varphi) \otimes K'_s$.

(3) *If $\rho : K_r \to K'_s \cup \infty$ is a place, which extends $\gamma : k \to L \cup \infty$, then $r \leq m$ and ρ can be further extended to a place from K_m to K'_s.*

Scholium 1.47. *Let L/k be an inessential field extension (see Definition 1.34 above) and φ again a form over k. Let $(L_i \mid 0 \leq i \leq h)$ be a generic splitting tower of $\varphi \otimes L$. It is then immediately clear from Definition 1.40 that this is also a generic splitting tower of φ. Hence $\mathrm{SP}(\varphi \otimes L) = \mathrm{SP}(\varphi)$, and the higher kernel forms of $\varphi \otimes L$ can also be interpreted as higher kernel forms of φ.*

Which strictly increasing sequences $(0, j_1, j_2, \ldots, j_n)$ occur as splitting patterns of anisotropic forms? What do anisotropic forms of given height h look like? These questions are difficult and at the moment the object of intense research. The first efforts towards answering them can be found in [32], [33], while more recent ones can be found in [22], [24], [25], [28] amongst others.

A complete answer is only known in the case $h = 1$. A form

$$\langle 1, a_1 \rangle \otimes \cdots \otimes \langle 1, a_d \rangle \quad (d \geq 1, \ a_i \in k)$$

is called a *d-fold Pfister form* over k.[8] If τ is a Pfister form of degree d, then the form τ', satisfying $\langle 1 \rangle \perp \tau' = \tau$, is called the *pure part* of τ.

Theorem 1.48. *An anisotropic form φ over k has height 1 if $\varphi \cong a\tau$ (dim φ even) or $\varphi \cong a\tau'$ (dim φ odd), with $a \in k^*$ and τ an anisotropic Pfister form of degree $d \geq 1$ in the first case and $d \geq 2$ in the second case.*

Note that therefore $\mathrm{SP}(\varphi) = (0, 2^{d-1})$ when dim φ is even and $\mathrm{SP}(\varphi) = (0, 2^{d-1} - 1)$ when dim φ is odd. Is $k = \mathbb{R}$ for example, then all $d \geq 1$, resp. $d \geq 2$ occur. One can take for instance $\varphi = 2^d \times \langle 1 \rangle$, resp. $\varphi = (2^d - 1) \times \langle 1 \rangle$.

A proof of Theorem 1.48 can be found in [33] and the books [55], [39]. In §3.6 and §3.9 we will prove two theorems for fields of arbitrary characteristic, from

---

[8] The form $\langle 1 \rangle$ also counts as a Pfister form, more precisely a 0-fold Pfister form.

which Theorem 1.48 can be obtained in characteristic $\neq 2$ (Theorem 3.45, Theorem 3.77).

For forms of height 2, the possible splitting patterns are known in even dimension [57]. Beyond that, little is known about such forms, but the known results are interesting and partly deep, see e.g. [34], [18], [28], [20], [21].

1.5 An Elementary Treatise on Quadratic Modules

We want to construct a specialization theory for *quadratic* forms, similar to the theory in §1.3 for symmetric bilinear forms. In order to do this we need some definitions and theorems about quadratic modules over rings, and in particular over valuation rings.

Let A be any ring (always commutative, with 1). We recall some fundamental definitions and facts about quadratic forms over A, cf. [50, Appendix 1].

Definition 1.49. Let M be an A-module. A *quadratic form on M* is a function $q : M \to A$, satisfying the following conditions:
(1) $q(\lambda x) = \lambda^2 q(x)$ for $\lambda \in A$, $x \in M$.
(2) The function $B_q : M \times M \to A$, $B_q(x, y) := q(x + y) - q(x) - q(y)$ is a bilinear form on M.

The pair (M, q) is then called a *quadratic module* over A.

Note that the bilinear form B_q is symmetric. Furthermore $B_q(x, x) = 2q(x)$ for all $x \in M$. If 2 is a unit in A, $2 \in A^*$, we can retrieve q from B_q. Also, every symmetric bilinear form B on M comes from a quadratic form q in this case, namely $q(x) = \frac{1}{2}B(x, x)$. So, if $2 \in A^*$, bilinear modules (see Definition 1.17) and quadratic modules over A are really the same objects.

If $2 \notin A^*$, and 2 is not a zero-divisor in A, we can still identify quadratic forms on an A-module M with special symmetric bilinear forms, namely the forms B with $B(x, x) \in 2A$ for all $x \in M$ ("even" bilinear forms). However, if 2 is a zero-divisor in A, then quadratic and bilinear modules over A are very different objects.

In what follows, primarily *free quadratic modules* will play a rôle, i.e. quadratic modules (M, q) for which the A-module is free and always of finite rank. If M is a free A-module with basis $e_1, \dots, e_n$ and (a_{ij}) a symmetric $(n \times n)$-matrix, then there exists exactly one quadratic form q on M with $q(e_i) = a_{ii}$ and $B_q(e_i, e_j) = a_{ij}$ for $i \neq j$ $(1 \leq i, j \leq n)$, namely

$$q\left(\sum_{i=1}^{n} x_i e_i\right) = \sum_{i=1}^{n} a_{ii} x_i^2 + \sum_{1 \leq i < j \leq n} a_{ij} x_i x_j.$$

We denote this quadratic module (M, q) by a symmetric matrix in square brackets, $(M, q) = [a_{ij}]$, and call $[a_{ij}]$ the *value matrix* of q with respect to the basis $e_1, \dots, e_n$. If (a_{ij}) is a diagonal matrix with coefficients $a_1, \dots, a_n$, then we write $(M, q) = [a_1, \dots, a_n]$.

Definition 1.50. Let (M_1, q_1) and (M_2, q_2) be quadratic A-modules. The *orthogonal sum* $(M_1, q_1) \perp (M_2, q_2)$ is the quadratic A-module

$$(M_1 \oplus M_2, q_1 \perp q_2),$$

consisting of the direct sum $M_1 \oplus M_2$ and the form

$$(q_1 \perp q_2)(x_1 + x_2) := q_1(x_1) + q_2(x_2)$$

for $x_1 \in M_1$, $x_2 \in M_2$.

If $(M_1, q_1) = [A_1]$ and $(M_2, q_2) = [A_2]$ are free quadratic modules with corresponding symmetric matrices A_1, A_2, then

$$(M_1, q_1) \perp (M_2, q_2) = \begin{bmatrix} A_1 & 0 \\ 0 & A_2 \end{bmatrix}.$$

Now it is also clear how to construct a multiple orthogonal sum $(M_1, q_1) \perp \cdots \perp (M_r, q_r)$. In particular we have for elements $a_1, \ldots, a_r \in A$ that

$$[a_1] \perp \cdots \perp [a_r] = [a_1, \ldots, a_r].$$

Let (M, q) be a quadratic $\mathfrak{o}$-module and suppose that M_1 and M_2 are submodules of the $\mathfrak{o}$-module M, then we write $M = M_1 \perp M_2$, when $B_q(M_1, M_2) = 0$ ("internal" orthogonal sum, in contrast with the "external" orthogonal sum of Definition 1.50). Clearly $(M, q) \cong (M_1, q|M_1) \perp (M_2, q|M_2)$ in this case.

If $\beta : M \times M \to A$ is a—*not necessarily symmetric*—bilinear form, then $q(x) := \beta(x, x)$ is a quadratic form on M. If M is free, one can easily verify that every quadratic form on M is of this form. Furthermore, two bilinear forms β, β' give rise to the same quadratic form q exactly when $\beta - \beta' = \gamma$ is an *alternating* bilinear form: $\gamma(x, x) = 0$ for all $x \in M$, and hence $\gamma(x, y) = -\gamma(y, x)$ for all $x, y \in M$.

Suppose now that (M_1, B_1) is a free bilinear module (cf. §1.3) and that (M_2, q_2) is a free quadratic module. We equip the free module $M := M_1 \otimes_A M_2$ with a quadratic form q as follows: first we choose a bilinear form β_2 on M_2 with $q_2(x) = \beta_2(x, x)$ for all $x \in M_2$. Next we form the tensor product $\beta := B_1 \otimes \beta_2 : M \times M \to A$ of the bilinear forms B_1, β_2. This bilinear form β is characterized by

$$\beta(x_1 \otimes x_2, y_1 \otimes y_2) = B_1(x_1, y_1)\beta_2(x_2, y_2)$$

for $x_1, y_1 \in M_1$ and $x_2, y_2 \in M_2$ (cf. [10, §1, No.9]). Finally we let $q(x) := \beta(x, x)$ for $x \in M_1 \otimes M_2$. This quadratic form q is *independent* of the choice of the bilinear form β_2, since if γ_2 is an alternating bilinear form on M_2, then $B_1 \otimes \gamma_2$ is an alternating bilinear form on M.

Definition 1.51. We call q the *tensor product* of the symmetric bilinear form B_1 and the quadratic form q_2, denoted $q = B_1 \otimes q_2$, and call the quadratic module (M, q) the tensor product of the bilinear module (M_1, B_1) and the quadratic module (M_2, q_2), $(M, q) = (M_1, B_1) \otimes (M_2, q_2)$.

The quadratic form $q = B_1 \otimes q_2$ is characterized by $B_q = B_1 \otimes B_{q_2}$ and

$$q(x_1 \otimes x_2) = B_1(x_1, x_1) \, q_2(x_2)$$

for $x_1 \in M_1$, $x_2 \in M_2$. For a one-dimensional bilinear module $\langle c \rangle$ and a quadratic free module (M, q) there is a natural isometry $\langle c \rangle \otimes (M, q) \cong (M, cq)$. In particular is for a symmetric $(n \times n)$-matrix A

$$\langle c \rangle \otimes [A] \cong [cA].$$

Later on we will often denote a quadratic module (M, q) by the single letter M and an (always symmetric) bilinear module (E, B) by the single letter E. The tensor product $E \otimes M$ is clearly additive in both arguments,

$$(E_1 \perp E_2) \otimes M \cong (E_1 \otimes M) \perp (E_2 \otimes M)$$
$$E \otimes (M_1 \perp M_2) \cong (E \otimes M_1) \perp (E \otimes M_2).$$

Consequently we have for example $(a_i, b_j \in A)$:

$$\langle a_1, \ldots, a_r \rangle \otimes [b_1, \ldots, b_s] \cong [a_1 b_1, a_1 b_2, \ldots, a_r b_s].$$

Let $M = (M, q)$ be a free quadratic A-module and $\alpha : A \to C$ a ring homomorphism. We associate to M a quadratic C-module $M' = (M', q')$ as follows: the C-module M' is the tensor product $C \otimes_A M$, formed by means of α. Choose a bilinear module $\beta : M \times M \to A$ with $q(x) = \beta(x, x)$ for all $x \in M$. Let $\beta' : M' \times M' \to C$ be the bilinear form over C associated to β, in other words

$$\beta'(c \otimes x, d \otimes y) = cd \, \alpha(\beta(x, y)) \tag{1.5}$$

for $x, y \in M$, $c, d \in C$. Let $q'(u) := \beta'(u, u)$ for $u \in M'$. This quadratic form q' is independent of the choice of β', since if γ is an alternating bilinear form on M, then the associated C-bilinear form γ' on M' is again alternating. The form q' can be characterized as follows:

$$q'(c \otimes x) = c^2 \alpha(q(x)), \quad B_{q'}(c \otimes x, d \otimes y) = cd \, \alpha(B_q(x, y)),$$

for $x \in M$, $y \in M$, $c \in C$, $d \in C$.

Definition 1.52. We say that the quadratic C-module (M', q') arises from $M = (M, q)$ by means of a *base extension* determined by α, and denote M' by $C \otimes_A M$ or by $C \otimes_\alpha M$ or, even more precisely, by $C \otimes_{A,\alpha} M$. We also use the notation $q' = q_C$.

If M is given by a symmetric matrix, $M = [a_{ij}]$ then $C \otimes_A M = [\alpha(a_{ij})]$.

Given a bilinear A-module $M = (M, B)$, we similarly define a bilinear C-module $C \otimes_A M = (C \otimes_A M, B_C)$, where B_C is determined in the obvious way by B, cf. (1.5) above.

If $M = (M, q)$ is a quadratic module over A and N a submodule of M, we also interpret N as a quadratic module, $N = (N, q|N)$ (quadratic submodule). Furthermore we denote by $N^\perp$ the submodule of M consisting of all elements $x \in M$ with $B_q(x, y) = 0$ for all $y \in N$. Of course, we also interpret $N^\perp$ as a quadratic module. In particular we can look at the quadratic submodule $M^\perp$ of M. If $M^\perp$ is free of finite rank r, then $M^\perp$ has the form $[a_1, \ldots, a_r]$ with elements $a_i \in A$.

Later on we will frequently use the following elementary lemma.

Lemma 1.53 ([50, p. 5]). *Let $M = (M, B)$ be a bilinear A-module and let N be a submodule of M. Suppose that the bilinear form B is nondegenerate on N, i.e. the homomorphism $x \mapsto B(x, -)|N$ from N to the dual module $\check{N} = \text{Hom}_A(N, A)$ is an isomorphism. Then $M = N \perp N^\perp$.*

Let $M = (M, q)$ be a quadratic module and let N be a submodule of M such that the bilinear form B_q is nondegenerate on N, then according to the lemma we also have that $M = N \perp N^\perp$.

To finish this section, we briefly examine *free hyperbolic* modules, being quadratic modules of the form $r \times \begin{bmatrix} 0 & 1 \\ 1 & 0 \end{bmatrix}$ with $r \in \mathbb{N}_0$, in other words direct sums of r copies of the quadratic module $\begin{bmatrix} 0 & 1 \\ 1 & 0 \end{bmatrix}$. {When $r = 0$, the zero module is meant.}

Lemma 1.54. *Let (M, q) be a quadratic A-module whose associated bilinear form B_q is nondegenerate. Let U be a submodule of M with $q|U = 0$. Suppose that U is free of rank r and that it is a direct summand of the A-module M. Then there exists a submodule $N \supset U$ of M with $N \cong r \times \begin{bmatrix} 0 & 1 \\ 1 & 0 \end{bmatrix}$ and $M = N \perp N^\perp$.*

Proof. (cf. [6, p.13 ff.]). We saw already in §1.3 (just after Definition 1.22) that there exists an exact sequence of A-modules

$$0 \longrightarrow U^\perp \longrightarrow M \overset{\alpha}{\longrightarrow} \check{U} \longrightarrow 0,$$

with $\check{U} := \text{Hom}_A(U, A)$, where α maps an element $z \in M$ to the linear form $B(z, -)|U$ on U. Since $\check{U}$ is free, this sequence splits. Choose a submodule V of M with $M = U^\perp \oplus V$. Then $\alpha|V : V \to \check{U}$ is an isomorphism. The modules U and V are therefore in perfect duality with respect to the pairing $U \times V \to A$, $(x, y) \mapsto B_q(x, y)$.

Now choose a—not necessarily symmetric—bilinear form $\beta : V \times V \to A$ with $\beta(x, x) = q(x)$ for $x \in V$. We then have an A-linear map $\varphi : V \to U$ such that $\beta(v, x) = B_q(v, \varphi(x))$ for $v \in V$, $x \in V$. Therefore

$$q(x - \varphi(x)) = q(x) - B_q(x, \varphi(x)) = q(x) - \beta(x, x) = 0$$

for every $x \in V$. Let $W := \{x - \varphi(x) \mid x \in V\}$. Since

$$B_q(u, x - \varphi(x)) = B_q(u, x)$$

for all $u \in U$ and $x \in V$, the modules W and U are also in perfect duality with respect to B_q. Furthermore is $q|W = 0$. Choose a basis $e_1, \ldots, e_r$ for U and let $f_1, \ldots, f_r$ be the dual basis of W with respect to B_q, i.e. $B_q(e_i, f_j) = \delta_{ij}$. Then $N := U \oplus W$ can be written as

$$N = \bigperp_{i=1}^{r} (Ae_i + Af_i) \cong r \times \begin{bmatrix} 0 & 1 \\ 1 & 0 \end{bmatrix}.$$

An appeal to Lemma 1.53 yields $M = N \perp N^{\perp}$.   □

Given a quadratic A-module $M = (M, q)$, we denote $(M, -q)$ by $-M$, as is already our practice for bilinear spaces.

Theorem 1.55. *Let* $M = (M, q)$ *be a quadratic A-module. Suppose that the bilinear form B_q is nondegenerate and that M is free of rank r. Then*

$$M \perp (-M) \cong r \times \begin{bmatrix} 0 & 1 \\ 1 & 0 \end{bmatrix}.$$

Proof. Let $(E, \tilde{q}) := (M, q) \perp (M, -q)$. We interpret the module $E = M \oplus M$ as the cartesian product $M \times M$. The diagonal $D := \{(x, x) \mid x \in M\}$ is a free E-submodule of rank r and D is a direct summand of E, for $E = D \oplus (M \times \{0\})$. Furthermore $q|D = 0$. The statement now follows from Lemma 1.54.   □

1.6 Quadratic Modules over Valuation Rings

In this section we let $\mathfrak{o}$ be a valuation ring with maximal ideal $\mathfrak{m}$, quotient field K and residue class field $k = \mathfrak{o}/\mathfrak{m}$. The case $\mathfrak{m} = 0$, i.e. $K = k = \mathfrak{o}$ is explicitly allowed. We shall present the theorems about quadratic modules over $\mathfrak{o}$ necessary for our specialization theory, and prove most of them.

Lemma 1.56. *Let (M, B) be a free bilinear module over $\mathfrak{o}$, and let N be a submodule of M. The submodule $N^{\perp}$ of M is free with finite basis and is a direct summand of M. Every submodule L of M with $M = N^{\perp} \oplus L$ is also free with finite basis.*

Proof. $M/N^{\perp}$ is torsion-free and finitely generated, hence free with finite basis. This follows from the fact that every finitely generated ideal of $\mathfrak{o}$ is principal, cf. [13, VII, §4]. (We used this already in the proof of Lemma 1.25.) If $\bar{x}_1, \ldots, \bar{x}_r$ is a basis of $M/N^{\perp}$ and $x_1, \ldots, x_r$ are the pre-images of the $\bar{x}_i$ in M, then $L_0 := \mathfrak{o}x_1 + \cdots + \mathfrak{o}x_r$ is free with basis $x_1, \ldots, x_r$ and $M = N^{\perp} \oplus L_0$. If $M = N^{\perp} \oplus L$ as well, then $L \cong M/N^{\perp} \cong L_0$, so that L is also free with finite basis. Finally, $N^{\perp}$ is torsion-free and finitely generated, hence free with finite basis.   □

In particular, if (M, B) is a free bilinear module over $\mathfrak{o}$, there is an orthogonal decomposition $M = M_0 \perp M^{\perp}$ with $B|M^{\perp} \times M^{\perp} = 0$.

Definition 1.57. $M^{\perp}$ is called the *radical* of the bilinear module (M, B). If (M, q) is a free quadratic $\mathfrak{o}$-module, then the radical $M^{\perp}$ of (M, B_q), equipped with the quadratic form $q|M^{\perp}$, is called the *quasilinear part* of M, and is denoted by $QL(M)$. If $M = M^{\perp}$, the quadratic module (M, q) is called *quasilinear*.

Definition 1.58. Let M be a free $\mathfrak{o}$-module with basis $e_1, \ldots, e_n$. We call a vector $x \in M$ *primitive in* M, if for the decomposition $x = \lambda_1 e_1 + \cdots + \lambda_n e_n$ ($\lambda_i \in \mathfrak{o}$) the ideal $\sum_1^m \lambda_i \mathfrak{o}$ is equal to $\mathfrak{o}$, i.e. at least one λ_i is a unit.

This property is independent of the choice of basis: x is primitive when $M/\mathfrak{o}x$ is torsion-free, hence when $\mathfrak{o}x$ is a direct summand of the module M.

Definition 1.59. We call a quadratic module (M, q) over $\mathfrak{o}$ *nondegenerate* if it satisfies the following conditions:

(Q0) M is free of finite rank.

(Q1) The bilinear form $\overline{B}_q$, induced by B_q on $M/M^{\perp}$ in the obvious way, is nondegenerate.

(Q2) $q(x) \in \mathfrak{o}^*$ for every vector x in $M^{\perp}$, which is primitive in $M^{\perp}$ (and hence in M).

If instead of (Q2), the following condition is satisfied

(Q2') $QL(M) = 0$ or $QL(M) \cong [\varepsilon]$ with $\varepsilon \in \mathfrak{o}^*$,

then we call (M, q) *regular*. In the special case $M^{\perp} = 0$, we call (M, q) *strictly regular*.

Remark 1.60. The strictly regular quadratic modules are identical with the "quadratic spaces" over $\mathfrak{o}$ in [32]. In case $\mathfrak{o} = K$, the term "nondegenerate" has the same meaning as in [32]. "Strictly regular" is a neologism, constructed with the purpose of avoiding collision with the confusingly many overlapping terms in the literature.

From a technical point of view, Definition 1.59 is the core definition of this book. The idea behind it is that the term "nondegenerate" captures a possibly large class of quadratic $\mathfrak{o}$-modules, which work well in the specialization theory (see §1.7 and following). {We will settle upon the agreement that a quadratic K-module E has "good reduction with respect to $\mathfrak{o}$" if $E \cong K \otimes_{\mathfrak{o}} M$, where M is a nondegenerate quadratic $\mathfrak{o}$-module, see Definition 1.84.} The requirements (Q0) and (Q1) are obvious, but (Q2) and (Q2') deserve some explanation.

If char $K \neq 2$, i.e. $2 \neq 0$ in $\mathfrak{o}$, then $q|M^{\perp} = 0$ and the requirement (Q2) implies that $M^{\perp} = 0$, hence implies—in conjunction with (Q0) and (Q1)—strict regularity. If $2 \in \mathfrak{o}^*$, then the nondegenerate quadratic $\mathfrak{o}$-modules are the same objects as the nondegenerate bilinear $\mathfrak{o}$-modules in the sense of our earlier definition.

Suppose now that char $K = 2$. The condition $M^{\perp} = 0$, in other words strict regularity, is very natural, but too limited for applications. Indeed, if $M^{\perp} = 0$, then the bilinear module (M, B_q) is nondegenerate and we have $B_q(x, x) = 2q(x) = 0$ for every $x \in M$. This implies that M has even dimension, as is well known. (To prove this, consider the vector space $K \otimes_{\mathfrak{o}} M$.) So if we insist on using strict regularity, we can only deal with quadratic forms of even dimension.

On the other hand, property (Q2) has an annoying defect: it is not always preserved under a basis extension. If $\mathfrak{o}' \supset \mathfrak{o}$ is another valuation ring, whose maximal ideal $\mathfrak{m}'$ lies over $\mathfrak{m}$, i.e. $\mathfrak{m}' \cap \mathfrak{o} = \mathfrak{m}$, and if M is nondegenerate, then $\mathfrak{o}' \otimes_{\mathfrak{o}} M$ can be degenerate.

However, if M satisfies (Q2'), this clearly cannot happen. Therefore we will limit ourselves later (from §2.1 onwards) in some important cases to regular modules.

The arguments will be clearer however, when we allow arbitrary nondegenerate quadratic modules as long as possible, and a lot of results in this generality are definitely important.

Our use of the terms "regular" and "nondegenerate" finds its justification in the requirements of the specialization theory presented in this book. Regular quadratic modules over valuation rings and fields (in the sense of Definition 1.59) are just those quadratic modules, for which the generic splitting theory in particular functions in a "regular way", as we are used to in the absence of characteristic 2 (cf. §1.4), see §2.1 below. The nondegenerate quadratic modules are those ones, for which the generic splitting theory still can get somewhere in a sensible way, see §2.2 and §2.4 below.

The author is aware of the fact that in other areas (number theory in particular) a different terminology is used. For example, Martin Kneser calls our regular modules "half regular" and our strictly regular modules "regular" in his lecture notes [42], and has a good reason to do so. In the literature, the word "nondegenerate" is almost exclusively used for the more restrictive class of quadratic modules, which we call "strictly regular", see Bourbaki [10, §3, No. 4] in particular. In our context, however, it would be a little bit silly to give the label "degenerate" to every quadratic module, which is not strictly regular.

We need some formulas, related to quadratic modules of rank 2.

Lemma 1.61. *Let $\alpha \in \mathfrak{o}, \beta \in \mathfrak{o}, \lambda \in \mathfrak{o}^*$. Then*

$$\begin{bmatrix} \alpha & \lambda \\ \lambda & \beta \end{bmatrix} \cong \begin{bmatrix} \alpha & 1 \\ 1 & \lambda^{-2}\beta \end{bmatrix}, \tag{1.6}$$

and furthermore

$$\langle \lambda \rangle \otimes \begin{bmatrix} \alpha & 1 \\ 1 & \beta \end{bmatrix} \cong \begin{bmatrix} \lambda\alpha & 1 \\ 1 & \lambda^{-1}\beta \end{bmatrix}. \tag{1.7}$$

Finally, if $\alpha \in \mathfrak{o}^*, \beta \in \mathfrak{o}$ then

$$\begin{bmatrix} \alpha & 1 \\ 1 & \beta \end{bmatrix} \cong \langle \alpha \rangle \otimes \begin{bmatrix} 1 & 1 \\ 1 & \alpha\beta \end{bmatrix}. \tag{1.8}$$

Proof. Let (M, q) be a free quadratic module with basis e, f and associated value matrix $\begin{bmatrix} \alpha & \lambda \\ \lambda & \beta \end{bmatrix}$. Then $e, \lambda^{-1}f$ is also a basis of M, whose associated value matrix equals $\begin{bmatrix} \alpha & 1 \\ 1 & \lambda^{-2}\beta \end{bmatrix}$. This settles (1). Furthermore $\langle \lambda \rangle \otimes \begin{bmatrix} \alpha & 1 \\ 1 & \beta \end{bmatrix}$ is determined by the quadratic module $(M, \lambda q)$, and so

$$\langle \lambda \rangle \otimes \begin{bmatrix} \alpha & 1 \\ 1 & \beta \end{bmatrix} \cong \begin{bmatrix} \lambda\alpha & \lambda \\ \lambda & \lambda\beta \end{bmatrix}.$$

Applying (1.6) to this, results in (1.7). Clearly (1.8) is a special case of (1.7).   □

Theorem 1.62. *Let $M = (M, q)$ be a nondegenerate quadratic module over $\mathfrak{o}$.*

(a) Then M is an orthogonal sum of modules $\begin{bmatrix} \alpha & 1 \\ 1 & \beta \end{bmatrix}$ with $1 - 4\alpha\beta \in \mathfrak{o}^*$ and modules $[\varepsilon]$ with $\varepsilon \in \mathfrak{o}^*$.

(b) If M is regular and $\dim M$ even (recall that $\dim M$ denotes the rank of the free module M), then M is strictly regular and equal to an orthogonal sum of modules $\begin{bmatrix} \alpha & 1 \\ 1 & \beta \end{bmatrix}$ with $1 - 4\alpha\beta \in \mathfrak{o}^*$.

Proof. First suppose that $2 \in \mathfrak{o}^*$. Then (M, B_q) is a bilinear space. Hence (M, B_q) has an orthogonal basis, as is well-known (cf. [50, I, Cor.3.4]), and so

$$(M, B_q) \cong \langle \varepsilon_1, \ldots, \varepsilon_n \rangle.$$

Therefore $(M, q) \cong \left[\frac{\varepsilon_1}{2}, \ldots, \frac{\varepsilon_n}{2} \right]$. For a binary quadratic module $N = [a, b]$ with units $a, b \in \mathfrak{o}^*$, we have

$$[a, b] \cong \begin{bmatrix} a & 2a \\ 2a & a + b \end{bmatrix},$$

since, if e, f is a basis of the module N with value matrix $\begin{bmatrix} a & 0 \\ 0 & b \end{bmatrix}$, then $e, e + f$ is a basis with value matrix $\begin{bmatrix} a & 2a \\ 2a & a+b \end{bmatrix}$. Then (1.6) yields

$$[a, b] \cong \begin{bmatrix} a & 1 \\ 1 & (2a)^{-1}(a + b) \end{bmatrix}.$$

Hence all the assertions of the theorem are clear when $2 \in \mathfrak{o}^*$.

Next suppose that $2 \in \mathfrak{m}$. The quasilinear part $M^\perp$ is of the form $[\varepsilon_1, \ldots, \varepsilon_r]$ with $\varepsilon_i \in \mathfrak{o}^*$. Let N be a module complement of $M^\perp$ in M. Then $(N, q|N)$ is strictly regular. We will show, by induction on $\dim N$, that N is the orthogonal sum of binary modules $\begin{bmatrix} \alpha & 1 \\ 1 & \beta \end{bmatrix}$.

If $N = 0$, nothing has to be done. So suppose that $N \neq 0$. We choose a primitive vector e in N. Since the bilinear form $B_q|N \times N$ is nondegenerate, there exists a vector $f \in N$ with $B_q(e, f) = 1$. Let $\alpha = q(e)$, $\beta = q(f)$. The determinant of the matrix $\begin{pmatrix} 2\alpha & 1 \\ 1 & 2\beta \end{pmatrix}$ is the unit $1 - 4\alpha\beta$. Hence the module $\mathfrak{o}e + \mathfrak{o}f$ is free with basis e, f. Using Lemma 1.53, we get an orthogonal decomposition $N = (\mathfrak{o}e + \mathfrak{o}f) \perp N_1$. Now $\mathfrak{o}e + \mathfrak{o}f \cong \begin{bmatrix} \alpha & 1 \\ 1 & \beta \end{bmatrix}$, and by our induction hypothesis, N_1 is also an orthogonal sum of binary quadratic modules of this form.

This establishes the proof of part *(a)*. In particular, N has even dimension. If M is regular, then $r \leq 1$. Is in addition $\dim M$ even, then $r = 0$, in other words, M has to be strictly regular.   □

If M is a free $\mathfrak{o}$-module, we interpret M as a subset of the K-vecor space $K \otimes_\mathfrak{o} M$, by identifying $x \in M$ with $1 \otimes x$. Then $K \otimes_\mathfrak{o} M = KM$.

Definition 1.63. A quadratic $\mathfrak{o}$-module (M, q) is called *isotropic* if there exists a vector $x \neq 0$ in M with $q(x) = 0$. Otherwise (M, q) is called *anisotropic*.

Remark. In this definition of "isotropic", we proceed in a different way, compared to Definition 1.23, since the bilinear form B_q can be degenerate. This would cause trouble over a local ring instead of over $\mathfrak{o}$.

Lemma 1.64. *Suppose that* (M, q) *is a quadratic* $\mathfrak{o}$*-module, and that* M *is free of finite rank over* $\mathfrak{o}$. *The following assertions are equivalent:*
(a) (M, q) *is isotropic.*
(b) $(K \otimes_{\mathfrak{o}} M, q_K)$ *is isotropic.*
(c) *The module* M *has a direct summand* $V \neq 0$ *with* $q|V = 0$.

Proof. The implications $(a) \Rightarrow (b)$ and $(c) \Rightarrow (a)$ are trivial. $(b) \Rightarrow (c)$: Let $E = K \otimes_{\mathfrak{o}} M = KM$. Now $q = q_K|M$. The K-vector space E contains a subspace $W \neq 0$ with $q_K|W = 0$. Let $V := W \cap M$. The $\mathfrak{o}$-module V is a direct summand of M, since M/V is torsion-free and finitely generated, hence free. We have $KV = W$. Therefore certainly $V \neq 0$ and $q|V = 0$. $\qquad\qquad\square$

Definition 1.65. Let (M, q) be a quadratic $\mathfrak{o}$-module. A pair of vectors e, f in M is called *hyperbolic* if $q(e) = q(f) = 0$ and $B_q(e, f) = 1$.

Note that according to Lemma 1.53, we then have that

$$M = (\mathfrak{o}e + \mathfrak{o}f) \perp N \cong \begin{bmatrix} 0 & 1 \\ 1 & 0 \end{bmatrix} \perp N$$

where N is a quadratic submodule of M, namely $N = (\mathfrak{o}e + \mathfrak{o}f)^{\perp}$.

Lemma 1.66. *Let* (M, q) *be a nondegenerate quadratic* $\mathfrak{o}$*-module, and let* e *be a primitive vector in* M *with* $q(e) = 0$. *Then* e *can be completed to a hyperbolic vector pair* e, f.

Proof. We choose a decomposition $M = N \perp M^{\perp}$ and write $e = x + y$ with $x \in N$, $y \in M^{\perp}$. Suppose for the sake of contradiction that the vector x is not primitive in N, hence not primitive in M. Then y is primitive in M. According to condition (Q2), we then have $q(y) \in \mathfrak{o}^*$. Hence also $q(x) = -q(y) \in \mathfrak{o}^*$. Therefore x has to be primitive: a contradiction.

Hence x is primitive in N. Since B_q is nondegenerate on N, there exists an element $z \in N$ with $B_q(x, z) = 1$. We also have $B_q(e, z) = 1$. Clearly $f := z - q(z)e$ completes the vector e to a hyperbolic pair. $\qquad\qquad\square$

Theorem 1.67 (Cancellation Theorem). *Let* M *and* N *be free quadratic* $\mathfrak{o}$*-modules. Let* G *be a strictly regular quadratic* $\mathfrak{o}$*-module, and suppose that* $G \perp M \cong G \perp N$. *Then* $M \cong N$.

For the proof we refer to [30], in which such a cancellation theorem is proved over local rings. An even more general theorem can be found in Kneser's works [41], [42, Ergänzung zu Kap. I]. A very accessible source for many aspects of the theory of quadratic forms over local rings is the book [6] of R. Baeza. Admittedly, Baeza only treats strictly regular quadratic modules, which he calls "quadratic spaces".

In the field case $\mathfrak{o} = K$, Theorem 1.67 was already demonstrated for $2 = 0$ by Arf [3] and for $2 \neq 0$—as is very well known—by Witt [58]. A nice discussion of the cancellation problem over fields, with a view towards the theory over local rings, can again be found in Kneser's work [40].

Theorem 1.68. *Let M be a nondegenerate quadratic module over $\mathfrak{o}$. There exists a decomposition*

$$M \cong M_0 \perp r \times \begin{bmatrix} 0 & 1 \\ 1 & 0 \end{bmatrix} \tag{1.9}$$

with M_0 nondegenerate and anisotropic, $r \geq 0$. The number $r \in \mathbb{N}_0$ and the isometry class of M_0 are uniquely determined by M.

Proof. Lemma 1.64 and Lemma 1.66 immediately imply the existence of a decomposition (1.9). Suppose now that

$$M \cong M_0' \perp r' \times \begin{bmatrix} 0 & 1 \\ 1 & 0 \end{bmatrix}$$

is another decomposition. Suppose without loss of generality that $r \leq r'$. Then Theorem 1.67 implies that

$$M_0 \cong M_0' \perp (r' - r) \times \begin{bmatrix} 0 & 1 \\ 1 & 0 \end{bmatrix}.$$

Since M_0 is anisotropic, we must have $r = r'$. $\qquad\square$

In the field case $\mathfrak{o} = K$, this theorem can again be found back in Arf [3] for $2 = 0$ and in Witt [58] for $2 \neq 0$.

Definition 1.69. We call a decomposition (1.9), as in Theorem 1.68, a *Witt decomposition* of M. We call M_0 the *kernel module* of the nondegenerate quadratic module M and r the *Witt index* of M and write $M_0 = \ker(M)$, $r = \operatorname{ind}(M)$.

Furthermore, we call two nondegenerate quadratic modules M and N over $\mathfrak{o}$ *Witt equivalent*, denoted by $M \sim N$, when $\ker(M) \cong \ker(N)$. We denote the Witt class of M, i.e. the equivalence class of M with respect to $\sim$, by $\{M\}$.

It is now tempting to define an addition of Witt classes $\{M\}$, $\{N\}$ of two nondegenerate quadratic modules M, N over $\mathfrak{o}$ in the same way as we did this for nondegenerate bilinear modules over fields in §1.2, namely $\{M\} + \{N\} := \{M \perp N\}$.

Although the orthogonal sum of two nondegenerate quadratic modules could be degenerate, the addition makes sense if one of the modules M, N is strictly regular, as we will show now.

Proposition 1.70. *Let M, M' be strictly regular quadratic $\mathfrak{o}$-modules with $M \sim M'$, and let N, N' be nondegenerate quadratic $\mathfrak{o}$-modules with $N \sim N'$. Then the modules $M \perp N$ and $M' \perp N'$ are nondegenerate and $M \perp N \sim M' \perp N'$.*

Proof. It follows immediately from Definition 1.59 that $M \perp N$ and $M' \perp N'$ are nondegenerate. Suppose without loss of generality that $\dim M' \geq \dim M$. Then $M' \cong M \perp s \times \begin{bmatrix} 0 & 1 \\ 1 & 0 \end{bmatrix}$ for an $s \geq 0$. Therefore $M' \perp N \cong (M \perp N) \perp s \times \begin{bmatrix} 0 & 1 \\ 1 & 0 \end{bmatrix}$, and so $M' \perp N \sim M \perp N$. A similar argument shows that $M' \perp N \sim M' \perp N'$. $\square$

Definition 1.71. We denote the set of Witt classes of nondegenerate quadratic $\mathfrak{o}$-modules by $\widetilde{Wq}(\mathfrak{o})$. We denote the subset of Witt classes of regular, resp. strictly regular quadratic $\mathfrak{o}$-modules by $Wqr(\mathfrak{o})$, resp. $Wq(\mathfrak{o})$.

Because of the proposition above, we now have a well-defined "addition" of classes $\{M\} \in Wq(\mathfrak{o})$ with classes $\{N\} \in \widetilde{Wq}(\mathfrak{o})$,

$$\{M\} + \{N\} := \{M \perp N\}.$$

The zero module $M = 0$ gives rise to a neutral element for this addition, $\{0\} + \{N\} = \{N\}$, $\{M\} + \{0\} = \{M\}$. If M and N are strictly regular, then $\{M\} + \{N\} = \{N\} + \{M\} = \{M \perp N\} \in Wq(\mathfrak{o})$. Therefore $Wq(\mathfrak{o})$ is an abelian semigroup with respect to this sum. Theorem 1.55 shows that $Wq(\mathfrak{o})$ is in fact an abelian group, since if (M, q) is a strictly regular quadratic $\mathfrak{o}$-module, then $\{(M, q)\} + \{(M, -q)\} = 0$.

Definition 1.72. We call the abelian group $Wq(\mathfrak{o})$ the *quadratic Witt group* of $\mathfrak{o}$, the set $\widetilde{Wq}(\mathfrak{o})$ the *quadratic Witt set* of $\mathfrak{o}$ and $Wqr(\mathfrak{o})$ the *regular quadratic Witt set* of $\mathfrak{o}$.

The group $Wq(\mathfrak{o})$ acts on the set $\widetilde{Wq}(\mathfrak{o})$ by means of the addition, introduced above. One can show easily that the isomorphism classes of nondegenerate quasi-linear quadratic $\mathfrak{o}$-modules (see Definition 1.57) form a system of representatives of the orbits of this action. $Wqr(\mathfrak{o})$ is a union of orbits.

If $\mathfrak{m} = 0$, hence $\mathfrak{o} = K$, then Witt classes of *degenerate* quadratic modules can be defined without difficulties. We will explain this next.

Definition 1.73. Let $M = (M, q)$ be a finite dimensional quadratic module over K. The *defect* $\delta(M)$ of M is the set of all $x \in M^\perp$ with $q(x) = 0$.

The defect $\delta(M)$ is clearly a subspace of M, and M is nondegenerate if and only if $\delta(M) = 0$. The form q induces a quadratic form $\overline{q} : M/\delta(M) \to K$ on the vector space $M/\delta(M)$ in the obvious way: $\overline{q}(\overline{x}) := q(x)$ for $x \in M$, where $\overline{x}$ is the image of x in $M/\delta(M)$. The quadratic K-module $(M/\delta(M), \overline{q})$ is clearly nondegenerate.

Definition 1.74. If $\delta(M) = 0$, i.e. if M is nondegenerate, then we call $M = (M, q)$ a *quadratic space* over K. In general we denote the quadratic space $(M/\delta(M), \overline{q})$ over K by $\widehat{M}$, and call $\widehat{M}$ the *quadratic space associated to* M.

We have

$$M \cong \widehat{M} \perp \delta(M) \cong \widehat{M} \perp s \times [0]$$

with $s := \dim \delta(M)$. Furthermore, given a quadratic space E over K, we obviously have

$$\delta(E \perp M) = \delta(M) \quad \text{and} \quad \widehat{E \perp M} \cong E \perp \widehat{M}.$$

Now it is clear that in the field case Theorem 1.68 can be expanded as follows:

Theorem 1.75. *Let M be a finite dimensional quadratic K-module. There exists a decomposition*

$$M \cong M_0 \perp r \times \begin{bmatrix} 0 & 1 \\ 1 & 0 \end{bmatrix}$$

with $\widehat{M_0}$ anisotropic and $r \geq 0$. The number $r \in \mathbb{N}_0$ and the isometry class of M_0 are uniquely determined by M.

Definition 1.76. We call M_0 the *kernel module* of M and r the *Witt index* of M, and write $M_0 = \ker M$, $r = \mathrm{ind}(M)$. Just as before, we call two finite dimensional quadratic K-modules M and N *Witt equivalent*, and write $M \sim N$, when $\ker(M) \cong \ker(N)$. We denote the Witt equivalence class of M by $\{M\}$.

Remark. It may seem more natural to call M and N equivalent when their "kernel spaces" $\widehat{\ker M}$, $\widehat{\ker N}$ are isometric. Our results in §1.7 and §2.3 are a bit stronger if we use the finer equivalence defined above.

We denote the set of Witt classes $\{M\}$ of finite dimensional quadratic K-modules M by $\widehat{Wq}(K)$ and call this the *defective quadratic Witt set* of K. The abelian group $Wq(K)$ acts on $\widehat{Wq}(K)$ in the usual way: $\{E\} + \{M\} := \{E \perp M\}$ for $\{E\} \in Wq(K)$, $\{M\} \in \widehat{Wq}(K)$. It even makes sense to add two arbitrary Witt classes together:

$$\{M\} + \{N\} := \{M \perp N\}.$$

The result is independent of the choice of representatives M, N, as can be shown by an argument, analogous to the proof Proposition 1.70. Therefore $\widehat{Wq}(K)$ is an abelian semigroup, having $Wq(K)$ as a subgroup.

Caution! If char $K = 2$, the semigroup $\widehat{Wq}(K)$ does not satisfy a cancellation rule. Is for example $a \in K^*$, then $[a] \perp [a] \cong [0] \perp [a]$, but it is *not* so that $[a] \sim [0]$.

To a field homomorphism $\alpha : K \to K'$ (i.e. a field extension), we can associate a well-defined map $\alpha_* : \widehat{Wq}(K) \to \widehat{Wq}(K')$, which sends the Witt class $\{M\}$ to the class $\{K' \otimes_\alpha M\}$. This map is a semigroup homomorphism. By restricting α_*, we can find maps from $Wq(K)$ to $Wq(K')$ and from $Wqr(K)$ to $Wqr(K')$, but (if char $K = 2$) in general *no* map from $\widetilde{Wq}(K)$ to $\widetilde{Wq}(K')$. This is the reason why we will sometimes extend the Witt set $\widetilde{Wq}(K)$ to the semigroup $\widehat{Wq}(K)$.

Let us now return to quadratic modules over arbitrary valuation rings.

At the end of §1.5 we introduced free hyperbolic modules. From now on, we call those modules simply "hyperbolic". {In this way, we stay in harmony with the terminology, used in the literature, for valuation rings (or local rings in general), cf. [8, V §2], [9].} Hence:

Definition 1.77. A quadratic o-module M is called *hyperbolic* if

$$M \cong r \times \begin{bmatrix} 0 & 1 \\ 1 & 0 \end{bmatrix}$$

for some $r \in \mathbb{N}_0$.

So, if M is not the zero module, this means that M contains a basis $e_1, f_1, e_2, f_2,$ $\dots, e_r, f_r$, where e_i, f_i are hyperbolic vector pairs, such that $\mathfrak{o}e_i + \mathfrak{o}f_i$ is orthogonal with $\mathfrak{o}e_j + \mathfrak{o}f_j$ for $i \neq j$. Clearly a hyperbolic module is strictly regular.

Lemma 1.78. *Let $M = (M, q)$ be a strictly regular quadratic $\mathfrak{o}$-module. M is hyperbolic if and only if M contains a submodule V with $q|V = 0$ and $V = V^{\perp}$.*

Proof. If M is hyperbolic and $e_1, f_1, \dots, e_r, f_r$ is a basis, consisting of pairwise orthogonal hyperbolic vector pairs, then the module $V := \mathfrak{o}e_1 + \mathfrak{o}e_2 + \dots + \mathfrak{o}e_r$ clearly has the properties $q|V = 0$ and $V = V^{\perp}$. Suppose now that M is strictly regular and that V is a submodule satisfying these properties. Since $M/V^{\perp} = M/V$ is torsion-free and finitely generated, it is free. Therefore V is a direct summand of the $\mathfrak{o}$-module M. V is torsion-free and finitely generated as well, and so also free. The form B_q gives rise to an exact sequence

$$0 \longrightarrow V^{\perp} \longrightarrow M \longrightarrow \check{V} \longrightarrow 0$$

(cf. the proof of Lemma 1.54), from which we deduce that $\dim M = \dim V + \dim V^{\perp} = 2 \dim V$.

According to Lemma 1.54 there is an orthogonal decomposition $M = N \perp N^{\perp}$ with $V \subset N$ and N hyperbolic. The module $V^{\perp}$ built in N instead of M also coincides with V. We conclude that $\dim N = 2 \dim V$ as well, so that $N = M$. $\qquad\square$

Theorem 1.79. *Let M be a strictly regular quadratic $\mathfrak{o}$-module with $K \otimes_{\mathfrak{o}} M$ hyperbolic. Then M itself is hyperbolic.*

Proof. We apply the criterion given by Lemma 1.78 twice. As before we interpret M as an $\mathfrak{o}$-submodule of $E := K \otimes_{\mathfrak{o}} M$, i.e. $M \subset E$, $E = KM$. Now E contains a subspace W with $q|W = 0$, $W^{\perp} = W$. The intersection $V := W \cap M$ is an $\mathfrak{o}$-submodule of M with $q|V = 0$ and $V^{\perp} = V$. (Here $V^{\perp}$ is considered in M.) $\qquad\square$

Definition 1.80. The valuation ring $\mathfrak{o}$ is called *quadratically henselian* if for every $\gamma \in \mathfrak{m}$, there exists a $\lambda \in \mathfrak{o}$ such that $\lambda^2 - \lambda = \gamma$.

Remark. One can convince oneself that this means that "Hensel's Lemma" is satisfied by monic polynomials of degree 2. Every henselian valuation ring is of course quadratically henselian (cf. [53, Chap. F], [16, §16]). In Definition 1.80 we isolated that part of the property "henselian", which is important for the theory of quadratic forms. Note that, if $\mathfrak{o}$ is quadratically henselian, then every $\varepsilon \in 1 + 4\mathfrak{m}$ is a square in $\mathfrak{o}$, since $\varepsilon = 1 + 4\gamma$ and $\gamma = \lambda^2 - \lambda$ imply that $\varepsilon = (1 - 2\gamma)^2$.

Lemma 1.81. *Let $\mathfrak{o}$ be quadratically henselian. Let $\alpha, \beta \in K$ and $\alpha\beta \in \mathfrak{m}$. Then we have that over K:*

$$\begin{bmatrix} \alpha & 1 \\ 1 & \beta \end{bmatrix} \cong \begin{bmatrix} 0 & 1 \\ 1 & 0 \end{bmatrix}.$$

Proof. According to Lemma 1.61, formula (1.8), we have

$$\begin{bmatrix} \alpha & 1 \\ 1 & \beta \end{bmatrix} \cong \langle \alpha \rangle \otimes \begin{bmatrix} 1 & 1 \\ 1 & \alpha\beta \end{bmatrix}.$$

There exists a $\lambda \in \mathfrak{o}$ with $\lambda^2 + \lambda + \alpha\beta = 0$. If e, f is a basis of the module $\begin{bmatrix} 1 & 1 \\ 1 & \alpha\beta \end{bmatrix}$, having the indicated value matrix, then $q(\lambda e + f) = \lambda^2 + \lambda + \alpha\beta = 0$. Therefore $\begin{bmatrix} \alpha & 1 \\ 1 & \beta \end{bmatrix}$ is isotropic. Applying Lemma 1.66, with $\mathfrak{o} = K$, gives $\begin{bmatrix} \alpha & 1 \\ 1 & \beta \end{bmatrix} \cong \begin{bmatrix} 0 & 1 \\ 1 & 0 \end{bmatrix}$. $\square$

Theorem 1.82. *Let $\mathfrak{o}$ be quadratically henselian. Let (M, q) be a nondegenerate anisotropic quadratic module over $\mathfrak{o}$. Then $q(e)$ is a unit for every vector $e \in M$ which is primitive in M.*

Proof. We choose a decomposition $M = N \perp M^\perp$. Let $e \in M$ be primitive. Suppose for the sake of contradiction that $q(e) \in \mathfrak{m}$. We write $e = x + y$ with $x \in N$, $y \in M^\perp$, and have $q(e) = q(x) + q(y)$.

Case 1: x is primitive in M, hence also in N. Then there exists an $f \in N$ with $B_q(x, f) = 1$, since B_q is nondegenerate on N. We have $B_q(e, f) = 1$. Therefore

$$\mathfrak{o}e + \mathfrak{o}f \cong \begin{bmatrix} \alpha & 1 \\ 1 & \beta \end{bmatrix}$$

with $\alpha = q(e) \in \mathfrak{m}$, $\beta = q(f) \in \mathfrak{o}$. According to Lemma 1.81, $\mathfrak{o}e + \mathfrak{o}f$ is isotropic. This is a contradiction, since M is anisotropic.

Case 2: x is not primitive in M. Now $x = \lambda x_0$ with x_0 primitive in M, $\lambda \in \mathfrak{m}$. The vector y has to be primitive in M. Since q is nondegenerate, it follows that $q(y) \in \mathfrak{o}^*$. We have

$$q(e) = \lambda^2 q(x_0) + q(y) \in \mathfrak{o}^*,$$

contradicting the assumption that $q(e) \in \mathfrak{m}$.

We conclude that $q(e) \in \mathfrak{o}^*$. $\square$

Later on, we will use Lemma 1.81 and Theorem 1.82 for a nondegenerate quadratic module M over an arbitrary valuation ring $\mathfrak{o}$, by going from M to the quadratic module $M^h := \mathfrak{o}^h \otimes_\mathfrak{o} M$ over the henselization $\mathfrak{o}^h$ of $\mathfrak{o}$ (see [53, Chap. F], [16, §17]). {Remark: One can also form a "quadratic henselization" $\mathfrak{o}^{qh}$ in the obvious way. It would suffice to take $\mathfrak{o}^{qh}$ instead of $\mathfrak{o}^h$.} In order to do so, the following lemma is important.

Lemma 1.83. *Let M be a nondegenerate quadratic module over $\mathfrak{o}$. Then the quadratic module M^h over $\mathfrak{o}^h$ is also nondegenerate.*

Proof. The properties (Q0) and (Q1) from Definition 1.59 stay valid for every basis extension. Hence we suppose that M satisfies (Q0) and (Q1), so that we only have to consider (Q2). We have $(M^h)^\perp = (\mathfrak{o}^h \otimes_\mathfrak{o} M)^\perp = \mathfrak{o}^h \otimes_\mathfrak{o} M^\perp = (M^\perp)^h$.

The property (Q2) for M says that the quasilinear space $k \otimes_\mathfrak{o} M^\perp = k \otimes_\mathfrak{o} QL(M)$ is anisotropic over k. To start with, we can identify $k \otimes_\mathfrak{o} M$ with the quadratic module $M/\mathfrak{m}M$ over $k = \mathfrak{o}/\mathfrak{m}$, whose quadratic form $\bar{q} : M/\mathfrak{m}M \to k$ is induced by the quadratic form $q : M \rightarrow o$ in the obvious way. In the same way we then have $k \otimes_o M^\perp = M^\perp/mM^\perp$. A vector $x \in M^\perp$ is primitive if and only if its image $\bar{x} \in M^\perp/mM^\perp$ is nonzero. (Q2) implies then that $\bar{q}(\bar{x}) \neq 0$.

Let m^h denote the maximal ideal of o^h. It is well-known that $m^h \cap o = m$ and that the residue class field o^h/m^h is canonically isomorphic to $o/m = k$. We identify o^h/m^h with k. Then $k \otimes_{o^h} M^h = k \otimes_o M$ and also $k \otimes_{o^h} QL(M^h) = k \otimes_o QL(M)$. Therefore (Q2) is satisfied for M if and only if (Q2) is satisfied for M^h. $\square$

1.7 Weak Specialization

In this section $\lambda : K \rightarrow L \cup \infty$ is a place, $o = o_\lambda$ is the valuation ring associated to λ, with maximal ideal m and residue class field $k = o/m$. Given a quadratic module $E = (E, q)$ over K, we want to associate a quadratic module $\lambda_*(E)$ over L to it, in so far this is possible in a meaningful way.

In order to do this, we limit ourselves to the case where E is nondegenerate (see Definition 1.59, but with K instead of o). This is not a real limitation, however. Since K is a field, every quadratic module $E = (E, q)$ of finite dimension over K has the form $E \cong F \perp [0, \ldots, 0] = F \perp s \times [0]$ with F nondegenerate (see §1.6). So, when we have associated a specialization $\lambda_*(F)$ to F in a satisfactory way, it is natural to put $\lambda_*(E) = \lambda_*(F) \perp s \times [0]$. We stay with nondegenerate quadratic K-modules, however, and now call them **quadratic spaces** over K, as agreed in Definition 1.74.

Definition 1.84. Let E be a quadratic space over K. We say that E has *good reduction with respect to λ* when $E \cong K \otimes_o M$, where M is a <u>nondegenerate</u> quadratic o-module.

This is the "good case". We would like to put $\lambda_*(E) := L \otimes_\lambda M$, where the tensor product is formed by means of the homomorphism $\lambda|o : o \rightarrow M$, see Definition 1.52. This tensor product can more easily be described as follows: First we go from the quadratic o-module $M = (M, q)$ to the quadratic space $M/mM = (M/mM, \bar{q})$ over $k = o/m$, where $\bar{q}$ is obtained from $q : M \rightarrow o$ in the obvious way. Next we extend scalars by means of the field extension $\bar{\lambda} : k \hookrightarrow L$, determined by λ, thus obtaining $\lambda_*(E) = L \otimes_k (M/mM)$.

Is this meaningful? With a similar argument as towards the end of §1.3 (second proof of Theorem 1.15), we now show that the answer is affirmative when E is *strictly regular* (which means that $E^\perp = 0$, cf. Definition 1.59).

Theorem 1.85. *Let E be strictly regular and let M, M' be nondegenerate quadratic o-modules with $E \cong K \otimes_o M \cong K \otimes_o M'$. Then $M \cong M'$, and hence $M/mM \cong M'/mM'$.*

Proof. Clearly $M^\perp = 0$ and $(M')^\perp = 0$. Hence M and M' are also strictly regular. By Theorem 1.55 the space $K \otimes_o [M' \perp (-M)] \cong E \perp (-E)$ is hyperbolic. So $M' \perp (-M)$ is itself hyperbolic, by Theorem 1.79. $M \perp (-M)$ is also hyperbolic, again by Theorem 1.55. The modules $M' \perp (-M)$ and $M \perp (-M)$ have the same rank. Hence $M' \perp (-M) \cong M \perp (-M)$, and so $M' \cong M$ by the Cancellation Theorem (Theorem 1.67).   □

The problem remains to see that $\lambda_*(E)$ does not depend on the choice of the module M, when E has good reduction with respect to λ, but is only nondegenerate. Furthermore, in case of bad reduction (i.e. not good reduction), we would neverthe-less like to associate a Witt class $\lambda_W(\{E\})$ over L to E in a meaningful way, just as for bilinear forms in §1.3 ("weak specialization"). Neither problem can be dealt with as in §1.3, in the first place because the Witt classes do not constitute an additive group this time. Thus we have to seek out a new path.

Lemma 1.86. *Let $\mathfrak{o}$ be quadratically henselian (see Definition 1.80) and let $V = (V, q)$ be an* anisotropic *quadratic module over K.*
(a) The sets

$$\mu(V) := \{x \in V \mid q(x) \in \mathfrak{o}\} \quad and \quad \mu_+(V) := \{x \in V \mid q(x) \in \mathfrak{m}\}$$

are $\mathfrak{o}$-submodules of V.
(b) For any $x \in \mu(V)$ and $y \in \mu_+(V)$, we have $q(x + y) - q(x) \in \mathfrak{m}$ and $B_q(x, y) \in \mathfrak{m}$.

Proof. (a) Let $x \in \mu(V)$ (resp. $x \in \mu_+(V)$) and $\lambda \in \mathfrak{o}$, then clearly $\lambda x \in \mu(V)$ (resp. $\mu_+(V)$). So we only have to show that for any $x, y \in \mu(V)$ (resp. $\mu_+(V)$), we have $x + y \in \mu(V)$ (resp. $\mu_+(V)$). Let $B := B_q$.

Let $x \in \mu(V)$, $y \in \mu(V)$. Suppose for the sake of contradiction that $x + y \notin \mu(V)$, i.e. $q(x + y) = q(x) + q(y) + B(x, y) \notin \mathfrak{o}$. Since $q(x) \in \mathfrak{o}$, $q(y) \in \mathfrak{o}$, we must have that $B(x, y) \notin \mathfrak{o}$. Hence $\lambda := B(x, y)^{-1} \in \mathfrak{m}$. The vectors $x, \lambda y$ give rise to the value matrix $\begin{bmatrix} q(x) & 1 \\ 1 & \lambda^2 q(y) \end{bmatrix}$. According to Lemma 1.81, the space $Kx + Ky = Kx + K\lambda y$ is isotropic. Contradiction, since V is anisotropic. Therefore $x + y \in \mu(V)$.

Next, let $x \in \mu_+(V)$, $y \in \mu_+(V)$. Suppose again for the sake of contradiction that $x + y \notin \mu_+(V)$. We just showed that $q(x + y) \in \mathfrak{o}$. By our assumption $q(x + y) \notin \mathfrak{m}$, so that $q(x+y) \in \mathfrak{o}^*$. Since $q(x) \in \mathfrak{m}$, $q(y) \in \mathfrak{m}$, we have $B(x, y) \in \mathfrak{o}^*$. Write $B(x, y) = \lambda^{-1}$ with $\lambda \in \mathfrak{o}^*$. Again, the vectors $x, \lambda y$ give rise to the value matrix $\begin{bmatrix} q(x) & 1 \\ 1 & \lambda^2 q(y) \end{bmatrix}$. Just as before, the space $Kx + Ky$ is isotropic, according to Lemma 1.81. Contradiction! We conclude that $x + y \in \mu_+(V)$.

(b) Suppose for the sake of contradiction that there exist vectors $x \in \mu(V)$, $y \in \mu_+(V)$ with $q(x + y) - q(x) \notin \mathfrak{m}$. We showed above that $q(x + y) \in \mathfrak{o}$. Hence $q(x + y) - q(x) = q(y) + B(x, y) \in \mathfrak{o}$. By assumption, this element doesn't live in $\mathfrak{m}$ and is thus a unit. Since $q(y) \in \mathfrak{m}$, we also have $B(x, y) \in \mathfrak{o}^*$. As before, we write $\lambda = B(x, y)^{-1}$ and observe that $Kx + Ky$ is isotropic, giving a contradiction. Therefore $q(x + y) - q(x) \in \mathfrak{m}$ and $B(x, y) = q(x + y) - q(x) - q(y) \in \mathfrak{m}$ as well.   □

We remain in the situation of Lemma 1.86. For $x \in \mu(V)$, $\lambda \in \mathfrak{m}$, we have $\lambda x \in \mu_+(V)$. Therefore

$$\rho(V) := \mu(V)/\mu_+(V)$$

is a k-vector space in a natural sense ($k = \mathfrak{o}/\mathfrak{m}$). We define a function $\overline{q} : \rho(V) \to k$ as follows:

$$\bar{q}(\bar{x}) := \overline{q(x)} \qquad (x \in \mu(V)),$$

where $\bar{x}$ denotes the image of $x \in \mu(V)$ in $\rho(V)$ and $\bar{a}$ denotes the image of $a \in \mathfrak{o}$ in k. Lemma 1.86 tells us that the map $\bar{q}$ is well-defined.

For $x \in \mu(V)$, $a \in \mathfrak{o}$, we have $\bar{q}(\bar{a}\,\bar{x}) = \bar{q}(\overline{ax}) = \overline{q(ax)} = \bar{a}^2\overline{q(x)} = \bar{a}^2\bar{q}(\bar{x})$. According to Lemma 1.86, the bilinear form $B := B_q$ has values in $\mathfrak{o}$ on $\mu(V) \times \mu(V)$ and values in $\mathfrak{m}$ on $\mu(V) \times \mu_+(V)$. Therefore, it induces on $\rho(V)$ a symmetric bilinear form $\bar{B}$ over k with $\bar{B}(\bar{x}, \bar{y}) = \overline{B(x, y)}$ for $x, y \in \mu(V)$. A very simple calculation now shows that

$$\bar{q}(\bar{x} + \bar{y}) - \bar{q}(\bar{x}) - \bar{q}(\bar{y}) = \bar{B}(\bar{x}, \bar{y})$$

for $x, y \in \mu(V)$. This furnishes the proof that $\bar{q}$ is a quadratic form on the k-vector space $\rho(V)$ with $B_{\bar{q}} = \bar{B}$. The quadratic k-module $(\rho(V), \bar{q})$ is clearly anisotropic.

Definition 1.87. ($\mathfrak{o}$ quadratically henselian.) We call the quadratic k-module $\rho(V) = (\rho(V), \bar{q})$ the *reduction* of the anisotropic quadratic K-module V with respect to the valuation ring $\mathfrak{o}$.

In order to associate to a quadratic space E over K, by means of the place $\lambda :$ $K \to L \cup \infty$, a Witt class $\lambda_W\{E\}$ over L, the following path presents itself now: Let $\mathfrak{o}^h$ be the henselization of the valuation ring $\mathfrak{o} = \mathfrak{o}_\lambda$, $\mathfrak{m}^h$ the maximal ideal of $\mathfrak{o}^h$ and K^h the quotient field of $\mathfrak{o}^h$. The residue class field $\mathfrak{o}^h/\mathfrak{m}^h$ is canonically isomorphic to $k = \mathfrak{o}/\mathfrak{m}$ and will be identified with k. Let

$$\lambda_W\{E\} = \{L \otimes_{\bar{\lambda}} \rho(\mathrm{Ker}\,(K^h \otimes E))\},$$

where, as before, $\bar{\lambda} : k \hookrightarrow L$ is the field embedding, determined by λ. The space over L on the right-hand side can then be considered to be a "weak specialization" of E with respect to λ.

All good and well, if only we knew whether the vector space

$$\rho(\mathrm{Ker}\,(K^h \otimes E))$$

has finite dimension! To guarantee this, we have to confine the class of allowed quadratic modules E.

In the following, $v : K \to \Gamma \cup \infty$ is a surjective valuation, associated to the valuation ring $\mathfrak{o}$, thus with Γ the value group of v, $\Gamma \cong K^*/\mathfrak{o}^*$. We use additive notation for Γ (so $v(xy) = v(x) + v(y)$ for $x, y \in K$).

Already in §1.3, we agreed to view the square class group $Q(\mathfrak{o})$ as a subgroup of $Q(K)$, and also to regard the elements of $Q(K)$ as one-dimensional bilinear spaces[9] over K.

Now we choose a complement Σ of $Q(\mathfrak{o})$ in $Q(K)$, in other words, a subgroup Σ of $Q(K)$ with $Q(K) = Q(\mathfrak{o}) \times \Sigma$. This is possible, since $Q(K)$ is elementary abelian of exponent 2, i.e. a vector space over the field with 2 elements. Further, we choose, for every square class $\sigma \in \Sigma$, an element $s \in \mathfrak{o}$ with $\sigma = \langle s \rangle$. For $\sigma = 1$ we choose the representative $s = 1$. Let S be the set of all elements s. For every $a \in K^*$ there exists

---

[9] Strictly speaking, square classes are *isomorphism classes* of one-dimensional spaces.

exactly one $s \in S$ and elements $\varepsilon \in \mathfrak{o}^*$, $b \in K^*$ with $a = s\varepsilon b^2$. Since $K^*/\mathfrak{o}^* \cong \Gamma$, it is clear that S (resp. Σ) is a system of representatives of $\Gamma/2\Gamma$ in K^* (resp. $Q(K)$) for the homomorphism from K^* to $\Gamma/2\Gamma$ (resp. from $Q(K)$ to $\Gamma/2\Gamma$), determined by $v : K^* \rightarrow \Gamma$.

Definition 1.88. A quadratic space E over K is called *obedient with respect to λ* (or *obedient with respect to $\mathfrak{o}$*) if there exists a decomposition

$$E \cong \perp_{s \in S} \langle s \rangle \otimes (K \otimes_{\mathfrak{o}} M_s), \tag{1.10}$$

where M_s is a nondegenerate quadratic $\mathfrak{o}$-module. (Of course $M_s \neq 0$ for finitely many s only.)

Clearly this property does not depend on the choice of system of representatives S.

Remark.

(1) Obedience is much weaker than the property "good reduction" (Definition 1.84).
(2) Let E and F be quadratic spaces over K, obedient with respect to $\mathfrak{o}$. If at least one of them is strictly regular, then $E \perp F$ is obedient.
(3) If char $K \neq 2$, then every quadratic space over K is strictly regular. Hence the orthogonal sum of two obedient quadratic spaces over K is again obedient.
(4) If $2 \in \mathfrak{o}^*$, i.e. char $L \neq 2$, then *every* quadratic space E over K is obedient with respect to λ, since K also has characteristic $\neq 2$ in this case. E has a decomposition in one-dimensional spaces, which are clearly obedient.
(5) If char $K = 0$, but char $L = 2$, then every quadratic space of odd dimension over K is *disobedient* ($=$ not obedient) with respect to λ. To see this, let E be an obedient quadratic space with orthogonal decomposition (1.10), as in Definition 1.88. Then the space $K \otimes_{\mathfrak{o}} M_i$ has quasilinear part $(K \otimes_{\mathfrak{o}} M_i)^{\perp} = 0$ for every $i \in \{1, \ldots, r\}$. Therefore M_i also has quasilinear part $M_i^{\perp} = 0$, and is thus strictly regular. Hence the space $M_i/\mathfrak{m}M_i$ over $k = \mathfrak{o}/\mathfrak{m}$ is strictly regular, and so must have even dimension. We conclude that E must have even dimension.

Let us now give an example of a two-dimensional disobedient space over a field K of characteristic 2.

First, recall the definition of the *Arf-invariant* $\mathrm{Arf}(\varphi)$ of a strictly regular form φ over a field k of characteristic 2. For $x \in k$, let $\wp(x) := x^2 + x$. Further, let $\wp(k)$ denote the set of all $\wp(x)$, $x \in k$. This is a subgroup of k^+, i.e. k regarded as an additive group. We choose a decomposition

$$\varphi \cong \begin{bmatrix} a_1 & 1 \\ 1 & b_1 \end{bmatrix} \perp \ldots \perp \begin{bmatrix} a_m & 1 \\ 1 & b_m \end{bmatrix} \tag{1.11}$$

and set

$$\mathrm{Arf}(\varphi) := a_1 b_1 + \cdots + a_m b_m + \wp(k) \in k^+/\wp(k).$$

It is well-known that $\mathrm{Arf}(\varphi)$ is independent of the choice of the decomposition (1.11) ([3], [55, Chap.IX. §4]).

Example 1.89. Let k be a field of characteristic 2 and $K = k(t)$ the rational function field in one variable t over k. Further, let $\mathfrak{o}$ be the discrete valuation ring of K with respect to the prime polynomial t in $k[t]$, i.e. $\mathfrak{o} = k[t]_{(t)}$.
Claim. The quadratic space $\begin{bmatrix} 1 & 1 \\ 1 & t^{-1} \end{bmatrix}$ over K is disobedient with respect to $\mathfrak{o}$.

Proof. Suppose for the sake of contradiction that this space is obedient. Then there exist a strictly regular space M over $\mathfrak{o}$ and an element $u \in K^*$ with $\begin{bmatrix} 1 & 1 \\ 1 & t^{-1} \end{bmatrix} \cong \langle u \rangle \otimes (K \otimes_{\mathfrak{o}} M)$. By Theorem 1.62 we have $M \cong \begin{bmatrix} \alpha & 1 \\ 1 & \beta \end{bmatrix}$ with $\alpha, \beta \in \mathfrak{o}$. Hence, by Lemma 1.61 (1.8), the following holds over K:

$$\begin{bmatrix} 1 & 1 \\ 1 & t^{-1} \end{bmatrix} \cong \langle \alpha u \rangle \begin{bmatrix} 1 & 1 \\ 1 & \alpha\beta \end{bmatrix}. \tag{1.12}$$

Comparing Arf-invariants of both sides, shows that there exists an element $a \in K$ with

$$t^{-1} = \alpha\beta + a^2 + a. \tag{1.13}$$

Hence there exists an element $a \in K$ with $t^{-1} + a^2 + a \in \mathfrak{o}$.

If we now move to the formal power series field $k((t)) \supset K$, we easily see that such an element a does not exist: for if $a = \sum_{r \geq d} c_r t^r$ with $d \in \mathbb{Z}$ and coefficients $c_r \in k$, $c_d \neq 0$, then

$$\sum_{r \geq d} c_r^2 t^{2r} + \sum_{r \geq d} c_r t^r + t^{-1} \in k[[t]],$$

and so $d < 0$. The term of lowest degree in the first sum on the left is $c_d^2 t^{2d}$. It cannot be compensated by other summands on the left. Therefore $t^{-1} + a^2 + a \notin \mathfrak{o}$, and the space $\begin{bmatrix} 1 & 1 \\ 1 & t^{-1} \end{bmatrix}$ is disobedient. $\square$

In this proof we used the Arf-invariant. We remark that we can use easier aids: anisotropic quadratic forms like $\begin{bmatrix} 1 & 1 \\ 1 & c \end{bmatrix}$ are norm forms of separable quadratic field extensions, and are as such "multiplicative". Therefore we can deduce immediately from (1.12) that $\begin{bmatrix} 1 & 1 \\ 1 & t^{-1} \end{bmatrix} \cong \begin{bmatrix} 1 & 1 \\ 1 & \alpha\beta \end{bmatrix}$ and then that there exists a relation (1.13).

Remark. Hitherto the quasilinear parts of free quadratic modules have played a predominantly negative role. This will continue to be the case. Rather often they cause a lot of complications compared with the theory of nondegenerate bilinear modules. But sometimes quasilinear quadratic modules can do good things. Suppose again that $K = k(t)$, $\mathfrak{o} = k[t]_{(t)}$ as in the example above. The space $\begin{bmatrix} 1 & 1 \\ 1 & t^{-1} \end{bmatrix}$ is disobedient with respect to $\mathfrak{o}$. However, $\begin{bmatrix} 1 & 1 \\ 1 & t^{-1} \end{bmatrix} \perp [t^{-1}]$ is obedient, since this space over K is isometric to $\begin{bmatrix} 1 & 1 \\ 1 & 0 \end{bmatrix} \perp [t^{-1}]$, and hence to $\begin{bmatrix} 0 & 1 \\ 1 & 0 \end{bmatrix} \perp [t]$.

In connection with the definition of obedience (Definition 1.88), we agree upon some more terminology. If $E = (E, q)$ is a quadratic space, obedient with respect to $\mathfrak{o}$, then we have an (internal) orthogonal decomposition

$$E = \perp_{s \in S} E_s$$

of the following kind: every vector space E_s contains a free $\mathfrak{o}$-submodule M_s with $E_s = KM_s$, such that $q|E_s = sq_s$ for a quadratic form $q_s : E_s \to K$, which takes values in $\mathfrak{o}$ on M_s and for which $(M_s, q_s|M_s)$ is a nondegenerate quadratic module. (Of course $E_s \neq 0$ for finitely many $s \in S$ only.)

Definition 1.90. We call such a decomposition $E = \underset{s \in S}{\perp} E_s$, together with a choice of modules $M_s \subset E_s$, a λ-modular, or also, an $\mathfrak{o}$-*modular representation* of the obedient quadratic space E.

Lemma 1.91. *Suppose again that $\mathfrak{o}$ is quadratically henselian and $E = (E,q)$ is a quadratic space over K, obedient with respect to $\mathfrak{o}$. Let $E = \underset{s \in S}{\perp} E_s$, $E_s = KM_s$ be an $\mathfrak{o}$-modular representation of E. Suppose furthermore that E is anisotropic. Then we have, with the notation of Lemma 1.86,[10]*

$$\mu(E) = M_1 \perp \underset{s \neq 1}{\perp} \mu_+(E_s),$$

$$\mu_+(E) = \mathfrak{m}M_1 \perp \underset{s \neq 1}{\perp} \mu_+(E_s).$$

Proof. As above, we set $q|E_s = sq_s$. Let x be a vector of E with $x \neq 0$. Then there exist finitely many pairwise distinct elements $s_1, \ldots, s_r$ in S, as well as *primitive* vectors $x_i \in M_i := M_{s_i}$ and scalars $a_i \in K^*$ $(1 \leq i \leq r)$, such that $x = \sum_{i=1}^{r} a_i x_i$. We have

$$q(x) = \sum_{i=1}^{r} a_i^2 s_i q_i(x_i),$$

where we used the abbreviation $q_i := q_{s_i}$. According to Theorem 1.82, we have $q_i(x_i) \in \mathfrak{o}^*$ for every $i \in \{1, \ldots, r\}$. Hence, for $i \neq j$, the values $v(a_i^2 s_i q_i(x_i)) = v(a_i^2 s_i)$ and $v(a_j^2 s_j q_j(x_j)) = v(a_j^2 s_j)$ are different. Let $v(a_k^2 s_k)$ be the smallest value among the $v(a_i^2 s_i)$, $1 \leq i \leq r$. Then $v(q(x)) = v(a_k^2 s_k) = v(q(a_k x_k))$. This shows that

$$\mu(E) = \underset{s \in S}{\perp} \mu(E_s), \quad \mu_+(E) = \underset{s \in S}{\perp} \mu_+(E_s).$$

For $s \neq 1$, $a \in K^*$ and primitive $y \in E_s$, we have $q(ay) = sa^2 q_s(y) \notin \mathfrak{o}^*$. Therefore, $\mu(E_s) = \mu_+(E_s)$ for every $s \in S \setminus \{1\}$. We still have to determine the modules $\mu(E_1)$ and $\mu_+(E_1)$. So, let $x \in E_1$, $x \neq 0$. We write $x = ay$ with $a \in K^*$ and primitive $y \in M_1$. Then $q(x) = a^2 q(y)$ and $q(y) \in \mathfrak{o}^*$. Hence $q(x) \in \mathfrak{o}$ exactly when $a \in \mathfrak{o}$, and $q(x) \in \mathfrak{m}$ exactly when $a \in \mathfrak{m}$. Therefore we conclude that $\mu(E_1) = M_1$ and $\mu_+(E_1) = \mathfrak{m}M_1$. $\square$

An immediate consequence of the lemma is

---

[10] In accordance with our earlier agreement, every E_s is considered as a quadratic subspace of E, $E_s = (E_s, q|E_s)$.

Theorem 1.92. *Let* $\mathfrak{o}$ *be quadratically henselian, and let* $E = (E, q)$ *be an anisotropic quadratic space over* K, *obedient with respect to* $\mathfrak{o}$. *Let* $E = \perp_{s \in S} E_s$, $E_s = KM_s$ *be an* $\mathfrak{o}$-*modular representation of* E. *Then we have for the reduction* $\rho(E) = (\rho(E), \overline{q})$ *of* E *with respect to* $\mathfrak{o}$:

$$(\rho(E), \overline{q}) \cong (M_1/\mathfrak{m}M_1, \overline{q}_1),$$

where $\overline{q}_1 : M_1/\mathfrak{m}M_1 \to k$ *is the quadratic form over* k, *determined by* $q|M_1 : M_1 \to \mathfrak{o}$ *in the obvious way.*   □

Furthermore, the lemma tells us that $M_1 = \mu(E_1)$ and $\mathfrak{m}M_1 = \mu_+(E_1)$. In particular we have

Theorem 1.93. *Let* $\mathfrak{o}$ *be quadratically henselian and let* $E = (E, q)$ *be an anisotropic quadratic space over* K, *having good reduction with respect to* $\mathfrak{o}$. *Then* $\mu(E) = (\mu(E), q|\mu(E))$ *is a nondegenerate quadratic* $\mathfrak{o}$-*module with* $E = K \otimes_{\mathfrak{o}} \mu(E)$, *and* $\mu_+(E) = \mathfrak{m}\mu(E)$. *Note that* $\mu(E)$ *is the only nondegenerate quadratic* $\mathfrak{o}$-*module* M *in* E, *with* $E = KM$.   □

We tone down Definition 1.90 as follows:

Definition 1.94. Let $E = (E, q)$ be a quadratic space over K, obedient with respect to λ. A λ-*modular* (or $\mathfrak{o}$-*modular*) *decomposition of* E is an orthogonal decomposition $E = \perp_{s \in S} E_s$, in which every space $\langle s \rangle \otimes E_s = (E_s, s \cdot (q|E_s))$ has good reduction with respect to λ.

If $\mathfrak{o}$ is quadratically henselian and E is anisotropic then, according to Theorem 1.93, every λ-modular decomposition corresponds to *exactly one* λ-modular representation of E.

Lemma 1.95. *Let* $\mathfrak{o}$ *be quadratically henselian. Let* $s_1, \ldots, s_r$ *be different elements of* S *and* $M_1, \ldots, M_r$ *anisotropic nondegenerate quadratic modules over* $\mathfrak{o}$. *Then*

$$E := \perp_{i=1}^{r} \langle s_i \rangle \otimes (K \otimes_{\mathfrak{o}} M_i)$$

is an anisotropic quadratic space over K.

Proof. Let a primitive vector $x_i \in M_i$ be given for every $i \in \{1, \ldots, r\}$. Also, let $a_1, \ldots, a_r \in K$ be scalars with

$$\sum_{i=1}^{r} s_i a_i^2 q_i(x_i) = 0, \tag{1.14}$$

where q_i denotes the quadratic form on M_i. We must show that all $a_i = 0$.

Suppose for the sake of contradiction that this is not so. After renumbering the M_i, we can assume without loss of generality that for a certain $t \in \{1, \ldots, r\}$, $a_i \neq 0$ for $1 \leq i \leq t$ and $a_i = 0$ for $t < i \leq r$. By Theorem 1.82 $q(x_i) \in \mathfrak{o}^*$ for every $i \in \{1, \ldots, t\}$. The values $v(s_i a_i^2 q_i(x)) = v(s_i a_i^2)$ with $1 \leq i \leq t$ are pairwise different. Since this contradicts equation (1.14), all $a_i = 0$.   □

We arrive at the main theorem of this chapter.

Theorem 1.96. *Let E be a quadratic space over K, obedient with respect to $\mathfrak{o}$. Let*

$$E = \underset{s \in S}{\perp} E_s = \underset{s \in S}{\perp} F_s$$

be two $\mathfrak{o}$-modular decompositions of E, and also let M_1, N_1 be nondegenerate quadratic modules with $E_1 \cong K \otimes_{\mathfrak{o}} M_1$, $F_1 \cong K \otimes_{\mathfrak{o}} N_1$. Then the quadratic spaces $M_1/\mathfrak{m}M_1$ and $N_1/\mathfrak{m}N_1$ over $k = \mathfrak{o}/\mathfrak{m}$ are Witt equivalent.

Proof. (a) For every $s \in S \setminus \{1\}$ we choose nondegenerate quadratic $\mathfrak{o}$-modules M_s, N_s with $E_s \cong \langle s \rangle \otimes (K \otimes_{\mathfrak{o}} M_s)$, $F_s \cong \langle s \rangle \otimes (K \otimes_{\mathfrak{o}} N_s)$. Suppose first that $\mathfrak{o}$ is quadratically henselian. In accordance with Theorem 1.68 we choose Witt decompositions

$$M_s = M_s^{\circ} \perp i_s \times \begin{bmatrix} 0 & 1 \\ 1 & 0 \end{bmatrix}, \quad N_s = N_s^{\circ} \perp j_s \times \begin{bmatrix} 0 & 1 \\ 1 & 0 \end{bmatrix},$$

with M_s°, N_s° anisotropic nondegenerate quadratic $\mathfrak{o}$-modules. By Lemma 1.95 above, the nondegenerate quadratic K-modules

$$U := \underset{s \in S}{\perp} \langle s \rangle \otimes (K \otimes_{\mathfrak{o}} M_s^{\circ}), \quad V := \underset{s \in S}{\perp} \langle s \rangle \otimes (K \otimes_{\mathfrak{o}} N_s^{\circ})$$

are anisotropic. Now,

$$E \cong U \perp \left(\sum_{s \in S} i_s \right) \times \begin{bmatrix} 0 & 1 \\ 1 & 0 \end{bmatrix} \cong V \perp \left(\sum_{s \in S} j_s \right) \times \begin{bmatrix} 0 & 1 \\ 1 & 0 \end{bmatrix}.$$

Hence $U \cong V$. According to Theorem 1.92, we have $\rho(U) \cong M_1^{\circ}/\mathfrak{m}M_1^{\circ}$, $\rho(V) \cong N_1^{\circ}/\mathfrak{m}N_1^{\circ}$. Therefore, $M_1^{\circ}/\mathfrak{m}M_1^{\circ} \cong N_1^{\circ}/\mathfrak{m}N_1^{\circ}$. Furthermore,

$$M_1/\mathfrak{m}M_1 \cong M_1^{\circ}/\mathfrak{m}M_1^{\circ} \perp i_1 \times \begin{bmatrix} 0 & 1 \\ 1 & 0 \end{bmatrix}, \quad N_1/\mathfrak{m}N_1 \cong N_1^{\circ}/\mathfrak{m}N_1^{\circ} \perp j_1 \times \begin{bmatrix} 0 & 1 \\ 1 & 0 \end{bmatrix}.$$

We conclude that $M_1/\mathfrak{m}M_1 \sim N_1/\mathfrak{m}N_1$.

(b) Suppose next that $\mathfrak{o}$ is arbitrary. We go from $(K, \mathfrak{o})$ to the henselization $(K^h, \mathfrak{o}^h)$. (By definition, K^h is the quotient field of the henselian valuation ring $\mathfrak{o}^h$.) As is well-known, the valuation $v : K \to \Gamma \cup \infty$ extends uniquely to a valuation $v^h : K^h \to \Gamma \cup \infty$, which therefore again has Γ as value group. Also, $\mathfrak{o}^h$ is the valuation ring of v^h. Consequently, S is also a system of representatives of $Q(K^h)/Q(\mathfrak{o}^h) \cong \Gamma/2\Gamma$.

Given a free quadratic module M over $\mathfrak{o}$, we set—as before—$M^h := \mathfrak{o}^h \otimes_{\mathfrak{o}} M$ and for a free quadratic module U over K, we set $U^h := K^h \otimes_K U$. By Lemma 1.83, the quadratic $\mathfrak{o}^h$-modules M_s^h, N_s^h are nondegenerate. Furthermore, $E^h = \underset{s \in S}{\perp} E_s^h = \underset{s \in S}{\perp} F_s^h$ and $E_s^h \cong K^h \otimes_{\mathfrak{o}^h} M_s^h$, $F_s^h \cong K^h \otimes_{\mathfrak{o}^h} N_s^h$ for every $s \in S$.

By the above, the quadratic spaces $M_1^h/\mathfrak{m}^h M_1^h$ and $N_1^h/\mathfrak{m}^h N_1^h$ over k are equivalent. However, these spaces can be canonically identified with $M_1/\mathfrak{m}M_1$ and

$N_1/\mathfrak{m}N_1$ (cf. the end of the proof of Lemma 1.83). We conclude that $M_1/\mathfrak{m}M_1 \sim N_1/\mathfrak{m}N_1$.　　　　　　　　　　　　　　　　　　　　　　　　　　　□

Definition 1.97. Let E be a quadratic space over K, obedient with respect to λ. If $E = \underset{s \in S}{\perp} E_s$ is a λ-modular decomposition of E and M_1 a nondegenerate quadratic $\mathfrak{o}$-module with $E_1 \cong K \otimes_\mathfrak{o} M_1$, then we call the quadratic space $L \otimes_\lambda M_1$ a *weak specialization of E with respect to λ.* (As before, $\otimes_\lambda$ denotes a base extension with respect to the homomorphism $\lambda|_\mathfrak{o} : \mathfrak{o} \to L$.) By Theorem 1.96, the space $L \otimes_\lambda M_1$ is uniquely determined by E and λ, up to Witt equivalence. We denote its Witt class by $\lambda_W(E)$, i.e.

$$\lambda_W(E) := \{L \otimes_\lambda M_1\}.$$

("W" as in "Witt" or "weak".)

If char $K = 2$, then $M_1/\mathfrak{m}M_1$ is a quadratic space over k, to be sure. Nevertheless $L \otimes_\lambda M_1 = L \otimes_{\overline{\lambda}} (M_1/\mathfrak{m}M_1)$ can be degenerate. In this case, we only have $\lambda_W(E) \in \widehat{Wq}(L)$ (see §1.6, from Definition 1.76 onwards). If char $K \neq 2$, this cannot happen since $M_1/\mathfrak{m}M_1$ is strictly regular in this case. So now, $\lambda_W(E) \in Wq(L)$.

Theorem 1.98. *If E and E' are quadratic spaces over K, obedient with respect to λ, and if $E \sim E'$, then $\lambda_W(E) = \lambda_W(E')$.*

Proof. Suppose without loss of generality that $\dim E \leq \dim E'$. Then $E' \cong E \perp r \times \begin{bmatrix} 0 & 1 \\ 1 & 0 \end{bmatrix}$ for a certain $r \in \mathbb{N}_0$. If we choose a nondegenerate $\mathfrak{o}$-module M_1 for E, as in Definition 1.97, then $M_1' := M_1 \perp r \times \begin{bmatrix} 0 & 1 \\ 1 & 0 \end{bmatrix}$ is a possibe choice for E'. (Here we consider $\begin{bmatrix} 0 & 1 \\ 1 & 0 \end{bmatrix}$ over $\mathfrak{o}$ instead of over K.) Therefore, $L \otimes_\lambda M_1' \sim L \otimes_\lambda M_1$.　　□

Example 1.99. Let $K = k(t_1, \ldots, t_n)$ be the rational function field in n variables $t_1, \ldots, t_n$ over an arbitrary field k. For every multi-index $\alpha = (\alpha_1, \ldots, \alpha_n) \in \mathbb{N}_0^n$, let t^α denote the monomial $t_1^{\alpha_1} \ldots t_n^{\alpha_n}$. We order the abelian group $\mathbb{Z}^n$ lexicographically and then have exactly one valuation $v : K \to \mathbb{Z}^n \cup \infty$ with $v(t^\alpha) = \alpha$ for every $\alpha \in \mathbb{N}_0^n$. Its valuation ring contains the field k, and the residue class field $\mathfrak{o}/\mathfrak{m}$ coincides with k. Let $\lambda : K \to k \cup \infty$ be the canonical place corresponding to $\mathfrak{o}$. Furthermore, let A be the set of all multi-indices $(\alpha_1, \ldots, \alpha_n)$ with $\alpha_i \in \{0, 1\}$ for every i, in other words $A = \{0, 1\}^n \subset \mathbb{N}_0^n$. As system of representatives S, in the above sense, we take the set of t^α with $\alpha \in A$.

Suppose now that we are given a family $(F_\alpha \mid \alpha \in A) = \mathcal{F}$ of 2^n quadratic spaces over k. We construct the space

$$E := \underset{\alpha \in A}{\perp} \langle t^\alpha \rangle \otimes (K \otimes_k F_\alpha)$$

over K. For every $\alpha \in A$, $\langle t^\alpha \rangle \otimes E$ is obedient with respect to λ and

$$\lambda_W(\langle t^\alpha \rangle \otimes E) = \{F_\alpha\}.$$

Hence the family $\mathcal{F}$ can be recovered from the space E, up to Witt equivalence. If all F_α are anisotropic, we even get the F_α back from E as kernel spaces of the Witt classes $\lambda_W(\langle t^\alpha \rangle \otimes E)$. In this case, E is anisotropic as well. This can easily be seen, using an argument similar to the one used in the proof of Lemma 1.95. One could say that the family $\mathcal{F}$ is "stored" in the space E over $k((t_1, \ldots, t_n))$.

This simple example is reminiscent of Springer's Theorem in §1.1. Without our theory of weak specialization, we can use Springer's Theorem and induction on n to show that the F_α are uniquely determined by E (at least when char $k \neq 2$). This is so, because the valuation v is "n-fold discrete". For more complicated value groups, we cannot fall back on Springer's Theorem for specialization arguments.

We return to the situation of arbitrary places $\lambda : K \to k \cup \infty$.

Definition 1.100.

(a) We denote the sets of Witt classes $\{E\}$ of obedient quadratic spaces (with respect to $\mathfrak{o}$) and obedient strictly regular quadratic spaces (with respect to $\mathfrak{o}$) by $\widetilde{Wq}(K, \mathfrak{o})$ and $Wq(K, \mathfrak{o})$ respectively. These are thus subsets of the sets $\widetilde{Wq}(K)$ and $Wq(K)$, introduced in Definition 1.72.

(b) We define a map $\lambda_W : \widetilde{Wq}(K, \mathfrak{o}) \to \widetilde{Wq}(L)$ by setting $\lambda_W(\{E\}) := \lambda_W(E)$. This map is well defined by Theorem 1.98.

Clearly $Wq(K, \mathfrak{o})$ is a subgroup of the abelian group $Wq(K)$. The group $Wq(K, \mathfrak{o})$ acts on the set $\widetilde{Wq}(K, \mathfrak{o})$ by restriction of the action of $Wq(K)$ on $\widetilde{Wq}(K)$, explained in §1.6. If char $K \neq 2$, then, of course, $\widetilde{Wq}(K, \mathfrak{o}) = Wq(K, \mathfrak{o})$, and λ_W maps $Wq(K, \mathfrak{o})$ to $Wq(L)$.

Remark. Let E and F be quadratic spaces over K, obedient with respect to λ and suppose that E is strictly regular. Obviously we then have

$$\lambda_W(E \perp F) = \lambda_W(E) + \lambda_W(F).$$

Therefore, restricting the map $\lambda_W : \widetilde{Wq}(K, \mathfrak{o}) \to \widehat{Wq}(L)$ gives rise to a homomorphism from $Wq(K, \mathfrak{o})$ to $Wq(L)$, which we also denote by λ_W. The map $\lambda_W : \widehat{Wq}(K, \mathfrak{o}) \to \widehat{Wq}(L)$ is equivariant with respect to the homomorphism $\lambda_W : Wq(K, \mathfrak{o}) \to Wq(L)$.

In §1.3, we got a homomorphism $\lambda_W : W(K) \to W(L)$, using a different method. Given a bilinear space $E = (E, B)$ over K, does there exist a description of the Witt class $\lambda_W(\{E\})$, analogous to our current Definition 1.94?

First of all, when $2 \in \mathfrak{o}^*$, i.e. char $L \neq 2$, we find complete harmony between §1.3 and §1.6. In this case, bilinear spaces over K and L are the same objects as quadratic spaces, and every such space over K is obedient with respect to λ.

Theorem 1.101. *If* char $L \neq 2$, *the homomorphism* $\lambda_W : W(K) \to W(L)$ *from §1.3 coincides with the homomorphism* $\lambda_W : Wq(K) \to Wq(L)$, *defined just now.*

Proof. It suffices to show that for a one-dimensional space $[a] = \langle a \rangle$, the element $\lambda_W(\{\langle 2a \rangle\})$ from §1.3 coincides with the currently defined Witt class $\lambda_W([a])$. If $a \in \mathfrak{o}^*$, then $\lambda_W(\{\langle 2a \rangle\}) = \{\langle 2\lambda(a) \rangle\} = \{[\lambda(a)]\}$, according to §1.3. However, if $\langle a \rangle \notin Q(\mathfrak{o})$, then $\langle 2a \rangle \notin Q(\mathfrak{o})$ and $\lambda_W(\{\langle 2a \rangle\}) = 0$. We obtain the same values for $\lambda_W([a])$, using Definition 1.97 above. $\square$

And so, when $\operatorname{char} L \neq 2$, the work performed hitherto gives us a new—more conceptual—proof of Theorem 1.12.

What is the situation when $\operatorname{char} L = 2$? First, we encounter in all generality definitions for bilinear spaces, analogous to Definitions 1.84 and 1.94.

Definition 1.102. Let $E = (E, B)$ be a bilinear space over K.

(a) E has *good reduction with respect to* λ (or *with respect to* $\mathfrak{o}$) when $E \cong K \otimes_{\mathfrak{o}} M$ for a bilinear space M over $\mathfrak{o}$. {Note: this is just a translation of the original definition of "good reduction" of §1.1, i.e. Definition 1.14, in geometric language.}
(b) A λ-*modular* (or $\mathfrak{o}$-*modular*) *decomposition of* E is an orthogonal decomposition $E = \underset{s \in S}{\perp} E_s$, in which every space $\langle s \rangle \otimes E_s$ has good reduction with respect to λ.

In contrast to the quadratic case, *every* bilinear space E over K has λ-modular decompositions. This is obvious, since E is the orthogonal sum of one-dimensional bilinear spaces and copies of the space $\left(\begin{smallmatrix} 0 & 1 \\ 1 & 0 \end{smallmatrix} \right)$, see §1.2.

Let $\lambda_W : W(K) \to W(L)$ be the map, introduced in §1.3. When E is a bilinear space over K, we set $\lambda_W(E) := \lambda_W(\{E\}) \in W(L)$.

Theorem 1.103. *Let E be a bilinear space over K and let $E = \underset{s \in S}{\perp} E_s$ be a λ-modular decomposition of E. Also, let M_1 be a bilinear space over $\mathfrak{o}$ with $E_1 \cong K \otimes_{\mathfrak{o}} M_1$. Then*

$$\lambda_W(E) = \{L \otimes_{\lambda} M_1\}.$$

Proof. We have $\lambda_W(E) = \sum_{s \in S} \lambda_W(E_s)$. By Theorem 1.92, we know that $\lambda_W(E_1) = \{L \otimes_{\lambda} M_1\}$. It remains to show that $\lambda_W(E_s) = 0$ for every $s \in S$ with $s \neq 1$. Hence we have to show that if $\langle s \rangle \in Q(K)$ is a square class, not in $Q(\mathfrak{o})$, and if N is a bilinear space over $\mathfrak{o}$, then $\lambda_W(\langle s \rangle \otimes (K \otimes_{\mathfrak{o}} N)) = 0$.

By Lemma 1.16, N has an orthogonal basis when the bilinear space $N/\mathfrak{m}N$ over k is not hyperbolic. In this case we have

$$\langle s \rangle \otimes (K \otimes_{\mathfrak{o}} N) \cong \langle s\varepsilon_1, \ldots, s\varepsilon_r \rangle$$

with units $\varepsilon_i \in \mathfrak{o}^*$. By definition of λ_W, we have $\lambda_W(\langle s\varepsilon_i \rangle) = 0$ for all $i \in \{1, \ldots, r\}$, hence $\lambda_W(\langle s \rangle \otimes (K \otimes_{\mathfrak{o}} N)) = 0$. If $N/\mathfrak{m}N$ is hyperbolic, we construct the space $N' := N \perp \langle 1 \rangle$ over $\mathfrak{o}$. Since N' has an orthogonal basis, we get $\lambda_W(\langle s \rangle \otimes (K \otimes_{\mathfrak{o}} N')) = 0$. Since $\lambda_W(\langle s \rangle \otimes \langle 1 \rangle) = 0$ as well, we conclude that $\lambda_W(\langle s \rangle \otimes (K \otimes N)) = 0$. $\square$

Theorem 1.103 allows us to speak of "weak specialization" of bilinear spaces, in analogy with Definition 1.97 for quadratic spaces.

Definition 1.104. Let E be a bilinear space over K. Also, let $E = \underset{s \in S}{\perp} E_s$ be a λ-modular decomposition of E and M_1 a bilinear space over $\mathfrak{o}$ with $E_1 \cong K \otimes_{\mathfrak{o}} M_1$. Then we call the space $L \otimes_{\lambda} M_1$ a *weak specialization* of the space E with respect to λ.

According to Theorem 1.103, a weak specialization of E with respect to λ is completely determined by E and λ up to Witt equivalence. One could ask if this can be shown in a direct—geometric—way, comparable to how we have done this for obedient quadratic spaces above. Reversing direction, this would give a new proof of Theorem 1.12, also when char $L = 2$.

We leave this question open. After all, we have a proof of Theorem 1.12, and with it a weak specialization theory for bilinear spaces. This theory is more satisfying than the corresponding theory for quadratic spaces, in that we do not have to demand obedience of the spaces.

We can now also make a statement about Problem 1.5(b). Let char $L = 2$ and let char $K = 0$. Let (E, q) be a quadratic space, obedient with respect to λ. Associated to it, we have the bilinear space (E, B_q). It may be that (E, q) and (E, B_q) are in principle the same object, but it does make a difference whether we weakly specialize E as a quadratic or a bilinear space. What is better?

Let (M, q) be a regular (= strictly regular) quadratic $\mathfrak{o}$-module. According to Theorem 1.62, we have a decomposition

$$(M, q) \cong \begin{bmatrix} a_1 & 1 \\ 1 & b_1 \end{bmatrix} \perp \cdots \perp \begin{bmatrix} a_r & 1 \\ 1 & b_r \end{bmatrix}.$$

(Note: $(M/\mathfrak{m}M, \overline{q})$ is strictly regular, and thus has even dimension.) Then

$$(M, B_q) \cong \begin{pmatrix} 2a_1 & 1 \\ 1 & 2b_1 \end{pmatrix} \perp \cdots \perp \begin{pmatrix} 2a_r & 1 \\ 1 & 2b_r \end{pmatrix}.$$

Hence,

$$L \otimes_\lambda (M, q) \cong \begin{bmatrix} \lambda(a_1) & 1 \\ 1 & \lambda(b_1) \end{bmatrix} \perp \cdots \perp \begin{bmatrix} \lambda(a_r) & 1 \\ 1 & \lambda(b_r) \end{bmatrix}.$$

However, $L \otimes_\lambda (M, B_q) \cong r \times \begin{pmatrix} 0 & 1 \\ 1 & 0 \end{pmatrix}$.

Therefore, a weak specialization of the space (E, B_q) is always hyperbolic and so gives hardly any information about (E, q). However, a weak specialization of (E, q) with respect to λ can give an interesting result.

Final Consideration. A last word about the central Definition 1.88 of an obedient quadratic space. It seems obvious to formally weaken it, by considering quadratic *modules* E instead of spaces E over K, subject only to the requirements that, as a K-vector space, E should have finite dimension and a decomposition (1.10) as presented there. But then E is already nondegenerate. We namely have

$$QL(E) \cong \underset{s \in S}{\perp} \langle s \rangle \otimes (K \otimes_\mathfrak{o} QL(M_s)),$$

and as in the proof of Lemma 1.95, we see that $QL(E)$ is anisotropic (without the requirement that $\mathfrak{o}$ is quadratically henselian). Thus these "obedient quadratic modules" are the same objects as those given by Definition 1.88.

1.8 Good Reduction

As before, we let $\lambda : K \to L \cup \infty$ be a place, $\mathfrak{o} = \mathfrak{o}_\lambda$ the valuation ring of K, $\mathfrak{m}$ its maximal ideal, $k = \mathfrak{o}/\mathfrak{m}$ its residue class field and v a valuation on K, associated to $\mathfrak{o}$, with value group Γ. Also, let $\Sigma \subset Q(K)$ be a complement of $Q(\mathfrak{o})$ in $Q(K)$ and S a system of representatives of Σ in K^* (with $1 \in S$), as introduced in §1.7.

From now on, we call a nondegenerate quadratic $\mathfrak{o}$-module a *quadratic space over* $\mathfrak{o}$. In case $\mathfrak{o} = K$, we already used this terminology in §1.7. Recall further the concept of a *bilinear space* over $\mathfrak{o}$ (Definition 1.21).

If E is a quadratic (resp. bilinear) space over K then, according to Definitions 1.84 and 1.102, E has *good reduction with respect to* λ (or: with respect to $\mathfrak{o}$) if there exists a quadratic (resp. bilinear) space M over $\mathfrak{o}$ with $E \cong L \otimes_\mathfrak{o} M$.

Theorem 1.105. *Let E be a quadratic or bilinear space over K, which has good reduction with respect to λ, and let M, M' be quadratic (resp. bilinear) spaces over $\mathfrak{o}$ with $E \cong K \otimes_\mathfrak{o} M \cong K \otimes_\mathfrak{o} M'$. Then the k-spaces $M/\mathfrak{m}M$ and $M'/\mathfrak{m}M'$ are isometric in the quadratic case and stably isometric[11] in the bilinear case. Therefore we also have, $L \otimes_\lambda M \cong L \otimes_\lambda M'$ resp. $L \otimes_\lambda M \approx L \otimes_\lambda M'$.*

Proof. By the theory of §1.7 (Theorems 1.96 and 1.103), $M/\mathfrak{m}M$ and $M'/\mathfrak{m}M'$ are Witt equivalent. Since these spaces have the same dimension, namely $\dim E$, they are isometric resp. stably isometric. $\square$

In the following, the words "good reduction" are used so often that it is appropriate to introduce an abbreviation. *From now on, we mostly write* GR *instead of "good reduction".*

Definition 1.106. Let E be a quadratic or bilinear space over K, which has GR with respect to λ, and let M be a quadratic, resp. bilinear, space over $\mathfrak{o}$ with $E \cong K \otimes_\mathfrak{o} M$. Then we denote the quadratic, resp. bilinear, module $L \otimes_\lambda M$ by $\lambda_*(E)$ and call it *"the" specialization of E with respect to* λ.

In the bilinear case, $\lambda_*(E)$ is nondegenerate, and thus a space. This is true in the quadratic case as well, as long as char $K \neq 2$. If char $K = 2$ however, $\lambda_*(E)$ can be degenerate (cf. our discussion about $\lambda_W(E)$ in §1.7 after Definition 1.97).

This terminology is convenient, but sloppy. According to Theorem 1.105, $\lambda_*(E)$ is only determined by E up to isometry in the quadratic case, and in the bilinear case even only up to stable isometry. Anyway, in what follows, we are interested in quadratic spaces (or modules) only up to isometry, and in bilinear spaces almost always only up to stable isometry.

Note that in the bilinear case, Definition 1.106 is only a translation of Definition 1.18 in geometric language.

Remark 1.107. Let F and G be bilinear spaces over K, which have GR with respect to λ. Then $F \perp G$ clearly also has GR with respect to λ, and

---

[11] See §1.2 for the term "stably isometric".

$$\lambda_*(F \perp G) \approx \lambda_*(F) \perp \lambda_*(G).$$

For quadratic spaces, we have to realize that (when char $K = 2$) the orthogonal sum of two spaces over $\mathfrak{o}$ can possibly be degenerate and thus need not again be a space. However, if F and G are quadratic spaces, which have GR with respect to λ, *and if F is strictly regular*, then $F \perp G$ has GR with respect to λ and

$$\lambda_*(F \perp G) \cong \lambda_*(F) \perp \lambda_*(G).$$

For further applications, the following theorem is of the utmost importance.

Theorem 1.108.

(a) *Let F and G be bilinear spaces over K. If the spaces F and $F \perp G$ have GR with respect to λ, then G also has GR with respect to λ.*

(b) *Let F and G be quadratic spaces over K. Suppose that F is strictly regular. If F and $F \perp G$ have GR with respect to λ, then G also has GR with respect to λ.*

Proof. Part (a) has already been proved in Theorem 1.26. Upon replacing every occurrence of "metabolic" by "hyperbolic", the argument there also yields a proof of part (b). (Use Theorem 1.68 and Lemma 1.64.) □

Corollary 1.109. *Let E and F be bilinear (resp. quadratic) spaces with $E \sim F$. If E has GR with respect to λ, then so has F, and $\lambda_*(E) \sim \lambda_*(F)$. Besides, if $E \approx F$ (resp. $E \cong F$), then $\lambda_*(E) \approx \lambda_*(F)$ (resp. $\lambda_*(E) \cong \lambda_*(F)$).*

This was already established in §1.3 for the bilinear case. The quadratic case can be proved similarly.

In part (b) of Theorem 1.108, the assumed strict regularity of F is essential.

Example 1.110. Let char $K = 2$ and let E be a three-dimensional quadratic space over K with basis e, f, g and corresponding value matrix $\begin{bmatrix} 1 & 1 \\ 1 & a \end{bmatrix} \perp [\varepsilon]$, where $a \in \mathfrak{o}$ and $\varepsilon \in \mathfrak{o}^*$. Let $F := E^\perp = Kg$ and $G := Ke + K(f + cg) \cong \begin{bmatrix} 1 & 1 \\ 1 & a+c^2\varepsilon \end{bmatrix}$ for some $c \in K$. We have $E = F \perp G$. The quadratic spaces E and F have good reduction, but G can have bad reduction because the possibility exists that the Arf-invariant $a + c^2\varepsilon + \wp K$, $\wp K := \{x^2 + x \mid x \in K\}$ does not contain an element of $\mathfrak{o}$.

For instance, let $K = k(t)$ where k is an imperfect field of characteristic 2 and t is an indeterminate. As in the example of a disobedient space, §1.7, let $\mathfrak{o} = k[t]_{(t)}$. We choose $\varepsilon \in k^*$ such that ε is not a square in k. Furthermore, we choose $a = 0$, $c = t^{-2}$. We have $\varepsilon t^{-2} + x^2 + x \notin \mathfrak{o}$ for *every* $x \in K$. This is immediately clear when we look at the power series expansions in $k((t))$: Let $x = \sum_{n \geq d} c_n t^n$ for some $d \in \mathbb{Z}$, all $c_n \in k$ and $c_d \neq 0$. If it was true that $\varepsilon t^{-2} + x^2 + x \in \mathfrak{o}$, then we would have

$$\varepsilon t^{-2} + \sum_{n \geq d} c_n^2 t^{2n} + \sum_{n \geq d} c_n t^n \in k[[t]].$$

But then $d < 0$, and so $d = -1$. It would follow that $\varepsilon = c_d^2$, an impossibility.

Theorem 1.111. *Let E be a quadratic space over K. E has* GR *with respect to λ if and only if QL(E) has* GR *with respect to λ and there exists a decomposition $E = F \perp QL(E)$ such that F has* GR *with respect to λ. In this case $\lambda_*(E) \cong \lambda_*(F) \perp \lambda_*(QL(E))$.*

Proof. Suppose first that we have a decomposition $E = F \perp QL(E)$, in which both F and $QL(E)$ have GR with respect to λ. Then, by Remark 1.107, E has GR with respect to λ and $\lambda_*(E) \cong \lambda_*(F) \perp \lambda_*(QL(E))$, because F is strictly regular.

Next, suppose that E has GR. We choose a quadratic space M over $\mathfrak{o}$ with $E \cong K \otimes_{\mathfrak{o}} M$ and a decomposition $M = N \perp QL(M)$. From the definition of spaces over $\mathfrak{o}$, in other words of nondegenerate quadratic $\mathfrak{o}$-modules (Definition 1.59), follows immediately that $QL(M)$ is nondegenerate and even that N is strictly regular. Now $E \cong K \otimes_{\mathfrak{o}} N \perp K \otimes_{\mathfrak{o}} QL(M)$, so that $QL(E) \cong K \otimes_{\mathfrak{o}} QL(M)$. Therefore $QL(E)$ has GR. Furthermore $F := K \otimes_{\mathfrak{o}} N$ has GR and is strictly regular. $\qquad\square$

If F and G are bilinear spaces over K, which have GR with respect to λ, then we know from §1.3 that $F \otimes G$ also has GR with respect to λ and that

$$\lambda_*(F \otimes G) \approx \lambda_*(F) \otimes \lambda_*(G),$$

see Theorem 1.30. As explained in §1.5, given a bilinear space F and a quadratic space G over K (or $\mathfrak{o}$), we can also construct a quadratic module $F \otimes G$ over K (resp. $\mathfrak{o}$), which can, however, be degenerate when char $K = 2$. For $F \otimes G$ to be again a space, we must require that G is strictly regular. Analogous to the above statement, we have:

Theorem 1.112. *Let F be a bilinear space over K, which has* GR *with respect to λ and let G be a quadratic space, having* GR *with respect to λ. In case* char $K = 2$, *suppose furthermore that G is strictly regular. Then $F \otimes G$ also has* GR *with respect to λ and*

$$\lambda_*(F \otimes G) \cong \lambda_*(F) \otimes \lambda_*(G).$$

Proof. We choose spaces M and N over $\mathfrak{o}$ with $F \cong K \otimes_{\mathfrak{o}} M$, $G \cong K \otimes_{\mathfrak{o}} N$. Then N is strictly regular. Therefore $M \otimes_{\mathfrak{o}} N$ is a strictly regular quadratic space over $\mathfrak{o}$. Furthermore, $F \otimes G \cong K \otimes_{\mathfrak{o}} (M \otimes_{\mathfrak{o}} N)$. Hence, $F \otimes G$ has GR and

$$\lambda_*(F \otimes G) \cong L \otimes_{\lambda} (M \otimes_{\mathfrak{o}} N) \cong (L \otimes_{\lambda} M) \otimes_L (L \otimes_{\lambda} N) \cong \lambda_*(F) \otimes \lambda_*(G). \quad \square$$

Remark. $\lambda_*(F)$ is only determined up to stable isometry and $\lambda_*(F \otimes G)$ only up to isometry. The theorem is valid—as the proof shows—for every choice of $\lambda_*(F)$. More generally we have: if E and E' are bilinear space over K with $E \approx E'$, and if F is a strictly regular quadratic space over K, then $E \otimes F \cong E' \otimes F$. This follows from the fact that there exists a bilinear space U over K with $E \perp U \cong E' \perp U$, implying that $(E \otimes F) \perp (U \otimes F) \cong (E' \otimes F) \perp (U \otimes F)$, and so $E \otimes F \cong E' \otimes F$ by the Cancellation Theorem.

The currently developed notion of the specialization $\lambda_*(E)$ of a space E which has GR allows us to complete our understanding of the results from §1.7 about obedience and weak specialization.

Theorem 1.113. *Let E be a quadratic module over K. Then E is nondegenerate and obedient with respect to λ if and only if*

$$E \cong \underset{s \in S}{\perp} \langle s \rangle \otimes F_s, \tag{1.15}$$

where the F_s are spaces over K, which have GR with respect to λ, only finitely many of them being nonzero. For every decomposition of the form (1.15), we have $\lambda_W(E) = \{\lambda_*(F_1)\}$.

Proof. If E is a space and is obedient with respect to λ, we clearly have a decomposition as above. The equality $\lambda_W(E) = \{\lambda_*(F_1)\}$ follows from the definition of $\lambda_W(E)$ in §1.7 and the definition of $\lambda_*(F_1)$.

Now let $(F_s \mid s \in S)$ be a family of spaces, having GR with respect to λ and $F_s = 0$ for almost all $s \in S$. Let $E := \underset{s \in S}{\perp} \langle s \rangle \otimes F_s$. Then the quadratic module E over K has finite dimension and $E^\perp = \underset{s \in S}{\perp} \langle s \rangle \otimes F_s^\perp$. According to the Final Consideration of §1.7, $E^\perp$ is anisotropic and therefore nondegenerate. It follows immediately from Definition 1.88 that E is obedient with respect to λ. □

We call every decomposition of the form (1.15), having the properties of Theorem 1.113, a λ-*modular decomposition* of E. This terminology is a bit sloppier than in §1.7, in the sense that we no longer discriminate between internal and external orthogonal sums. We talk about λ-modular decompositions of bilinear spaces in a similar fashion.

Theorem 1.114. *Let F be a bilinear space over K, which has GR with respect to λ.*
(i) If G is a bilinear space over K, then

$$\lambda_W(F \otimes G) = \lambda_W(F)\lambda_W(G) = \{\lambda_*(F)\}\lambda_W(G). \tag{1.16}$$

(ii) If G is a strictly regular quadratic space over K, obedient with respect to λ, then $F \otimes G$ is also obedient with respect to λ, and (1.15) holds again.

Proof. *(ii)* Let $G \cong \underset{s \in S}{\perp} \langle s \rangle \otimes G_s$ be a λ-modular decomposition of G. Every G_s has GR with respect to λ and is strictly regular. By Theorem 1.112, $F \otimes G_s$ has GR with respect to λ and $\lambda_*(F \otimes G_s) \cong \lambda_*(F) \otimes \lambda_*(G_s)$. Hence,

$$F \otimes G \cong \underset{s \in S}{\perp} \langle s \rangle \otimes (F \otimes G_s)$$

is a λ-modular decomposition of $F \otimes G$ and $\lambda_W(F \otimes G) = \{\lambda_*(F \otimes G_1)\} = \{\lambda_*(F) \otimes \lambda_*(G_1)\} = \{\lambda_*(F)\}\{\lambda_*(G_1)\} = \lambda_W(F)\lambda_W(G)$.

The proof of *(i)* is analogous. Less care is needed here than for *(ii)*. □

Similarly we can show:

Theorem 1.115. *Let F be a bilinear space and G a strictly regular quadratic space over K. Suppose that the space G has* GR *with respect to* λ. *Then* $F \otimes G$ *is obedient with respect to* λ, *and*

$$\lambda_W(F \otimes G) = \lambda_W(F)\lambda_W(G) = \lambda_W(F)\{\lambda_*(G)\}.$$

To conclude this section, we have a look at the reduction behaviour of quadratic spaces under base field extensions. The following—almost banal—theorem, together with Theorem 1.108(b) above, lays the foundations upon which we will build the generic splitting theory of regular quadratic spaces in the next section.

Theorem 1.116. *Let* $K' \supset K$ *be a field extension and* $\mu : K' \to L \cup \infty$ *an extension of the place* $\lambda : K \to L \cup \infty$. *Let E be a regular quadratic space over K, which has* GR *with respect to* λ. *The space* $K' \otimes_K E$ *has* GR *with respect to* μ *and* $\mu_*(K' \otimes_K E) \cong \lambda_*(E)$.

Proof. Let $\mathfrak{o}'$ be the valuation ring associated to μ. Furthermore, let M be a regular quadratic space over $\mathfrak{o}$ with $E \cong K \otimes_{\mathfrak{o}} M$. Then $\mathfrak{o}' \otimes_{\mathfrak{o}} M$ is a regular quadratic space over $\mathfrak{o}'$ (sic!), and $K' \otimes_K E \cong K' \otimes_{\mathfrak{o}} M = K' \otimes_{\mathfrak{o}'} (\mathfrak{o}' \otimes_{\mathfrak{o}} M)$. Therefore, $K' \otimes_K E$ has GR with respect to μ and

$$\mu_*(K' \otimes_K E) \cong L \otimes_\mu (\mathfrak{o}' \otimes_{\mathfrak{o}} M) = L \otimes_\lambda M \cong \lambda_*(E). \qquad \square$$

If we only require that E is nondegenerate instead of regular, the statement of Theorem 1.116 becomes false. Looking for a counterexample, we can restrict ourselves to quasilinear spaces, by Theorems 1.111 and 1.116.

Example. Let k be an imperfect field of characteristic 2 and let a be an element of k which is not a square. Consider the power series field $K = k((t))$ in one indeterminate t. Let $\lambda_0 : K \to k \cup \infty$ be the place with valuation ring $\mathfrak{o} = k[[t]]$, which maps every power series $f(t) \in \mathfrak{o}$ to its constant term $f(0)$. The quasilinear quadratic $\mathfrak{o}$-module $M = [1, a + t]$ is nondegenerate. For, if $f(t) = \sum_{i \geq 0} b_i t^i$ and $g(t) = \sum_{i \geq 0} c_i t^i$ are elements of $\mathfrak{o}$, whose constant terms b_0, c_0 are not both zero, then $f(t)^2 + (a + t)g(t)^2$ has nonzero constant term $b_0^2 + ac_0^2$, and is thus a unit in $\mathfrak{o}$. Therefore, axiom (QM2) of Definition 1.59 is satisfied.

Hence the space $E := K \otimes_{\mathfrak{o}} M$ has GR with respect to λ_0, and $(\lambda_0)_*(E) \cong [1, a]$. Consider now $L := k(\sqrt{a})$ and the place $\lambda := j \circ \lambda_0 : K \to L \cup \infty$, obtained by composing λ_0 with the inclusion $j : k \hookrightarrow L$. Then E also has GR with respect to λ, and over L we have,

$$\lambda_*(E) = L \otimes_k (\lambda_0)_*(E) \cong [1, a] \cong [1, 1] \cong [1, 0].$$

Finally, let $K' := K(\sqrt{a}) = L((t))$ and let $\mu : K' \to L \cup \infty$ be the place with valuation ring $\mathfrak{o}' := L[[t]]$, which again maps every power series $f(t) \in \mathfrak{o}'$ to its constant term $f(0)$. Then μ extends the place λ. Over K', we have

$$K' \otimes_K E = [1, (\sqrt{a})^2 + t] \cong [1, t].$$

Therefore, $K' \otimes_K E$ is obedient with respect to μ and $\mu_W(K' \otimes_K E) = \{[1]\}$. It is now clear that $K' \otimes_K E$ has bad reduction with respect to μ, since the Witt class $\mu_W(K' \otimes_K E)$ would contain a two-dimensional module otherwise.

Remark. This argument shows nicely that it is profitable to give such an elaborate definition of Witt equivalence of degenerate quadratic modules over fields, as done in Definition 1.76.

Chapter 2
Generic Splitting Theory

2.1 Generic Splitting of Regular Quadratic Forms

From now on we leave the geometric arena behind, and mostly talk of quadratic and bilinear *forms*, instead of *spaces*, over fields. The importance of the geometric point of view was to bring quadratic and bilinear modules over valuation rings into the game. For our specialization theory, these modules were merely an aid however, and their rôle has now more or less ended.

We should indicate one problem though, which occurs when we make the transition from quadratic spaces to quadratic forms: If $\varphi(x_1, \ldots, x_n)$ and $\psi(x_1, \ldots, x_n)$ are two quadratic forms over a field K, they are isometric, $\psi \cong \varphi$, if the polynomial ψ emerges from φ through a linear coordinate transformation. Now, if φ—i.e. the space (E, q) associated to φ—is degenerate, then it is possible that not all of the coordinates $x_1, \ldots, x_n$ ($n := \dim E$) occur in the polynomial ψ (and possibly neither in φ). The dimension of ψ in the naive sense, i.e. the number of occurring variables, can be smaller than the dimension of E. In order to recover the space (E, q) from ψ, one would have to attach a "virtual dimension" n to ψ, pretty horrible!

As soon as one allows degenerate forms, the geometric language is more precise—and thus more preferred—than the "algebraic" language. But now we only consider *regular* quadratic forms ($\hat{=}$ regular quadratic spaces) over fields, so that the forms are guaranteed to remain nondegenerate under base field extensions.

Many concepts which we introduced for quadratic spaces over fields (§1.5–§1.8) will now be used for quadratic forms, usually without any further comments. *In what follows, a "form" is always a regular quadratic form over a field.* We recall once more the concepts GR (= good reduction) and specialization of forms and try to be as down to earth as possible.

So, let φ be a form over a field K. If $\dim \varphi$ is even (resp. odd), then $\varphi = [a_{ij}]$ (resp. $\varphi \cong [a_{ij}] \perp [c]$), where (a_{ij}) is a symmetric $(2m) \times (2m)$-matrix, such that $(r := 2m)$

$$
\det \begin{pmatrix}
2a_{11} & a_{12} & \cdots & a_{1r} \\
a_{21} & 2a_{22} & \cdots & a_{2r} \\
\vdots & \vdots & & \vdots \\
a_{r1} & a_{r2} & \cdots & 2a_{rr}
\end{pmatrix} \neq 0
$$

and $c \neq 0$. (See the beginning of §1.5 for the notation used here.)

Let $\lambda : K \to L \cup \infty$ be a place and $\mathfrak{o} = \mathfrak{o}_\lambda$ its valuation ring. The form φ has GR with respect to λ when $\varphi \cong [a_{ij}]$, resp. $\varphi \cong [a_{ij}] \perp [c]$, such that all a_{ij} are in $\mathfrak{o}$ and c, as well as the determinant above, are units in $\mathfrak{o}$. We then have $\lambda_*(\varphi) \cong [\lambda(a_{ij})]$ resp. $\lambda_*(\varphi) \cong [\lambda(a_{ij})] \perp [\lambda(c)]$. By Theorem 1.62(b), $[a_{ij}]$ is the orthogonal sum of m binary forms $\begin{bmatrix} a_i & 1 \\ 1 & b_i \end{bmatrix}$ $(1 \le i \le m)$ with $a_i \in \mathfrak{o}$, $b_i \in \mathfrak{o}$, $1 - 4a_i b_i \in \mathfrak{o}^*$. Thus, after a coordinate transformation, we have

$$
\varphi(x_1, \ldots, x_n) = \sum_{i=1}^{m} (a_i x_{2i}^2 + x_{2i} x_{2i+1} + b_i x_{2i+1}^2) \qquad (\ + c x_{2m+1}^2). \tag{2.1}
$$

We obtain $(\lambda_* \varphi)(x_1, \ldots, x_n)$ from this form, upon replacing the coefficients a_i, b_i, c by $\lambda(a_i)$, $\lambda(b_i)$, $\lambda(c)$.

Let $K' \supset K$ be a field extension. If $\varphi = \varphi(x_1, \ldots, x_n)$ is a form over K, we write $\varphi \otimes K'$ or $\varphi \otimes_K K'$ for this form considered over K'. If (E, q) is the quadratic space associated to φ, i.e. $\varphi \,\hat{\cong}\, (E, q)$, then the basis extension $(K' \otimes_K E, q_{K'})$ of (E, q) is the quadratic space associated to $\varphi \otimes K'$.[1]

Let $\mu : K' \to L \cup \infty$ be an extension of the place $\lambda : K \to L \cup \infty$. If φ has GR with respect to λ, then $\varphi \otimes K'$ has GR with respect to μ and $\mu_*(\varphi \otimes K') \cong \lambda_*(\varphi)$. This is clear now and was already established in Theorem 1.116 anyway.

We can extend the observation we made at the start of our treatment of generic splitting in §1.4 to regular quadratic forms. Thus let k be a field and φ a form over k. Furthermore, let K and L be fields, containing k, and let $\lambda : K \to L \cup \infty$ be a place over k. Then $\varphi \otimes K$ has GR with respect to λ and $\lambda_*(\varphi \otimes K) \cong \varphi \otimes L$.

We use H to denote the quadratic form $\begin{bmatrix} 0 & 1 \\ 1 & 0 \end{bmatrix}$, regardless of the field we are working in. The bilinear form B_φ, associated to $\varphi = \begin{bmatrix} 0 & 1 \\ 1 & 0 \end{bmatrix}$, is $\begin{pmatrix} 0 & 1 \\ 1 & 0 \end{pmatrix}$. We denote this bilinear form henceforth by $\widetilde{H}$. (This notation differs from §1.4!) If

$$
\varphi \otimes K \cong \varphi_1 \perp r_1 \times H
$$

is the Witt decomposition of $\varphi \otimes K$, then the form φ_1 has GR with respect to λ by Theorem 1.108(b) and we have

$$
\varphi \otimes L \cong \lambda_*(\varphi_1) \perp r_1 \times H.
$$

Therefore, $\operatorname{ind}(\varphi \otimes L) \ge \operatorname{ind}(\varphi \otimes K)$. If K and L are specialization equivalent over k (see Definition 1.32), then

---

[1] We write $\varphi \otimes K'$ instead of $K' \otimes \varphi$ in order to be in harmony with the notation in §1.3 and in the literature. {Many authors more briefly write $\varphi_{K'}$. Starting with §3.2 we will occasionally do this too.}

$$\text{ind}(\varphi \otimes L) = \text{ind}(\varphi \otimes K) \quad \text{and} \quad \ker(\varphi \otimes L) \cong \lambda_*(\varphi_1).$$

We transfer the definition of generic zero field (Definition 1.34) literally to the current situation. Just as in §1.4 we define, for $n := \dim \varphi \geq 2$, a field extension $k(\varphi)$ of k as follows: if $n > 2$ or $n = 2$ and $\varphi \not\cong H$, let $k(\varphi)$ be the quotient field of the integral domain $k[X_1, \ldots, X_n]/(\varphi(X_1, \ldots, X_n))$. {It is easy to see that the polynomial $\varphi(X_1, \ldots, X_n)$ is irreducible.} If $\varphi \cong H$, however, i.e. $\varphi(X_1, X_2) \cong X_1 X_2$, let $k(\varphi) = k(t)$ for an indeterminate t.

We want to show that $k(\varphi)$ is a generic zero field of φ. This is very simple when φ is isotropic: Obviously k itself is a generic zero field of φ in this case. We thus have to show that $k(\varphi)$ is specialization equivalent to k over k, which now means that there exists a place from $k(\varphi)$ to k over k.

Lemma 2.1. *If φ is an isotropic form over k, then $k(\varphi)$ is a purely transcendental field extension of k.*

Proof. This is by definition clear when $\dim \varphi = 2$, i.e. when $\varphi \cong H$. Now let $n := \dim \varphi > 2$. We have a decomposition $\varphi \cong H \perp \psi$ and may suppose without loss of generality that $\varphi = H \perp \psi$. So, if $X_1, \ldots, X_n$ are indeterminates, we have

$$\varphi(X_1, \ldots, X_n) = X_1 X_2 + \psi(X_3, \ldots, X_n).$$

Therefore,[2]

$$k(\varphi) = \text{Quot}(k[X_1, \ldots, X_n]/(X_1 X_2 + \psi(X_3, \ldots, X_n)) = k(x_1, \ldots, x_n),$$

where x_i of course denotes the image of X_i in $k(\varphi)$. The elements $x_2, x_3, \ldots, x_n$ are algebraically independent over k and $x_1 = -x_2^{-1}\psi(x_3, \ldots, x_n)$. Therefore, $k(\varphi) = k(x_2, \ldots, x_n)$ is purely transcendental over k.   □

Since $k(\varphi)/k$ is purely transcendental, there are many places from $k(\varphi)$ to k over k, cf. [11, §10, Prop. 1].

We will move on to prove that also for anisotropic φ with $\dim \varphi \geq 2$, $k(\varphi)$ is a generic zero field of φ. *De facto* we will obtain a stronger result (Theorem 2.5 below), and we will need its full strength later on as well. We require a lemma about the extension of places to quadratic field extensions.

Lemma 2.2. *Let E be a field, K a quadratic extension of E and α a generator of K over E. Let $p(T) = T^2 - aT + b \in E[T]$ be the minimal polynomial of α over E. Furthermore, let $\rho : E \to L \cup \infty$ be a place with $\rho(b) \neq \infty$ and $\rho(a) \neq \infty$, $\rho(a) \neq 0$. Finally, let β be an element of L such that*

$$\beta^2 - \rho(a)\beta + \rho(b) = 0.$$

Then there exists a unique place $\lambda : K \to L \cup \infty$ with $\lambda(\alpha) = \beta$, which extends ρ.

---

[2] If A is an integral domain, $\text{Quot}(A)$ denotes the quotient field of A.

Proof. Let $\mathfrak{o}$ denote the valuation ring of ρ, $\mathfrak{m}$ its maximal ideal and k the residue class field $\mathfrak{o}/\mathfrak{m}$. Let $\gamma : E \to k \cup \infty$ be the canonical place associated to $\mathfrak{o}$. Then $\rho = \overline{\rho} \circ \gamma$, where $\overline{\rho} : k \hookrightarrow L$ is a field embedding. We may assume without loss of generality that k is a subfield of L and that $\overline{\rho}$ is the inclusion map $k \hookrightarrow L$. Every place $\lambda : K \to L \cup \infty$ which extends ρ, has as image a field which is algebraic over k. Therefore we may replace L by the algebraic closure of k in L and thus assume without loss of generality that L is algebraic over k. If c is an element of $\mathfrak{o}$, then $\overline{c}$ will denote the image of c in $\mathfrak{o}/\mathfrak{m} = k$, i.e. $\overline{c} = \gamma(c)$. We have $\beta^2 - \overline{a}\beta + \overline{b} = 0$.

By the general extension theorem for places [11, §2, Prop. 3], there exists a place from K to the algebraic closure $\tilde{k}$ of k, which extends γ. We choose such a place $\delta : K \to \tilde{k} \cup \infty$. Let $\mathfrak{O}$ be the valuation ring of K, associated to δ, $\mathfrak{M}$ its maximal ideal and $F := \mathfrak{O}/\mathfrak{M}$ its residue class field. Finally, let $\sigma : K \to F \cup \infty$ be the canonical place of $\mathfrak{O}$. Then σ extends the place $\gamma : E \to k \cup \infty$. By general valuation theory [11, §8, Th. 1], we have $[F : k] \le 2$. We may envisage F and L as subfields of $\tilde{k}$, which both contain k. One of the things we will show, is that F is equal to the subfield $L' := k(\beta)$ of L.

The field extension K/L is separable since the coefficient a in the minimal polynomial $p(T)$ is different from zero. Let j be the involution of K over E, i.e. the automorphism of K with fixed field E.

Case 1: $L' \ne k$. So $\beta \notin k$ and $T^2 - \overline{a}T + \overline{b}$ is the minimal polynomial of β over k. Now $\alpha^2 - \alpha a + b = 0$ implies $\sigma(\alpha)^2 - \sigma(\alpha)\overline{a} + \overline{b} = 0$. Thus $\sigma(\alpha) = \beta$ or $\sigma(\alpha) = \overline{a} - \beta$. In particular, $L' = k(\sigma(\alpha)) \subset F$. Since $[L' : k] = 2$ and $[F : k] \le 2$, we have $L' = F$ and $F \subset L$.

We have $j(\alpha) = a - \alpha$, and thus $\sigma j(\alpha) = \overline{a} - \sigma(\alpha)$. Since $\overline{a} \ne 0$ this implies $\sigma j(\alpha) \ne \sigma(\alpha)$. Therefore the places σ and $\sigma \circ j$ from K to L are different. They both extend the place ρ. By general valuation theory [11, §8, Th. 1], there can be no other places from K to $\tilde{k}$ which extend ρ. Now take $\lambda = \sigma$ in case $\sigma(\alpha) = \beta$, and $\lambda = \sigma \circ j$ in case $\sigma(\alpha) = \overline{a} - \beta$. Then $\lambda : K \to L \cup \infty$ is the only place which extends ρ and maps α to β.

Case 2: $L' = k$. This time $\beta \in k$. Since β is a root of $T^2 - \overline{a}T + \overline{b}$, we have $T^2 - \overline{a}T + \overline{b} = (T - \beta)(T - \overline{a} + \beta)$ and $\alpha^2 - \alpha\overline{a} + \overline{b} = 0$ implies again that $\sigma(\alpha)^2 - \sigma(\alpha)\overline{a} + \overline{b} = 0$. Therefore,

$$(\sigma(\alpha) - \beta)(\sigma(\alpha) - \overline{a} + \beta) = 0.$$

Hence $\sigma(\alpha) = \beta$ or $\sigma(\alpha) = \overline{a} - \beta$. In particular we have $\sigma(\alpha) \in k$.

Again the places σ and $\sigma \circ j$ are different. By general valuation theory, they are exactly all the places from K to $\tilde{k}$ which extend ρ and $[F : k] = 1$, i.e. $F = k$. As in Case 1, we take $\lambda = \sigma$ in case $\sigma(\alpha) = \beta$, and $\lambda = \sigma \circ j$ in case $\sigma(\alpha) = \overline{a} - \beta$. Again λ is the only place from K to L (even the only place from K to $\tilde{k}$) which extends ρ and maps α to β. $\square$

Theorem 2.3. *Let $\lambda : K \to L \cup \infty$ be a place and φ a form over K with GR with respect to λ and $\dim \varphi \ge 2$. Let $\overline{\varphi} := \lambda_*(\varphi)$. Then λ can be extended to a place $\mu : K(\varphi) \to L(\overline{\varphi}) \cup \infty$.*

Proof. If φ is isotropic, then $K(\varphi)/K$ is purely transcendental by Lemma 2.1. In this case λ can be extended in many ways to a place from $K(\varphi)$ to L. So suppose that φ is anisotropic.

Let $n := \dim \varphi$ and let $\mathfrak{o}$ be the valuation ring of λ. If $n = 2$ we assume in addition that $\overline{\varphi} \not\cong H$, deferring the case $\overline{\varphi} \cong H$ to the end of the proof. After a linear transformation of K^n we may assume without loss of generality that φ is of the form (2.1) as in the beginning of this section (thus with $a_i, b_i \in \mathfrak{o}, c \in \mathfrak{o}^*$ etc.). We denote by $\overline{\varphi}$ the form over L, obtained from φ by replacing the coefficients a_i, b_i and—if n is odd—c by $\overline{a}_i := \lambda(a_i), \overline{b}_i := \lambda(b_i), \overline{c} := \lambda(c)$. Let $X_1, \ldots, X_n$, resp. $U_1, \ldots, U_n$ be indeterminates, then we also write

$$\varphi(X_1, \ldots, X_n) = a_1 X_1^2 + X_1 X_2 + b_1 X_2^2 + \psi(X_3, \ldots, X_n),$$

and accordingly

$$\overline{\varphi}(U_1, \ldots, U_n) = \overline{a}_1 U_1^2 + U_1 U_2 + \overline{b}_1 U_2^2 + \overline{\psi}(U_3, \ldots, U_n).$$

Furthermore we write

$$K(\varphi) = \text{Quot}\frac{K[X_1, \ldots, X_n]}{(\varphi(X_1, \ldots, X_n))} = K(x_1, \ldots, x_n),$$

$$L(\overline{\varphi}) = \text{Quot}\frac{L(U_1, \ldots, U_n)}{(\overline{\varphi}(U_1, \ldots, U_n))} = L(u_1, \ldots, u_n),$$

where x_i denotes of course the image of X_i in $K(\varphi)$ and u_i the image of U_i in $L(\overline{\varphi})$. We then have the relations

$$a_1 x_1^2 + x_1 x_2 + b_1 x_2^2 + \psi(x_3, \ldots, x_n) = 0,$$

$$\overline{a}_1 u_1^2 + u_1 u_2 + \overline{b}_1 u_2^2 + \overline{\psi}(u_3, \ldots, u_n) = 0.$$

{If $n = 2$, the last summands on the left should be read as zero.} It is easy to see that the space $N := \begin{bmatrix} a_1 & 1 \\ 1 & b_1 \end{bmatrix}$ over $\mathfrak{o}$ is isometric to a space $\begin{bmatrix} a_1' & 1 \\ 1 & b_1' \end{bmatrix}$ with $a_1' \in \mathfrak{o}^*$. {Lift a suitable basis of $N/\mathfrak{m}N$ to a basis of N.} Therefore we additionally assume, without loss of generality, that $a_1 \in \mathfrak{o}^*$, i.e. $\overline{a}_1 \neq 0$.

The elements $x_2, \ldots, x_n$ are algebraically independent over K and likewise the elements $u_2, \ldots, u_n$ are algebraically independent over L. Let

$$E := K(x_2, \ldots, x_n) \subset K(\varphi)$$

and

$$F := L(u_2, \ldots, u_n) \subset L(\overline{\varphi}).$$

Our place λ has exactly one extension $\tilde{\lambda} : E \to F \cup \infty$ with $\tilde{\lambda}(x_i) = u_i$ ($2 \leq i \leq n$) (cf. [11, §10, Prop. 2]).

Let $\tilde{o}$ be the valuation ring of $\tilde{\lambda}$. We have $K(\varphi) = E(x_1)$, $L(\overline{\varphi}) = F(u_1)$ with relations

$$x_1^2 + ax_1 + b = 0, \quad u_1^2 + \overline{a}u_1 + \overline{b} = 0,$$

where $a := a_1^{-1}x_2$, $b := a_1^{-1}(b_1 x_2^2 + \psi(x_3, \ldots, x_n))$ are in $\tilde{o}$ and $\overline{a} = \tilde{\lambda}(a)$, $\overline{b} = \tilde{\lambda}(b)$.

Surely $K(\varphi) \neq E$, because $\varphi \otimes K(\varphi)$ is isotropic, but $\varphi \otimes E$ is anisotropic since the extension E/K is purely transcendental. Therefore $K(\varphi)$ is a quadratic field extension of E. Furthermore $\overline{a} = \overline{a}_1^{-1}u_2 \neq 0$. Hence we can apply Lemma 2.2 to the place $\tilde{\lambda} : E \to F \cup \infty$ and the field extensions $K(\varphi)/E$, $L(\overline{\varphi})/F$. According to this lemma, there exists a (unique) place $\mu : K(\varphi) \to L(\overline{\varphi})$ which extends $\tilde{\lambda}$ and maps x_1 to u_1. This proves the theorem in case $\overline{\varphi} \not\approx H$.

Finally, let $n = 2$, φ anisotropic, but $\overline{\varphi}$ isotropic. We can still use the description of $K(\varphi)$ given above (this time with $\psi = 0$), i.e. $\varphi = \begin{bmatrix} a_1 & 1 \\ 1 & b_1 \end{bmatrix}$ with $a_1 \in o^*$, $b_1 \in o$, $K(\varphi) = K(x_1, x_2)$ with x_2 transcendental over K and

$$a_1 x_1^2 + x_1 x_2 + b_1 x_2^2 = 0. \tag{2.2}$$

We want to extend λ to a place $\mu : K(\varphi) \to L \cup \infty$. In order to do this, we choose a nontrivial zero $(c_1, c_2) \in L^2$ of $\overline{\varphi}$. Then we choose an extension $\tilde{\lambda} : K(x_2) \to L \cup \infty$ of λ with $\tilde{\lambda}(x_2) = c_2$, cf. [11, §10, Prop. 1]. {Take the "variable" $x_2 - c_2$ there.} We have

$$\overline{a}_1 c_1^2 + c_1 c_2 + \overline{b}_1 c_2^2 = 0. \tag{2.3}$$

If $c_2 = 0$, then $c_1 = 0$ by (2). Our zero is nontrivial, however, so $c_2 \neq 0$.

Equation (2.2) shows that $x_1^2 + ax_1 + b = 0$ with $a := a_1^{-1}x_2$, $b := a_1^{-1}b_1 x_2^2$. We have $\tilde{\lambda}(a) = \overline{a}_1^{-1}c_2 \neq 0$, $\tilde{\lambda}(b) = \overline{a}_1^{-1}\overline{b}_1 c_2^2$. Equation (2.3) shows that $c_1^2 + \tilde{\lambda}(a)c_1 + \tilde{\lambda}(b) = 0$. Since φ is anisotropic, $K(\varphi) \neq K(x_1)$, thus $[K(\varphi) : K(x_1)] = 2$. Lemma 2.2 tells us that there exists a (unique) place $\mu : K(\varphi) \to L \cup \infty$ which extends $\tilde{\lambda}$. $\square$

Remark 2.4. In the proof we did not need the uniqueness statement of Lemma 2.2. Nonetheless, it deserves some attention. For example, one can use it to deduce from our proof that, given an anisotropic $\overline{\varphi}$, there is *exactly one* place $\mu : K(\varphi) \to L(\overline{\varphi}) \cup \infty$ which extends λ and maps x_i to u_i ($1 \leq i \leq n$). Now, this holds for the generators $x_1, \ldots, x_n$ of $K(\varphi)$ and $u_1, \ldots, u_n$ of $L(\overline{\varphi})$, associated to the special representation (2.1) of φ above (still with $a_1 \in o^*$) and can, by means of a coordinate transformation (with coefficients in o, etc.), be transferred to the case where $\varphi = [a_{ij}]$ for an arbitrary symmetric matrix (a_{ij}) over o with

$$\det \begin{pmatrix} 2a_{11} & \cdots & a_{1n} \\ \vdots & & \vdots \\ a_{n1} & \cdots & 2a_{nn} \end{pmatrix} \in o^*.$$

Theorem 2.5. *Let $\lambda : K \to L \cup \infty$ be a place. Let φ be a form over k with $\dim \varphi \geq 2$, which has GR with respect to λ. Then $\lambda_*(\varphi)$ is isotropic if and only if λ can be extended to a place $\mu : K(\varphi) \to L \cup \infty$.*

Proof. If λ can be extended to a place μ from $K(\varphi)$ to L, then $\lambda_*(\varphi)$ has to be isotropic by the standard argument, which we already used in §1.4, just after Theorem 1.41.

Conversely, suppose that $\overline{\varphi} := \lambda_*(\varphi)$ is isotropic. By Theorem 2.3, there exists a place

$$\mu : K(\varphi) \to L(\overline{\varphi}) \cup \infty$$

which extends λ. By Lemma 2.1, $L(\overline{\varphi})/L$ is a purely transcendental extension. Therefore there exists a place

$$\rho : L(\overline{\varphi}) \to L \cup \infty$$

over L. Now $\rho \circ \mu : K(\varphi) \to L \cup \infty$ is a place which extends λ.   □

Remark. We could have adapted the proof of Theorem 2.3 in such a way that we would have obtained Theorem 2.5 immediately (see in particular our argument there in the case $\overline{\varphi} \cong H$). On the other hand, we can obtain Theorem 2.3 from Theorem 2.5 by applying the latter to the place

$$j \circ \lambda : K(\varphi) \to L(\overline{\varphi}) \cup \infty,$$

where j is the inclusion $L \hookrightarrow L(\overline{\varphi})$. For our further investigations (§2.2, §2.4, §2.5) it is better, however, to isolate Theorem 2.3 and its proof as a stopover to Theorem 2.5.

Corollary 2.6. *Let φ be a form over a field k with* $\dim \varphi \geq 2$. *Then $k(\varphi)$ is a generic zero field of φ.*

Proof. Clearly $\varphi \otimes k(\varphi)$ is isotropic. Now let $L \supset k$ be a field extension with $\varphi \otimes L$ isotropic. Applying Theorem 2.5 to the trivial place $k \hookrightarrow L$, we see that there exists a place $\mu : k(\varphi) \to L \cup \infty$ over k.   □

Since Theorems 1.39 and 1.41 of §1.4 are subcases of our current Theorem 2.5, they are now proved. The definitions and theorems following these two theorems remain valid. Later on they will be used without further comments in renewed generality, namely for arbitrary characteristic instead of characteristic $\neq 2$, regular quadratic forms instead of the nondegenerate symmetric bilinear forms which occur there. In particular, for every regular quadratic form φ over k, we have a generic splitting tower $(K_i \mid 0 \leq i \leq h)$ with higher indices i_r and higher kernel forms φ_r.

Comments. It should be noted that I. Kersten and U. Rehmann have used a different route in [29, §6] to construct, for every form φ over k and every $r \leq \left[\frac{\dim \varphi}{2} \right]$, a field extension of k which is generic for the splitting off of r hyperbolic planes. In particular one can already find a generic splitting tower of φ in their work. In [37] the generic splitting of regular quadratic forms in arbitrary characteristic is based on these foundations.

The results about forms of height 1 and 2, cited at the end of our §1.4, have so far only been established for characteristic $\neq 2$ in the literature.

2.2 Separable Splitting

All fields occurring in this section are supposed to have characteristic 2. If φ is a nondegenerate quadratic form over such a field k, then its quasilinear part (cf. Definition 1.57), which we denote by $QL(\varphi)$, is anisotropic. Now, if $K \supset k$ is a field extension, then it can happen that $QL(\varphi \otimes K) = QL(\varphi) \otimes K$ is isotropic, and thus that $\varphi \otimes K$ is degenerate. This possibility prompted us, when dealing with the theory of generic splitting in the previous sections, to allow only regular forms, i.e. quadratic forms φ with $\dim QL(\varphi) \leq 1$.

We will now make a course change and allow arbitrary nondegenerate quadratic forms. We will however only tolerate a restricted class of field extensions, the so-called separable field extensions. We will see that a reasonably satisfying theory of generic splitting is still possible in this case.

A field extension $K \supset k$ is called *separable* if every finitely generated subextension $E \supset k$ has a *separating transcendence basis*, i.e. a transcendence basis $t_1, \ldots, t_n$, such that E is separably algebraic over $k(t_1, \ldots, t_n)$, cf. [12], [46, X, §6], [27, IV, §5]. If K is already finitely generated over k, then it suffices to check if K itself contains a separating transcendence basis over k (*loc. cit.*).

In what follows, a "form" will always be understood to be a quadratic form. The foundation for the rest of this section is

Theorem 2.7. *Let φ be a nondegenerate form over k, and let $K \supset k$ be a separable field extension. Then $\varphi \otimes K$ is also nondegenerate.*

Proof. We work in the algebraic closure $\widetilde{K}$ of K. We have $QL(\varphi) = [a_1, \ldots, a_r]$ with $a_i \in K$. This form is anisotropic. We have to show that $QL(\varphi) \otimes K$ is also anisotropic. Each a_i has exactly one square root $\sqrt{a_i} \in \widetilde{K}$, which is already in the radical closure $k^{1/2^\infty} \subset K^{1/2^\infty} \subset \widetilde{K}$ of k. Now, since $QL(\varphi)$ is anisotropic, the elements $a_1, \ldots, a_n$ are linearly independent over the subfield $k^2 = \{x^2 \mid x \in k\}$ of k, or equivalently, the elements $\sqrt{a_1}, \ldots, \sqrt{a_n}$ are linearly independent over k. By an important theorem about separable field extensions ("MacLane's Criterion", *loc. cit.*), the fields K and $k^{1/2^\infty}$ are linearly disjoint over k. Therefore, the elements $\sqrt{a_1}, \ldots, \sqrt{a_r}$ are linearly independent over K. This shows that $QL(\varphi) \otimes K$ is anisotropic. $\square$

If $\mathfrak{o}$ is a valuation ring with maximal ideal $\mathfrak{m}$, we denote the residue class field $\mathfrak{o}/\mathfrak{m}$ from now on by $\kappa(\mathfrak{o})$.

Theorem 2.8. *Let $\lambda : K \to L \cup \infty$ be a place and φ a form which has GR with respect to λ.[3] Suppose that the form $\lambda_*(\varphi)$ is nondegenerate. Suppose further that $K' \supset K$ is a field extension and that $\mu : K' \to L \cup \infty$ is an extension of the place λ. Then the form $\varphi \otimes K'$ has GR with respect to μ and $\mu_*(\varphi \otimes K') \cong \lambda_*(\varphi)$.*

Proof. Let $\mathfrak{o} := \mathfrak{o}_\lambda$, $\mathfrak{o}' := \mathfrak{o}_\mu$. The field extension $\overline{\lambda} : \kappa(\mathfrak{o}) \hookrightarrow L$ is a combination of the extensions $\kappa(\mathfrak{o}) \hookrightarrow \kappa(\mathfrak{o}')$ and $\overline{\mu} : \kappa(\mathfrak{o}') \hookrightarrow L$, where the first extension is induced by the inclusion $\mathfrak{o} \hookrightarrow \mathfrak{o}'$.

---

[3] This assumption presupposes that φ is nondegenerate (cf. Definition 1.84).

Let E be a quadratic space for φ and M a nondegenerate quadratic $\mathfrak{o}$-module with $E \cong K \otimes_{\mathfrak{o}} M$. Then $K' \otimes E = K' \otimes_{\mathfrak{o}'} M'$ with $M' := \mathfrak{o}' \otimes_{\mathfrak{o}} M$. The quasilinear quadratic $\kappa(\mathfrak{o})$-module $G := \kappa(\mathfrak{o}) \otimes_{\mathfrak{o}} QL(M)$ is anisotropic. By assumption, $L \otimes_{\bar{\lambda}} G = QL(L \otimes_{\lambda} M)$ is also anisotropic. Therefore

$$\kappa(\mathfrak{o}') \otimes_{\kappa(\mathfrak{o})} G = \kappa(\mathfrak{o}') \otimes_{\mathfrak{o}'} QL(M')$$

is anisotropic. This proves that M' is a nondegenerate quadratic $\mathfrak{o}'$-module. Hence $\varphi \otimes K'$ is nondegenerate and has GR with respect to μ. Furthermore, $\mu_*(\varphi \otimes K')$ corresponds to the quadratic space

$$L \otimes_{\mu} M' = L \otimes_{\mu} (\mathfrak{o}' \otimes_{\mathfrak{o}} M) = L \otimes_{\lambda} M.$$

Hence $\mu_*(\varphi \otimes K') \cong \lambda_*(\varphi)$. $\square$

Now we can faithfully repeat the observation, made at start of our treatment of the theory of generic splitting in §1.4.

So, let φ be a nondegenerate form over a field k (of characteristic 2), and let $K \supset k$ and $L \supset k$ be field extensions of k with $L \supset k$ separable. Let $\lambda : K \to L \cup \infty$ be a place over k. On the basis of Theorem 2.7, we can apply Theorem 2.8 to the trivial place $k \hookrightarrow L$ and its extension λ. We see that $\varphi \otimes K$ is nondegenerate and has GR with respect to λ, and that $\lambda_*(\varphi \otimes K) \cong \varphi \otimes L$.

Let $\varphi \otimes K \cong \varphi_1 \perp r_1 \times H$ be the Witt decomposition of φ. By an established argument (Theorem 1.108(b)), φ_1 has GR with respect to λ, and $\varphi \otimes L \cong \lambda_*(\varphi_1) \perp r_1 \times H$. Hence, $\mathrm{ind}(\varphi \otimes L) \geq \mathrm{ind}(\varphi \otimes K)$. If K is also separable over k and if K and L are specialization equivalent over k, it follows again that $\mathrm{ind}(\varphi \otimes L) = \mathrm{ind}(\varphi \otimes K)$ and $\ker(\varphi \otimes L) \cong \lambda_*(\varphi_1)$.

Definition 2.9. A *generic separable zero field of* φ is a separable field extension $K \supset k$ which has the following properties:

(a) $\varphi \otimes K$ is isotropic.

(b) If $L \supset k$ is a separable field extension with $\varphi \otimes L$ isotropic, then there exists a place $\lambda : K \to L \cup \infty$ over k.

According to Theorem 2.7, φ can become isotropic over a separable field extension L of k only if $\varphi \neq QL(\varphi)$, i.e. if $\dim \varphi - \dim QL(\varphi) \geq 2$. If φ is such a form with $\dim \varphi \neq 2$, then the polynomial $\varphi(X_1, \ldots, X_n)$ is irreducible over the algebraic closure $\tilde{k}$ of k. This can easily be seen by writing $\varphi(X_1, \ldots, X_n) = X_1 X_2 + \psi(X_3, \ldots, X_n)$ over $\tilde{k}$, after a coordinate transformation. Here ψ is a quadratic polynomial which is not the zero polynomial. Therefore, we can again construct the field

$$k(\varphi) = \mathrm{Quot}\, \frac{k[X_1, \ldots, X_n]}{(\varphi(X_1, \ldots, X_n))}.$$

This extends our definition of $k(\varphi)$ in §2.1 from regular forms φ to nondegenerate forms φ. {Obviously $k(\varphi)$ will have its original meaning when $\dim \varphi = 2$, $QL(\varphi) = 0$.}

It is now easy to see that $k(\varphi)$ is separable over k: Let $x_1, \ldots, x_n$ be the images of $X_1, \ldots, X_n$ in $k(\varphi)$, i.e. $k(\varphi) = k(x_1, \ldots, x_n)$. After a coordinate transformation we may suppose, without loss of generality, that

$$\varphi(X_1, \ldots, X_n) = a_1 X_1^2 + X_1 X_2 + b_1 X_2^2 + \psi(X_3, \ldots, X_n).$$

The elements $x_2, \ldots, x_n$ form a transcendence basis of $k(\varphi)$ over k. If $x_1 \notin k(x_2, \ldots, x_n)$, then $k(\varphi)$ is a separable quadratic extension of $k(x_2, \ldots, x_n)$. If the form φ is isotropic, we can make it so that $a_1 = 0$. Then $k(\varphi) = k(x_2, \ldots, x_n)$ is purely transcendental over k.

Theorem 2.10. *Let φ be a nondegenerate form over a field K with $\varphi \neq QL(\varphi)$ and let $\lambda : K \to L \cup \infty$ be a place such that φ has GR with respect to λ. Suppose that also the form $\lambda_*(\varphi)$ is nondegenerate. Then $\lambda_*(\varphi)$ is isotropic if and only if λ can be extended to a place $\mu : K(\varphi) \to L \cup \infty$.*

Proof. If there exists such a place μ, then $\lambda_*(\varphi)$ is isotropic by an established argument, using Theorem 2.8 above.

Suppose now that $\overline{\varphi} := \lambda_*(\varphi)$ is isotropic. By the theory in §2.1, we may suppose that $\dim QL(\varphi) \geq 2$, thus $\dim \varphi \geq 4$. Just as in the proof of Theorem 2.3, we see that λ can be extended to a place $\mu : K(\varphi) \to L(\overline{\varphi}) \cup \infty$. {Note that this is also true in case $\overline{\varphi}$ is anisotropic.} Since $\overline{\varphi}$ is isotropic, $L(\overline{\varphi})$ is a purely transcendental extension of L, as established above. Hence there exists a place ρ from $L(\overline{\varphi})$ to L over L. The place $\rho \circ \mu : K(\varphi) \to L \cup \infty$ extends λ.   □

Corollary 2.11. *Let φ be a nondegenerate form over a field k with $\varphi \neq QL(\varphi)$. Then $k(\varphi)$ is a generic separable zero field of φ.*

Since we have secured the existence of a generic separable zero field, we obtain for every form φ over k a *generic separable splitting tower*

$$(K_r \mid 0 \leq r \leq h)$$

with higher indices i_r and higher kernel forms φ_r ($0 \leq r \leq h$), in complete analogy with the construction of generic splitting towers in §1.4 (just before Definition 1.40). Thus K_0/k is a *separable* inessential field extension, $\varphi_r = \ker(\varphi \otimes K_r)$, and K_{r+1} is a generic separable zero field of φ_r for $r < h$. The height h satisfies $h \leq \frac{1}{2}(\dim \varphi - \dim QL(\varphi))$.

In analogy with Theorem 1.42, we have the following theorem, with *mutatis mutandis* the same proof.

Theorem 2.12. *Let φ be a nondegenerate form over k. Let $(K_r \mid 0 \leq r \leq h)$ be a generic separable splitting tower of φ with associated higher kernel forms φ_r and indices i_r. Let $\gamma : k \to L \cup \infty$ be a place such that φ has GR with respect to γ. Suppose that the form $\gamma_*(\varphi)$ is nondegenerate. Finally, for an m with $0 \leq m \leq h$, let a place $\lambda : K_m \to L \cup \infty$ be given, which extends γ and which cannot be extended to K_{m+1} in case $m < h$. Then φ_m has GR with respect to λ. The form $\gamma_*(\varphi)$ has kernel form $\lambda_*(\varphi_m)$ and Witt index $i_0 + \cdots + i_m$.*

Now we can faithfully repeat Scholium 1.43 to 1.47 from §1.4, but this time with generic separable splitting towers and separable field extensions. We leave this to the reader. {In Scholium 1.46, one should of course assume—as in Theorem 2.12 above—that $\gamma_*(\varphi)$ is nondegenerate.}

In particular, the generic separable splitting tower $(K_r \mid 0 \le r \le h)$ regulates the splitting behaviour of φ with respect to *separable* field extensions $L \supset k$, as described in Scholium 1.43.

2.3 Fair Reduction and Weak Obedience

Our specialization theory of quadratic forms, developed in §1.6–§1.8, gave a satisfying basis for understanding the splitting behaviour of quadratic forms under field extensions (if the forms were not regular, we had to limit ourselves to separable field extensions). There is one important point, however, where the specialization theory is disappointing.

For instance, let $\lambda : K \to L \cup \infty$ be a place from a field K of characteristic 0 to a field L of characteristic 2. If φ is a quadratic form over K which has good reduction with respect to λ, then $\lambda_*(\varphi)$ is automatically a strictly regular form. Conversely, given a strictly regular form ψ over L, one can easily find a strictly regular (= nondegenerate) form φ over K which has good reduction with respect to λ and such that $\lambda_*(\varphi) \cong \psi$. One could then try to deduce properties of ψ from properties of the "lifting" φ of ψ, in the hope that φ is easier to deal with than ψ because char $K = 0$. Good examples can be obtained from the generalization of Scholium 1.46, at the end of §2.1, but we will not carry this out.

It is furthermore desirable to lift a nondegenerate form ψ over L, with quasilinear part $QL(\psi) \ne 0$, to a form φ over K. This is not possible with our specialization theory as it stands.

Let us review the foundations of the current theory! We return to the use of geometric language. Let $\lambda : K \to L \cup \infty$ be a place with associated valuation ring $\mathfrak{o}$. Let $\mathfrak{m}$ be the maximal ideal of $\mathfrak{o}$ and $k = \mathfrak{o}/\mathfrak{m}$ its residue class field. Finally, let $\bar{\lambda} : k \hookrightarrow L$ be the field extension determined by λ.

Given a suitable quadratic space E over K, the idea was to attach a space F over k to E and then to define the specialization $\lambda_*(E)$ of E with respect to λ as the space $L \otimes_{\bar{\lambda}} F$. We required E to have good reduction, i.e. $E \cong K \otimes_{\mathfrak{o}} M$ for a nondegenerate quadratic module M over $\mathfrak{o}$. Then we could choose our space F over k to be $F = M/\mathfrak{m}M$.

The preceding is analogous to the specialization theory for varieties in algebraic and arithmetic geometry (and related areas, such as rigid analysis): one defines a class of "nondegenerate" objects X over the valuation ring $\mathfrak{o}$ and associates to it, by means of a base extension, the objects $K \otimes_{\mathfrak{o}} X$ and $k \otimes_{\mathfrak{o}} X$ over K and k. Finally one decrees that $L \otimes_{\bar{\lambda}} (k \otimes_{\mathfrak{o}} X) = L \otimes_{\lambda} X$ is the specialization of $K \otimes_{\mathfrak{o}} X$ with respect to λ.

Without any doubt, we have found in the nondegenerate quadratic spaces, as defined in Definition 1.59, a respectable class X of such objects which easily comes to mind. Nevertheless we are a bit unfair to all the quadratic spaces over K, waiting to be specialized. It would be fairer to require only that $E \cong K \otimes_o M$ with M a free quadratic o-module such that $M/\mathfrak{m}M$ is nondegenerate, and then to define $\lambda_*(E) := L \otimes_\lambda M = L \otimes_{\bar{\lambda}} (M/\mathfrak{m}M)$. But then one should show, that up to isometry, $\lambda_*(E)$ depends only on λ and E and not on the choice of M.

This is possible, as we shall show now. In the following a *quadratic module* over a valuation ring (in particular a field) will always be understood to be a *free quadratic module of finite rank*.

Let o be a valuation ring with maximal ideal $\mathfrak{m}$, quotient field K and residue class field $k = o/\mathfrak{m}$. If $2 \notin \mathfrak{m}$, then we already know all that follows. On formal grounds, we allow nevertheless the uninteresting case that char $k \neq 2$.

Definition 2.13.

(a) A quadratic module M over o is called *reduced nondegenerate* if the quadratic module $M/\mathfrak{m}M$ over k is nondegenerate. (Note: if char $k \neq 2$, this implies that M itself is nondegenerate.)

(b) A quadratic module E over K has *fair reduction* (or: FR for short) *with respect to* λ (or: *with respect to* o), if there exists a reduced nondegenerate quadratic module M over o such that $E \cong K \otimes_o M$.[4]

It is our task to prove that in the situation of Definition 2.13(b), the space $M/\mathfrak{m}M$ is independent of the choice of reduced nondegenerate module M up to isometry. We will however proceed in a more general direction than necessary, in order to develop at the same time the equipment necessary to obtain a generalization of the important Theorem 1.108(b) for fair reduction instead of good reduction. First a very general definition.

Definition 2.14. Let $M = (M, q)$ and $M' = (M', q')$ be quadratic modules over a ring A. We say that M *represents* the quadratic module M', and write $M' < M$, if the A-module M has a direct sum decomposition $M = M_1 \oplus M_2$ with $(M_1, q|M_1) \cong (M', q')$.

Now, if M is a quadratic o-module, we always regard M as an o-submodule of the K-vector space $K \otimes_o M$. So we have $K \otimes_o M = KM$. In the following, we will almost always denote the quadratic form on M by q, and its associated bilinear form B_q by B. We will also use q to denote the quadratic form on $K \otimes_o M$ with values in K, obtained from q. We will often denote the module $M/\mathfrak{m}M$ over k by $\overline{M}$, and the image of a vector $x \in M$ in $\overline{M}$ with $\bar{x}$. Likewise we denote the image of a scalar $a \in o$ in k by $\bar{a}$. Finally, we denote by $\bar{q}$ the quadratic form on $\overline{M}$ with values in k, induced by q, i.e. $\bar{q}(\bar{x}) = \overline{q(x)}$ for $x \in M$. Furthermore, $\overline{B}$ will stand for the associated bilinear form $B_{\bar{q}}$, i.e. $\overline{B}(\bar{x}, \bar{y}) = \overline{B(x, y)}$ for $x, y \in M$.

As before, we call a nondegenerate quadratic module over a field or valuation ring a *quadratic space*, or just a *space*. If M is a reduced nondegenerate quadratic o-module, then $\overline{M} = M/\mathfrak{m}M$ is a space over k. It is easy to show that $K \otimes M$ is a space over K too (see Remark 2.22 below), but that is not so important at the moment.

---

[4] In [32] the words "nearly good reduction" are used instead of "fair reduction". We avoid this terminology here because we will talk about a different concept, "almost good reduction", later (Chapter 4).

Lemma 2.15. *Let M be a reduced nondegenerate quadratic module over $\mathfrak{o}$. There exists a decomposition $M = M_1 \perp M_2$ with M_1 strictly regular and $B(M_2 \times M_2) \subset \mathfrak{m}$. If such a decomposition is given and x is a primitive vector of M_2, then $q(x) \in \mathfrak{o}^*$.*

Proof. We have a decomposition $\overline{M} = U \perp QL(\overline{M})$ with U strictly regular.[5] Let $x_1, \ldots, x_r$ be vectors of M, such that the images $\overline{x}_1, \ldots, \overline{x}_1$ in $\overline{M}$ form a basis of the k-vector space U. Then the determinant of the value matrix $(B(x_i, x_j))$ with $1 \leq i$, $j \leq r$ is a unit of $\mathfrak{o}$. Therefore, $M_1 := \sum_{i=1}^{r} \mathfrak{o} x_i$ is a strictly regular space. Let M_2 be the orthogonal complement $M_1^{\perp}$ of M_1 in M. Then $M = M_1 \perp M_2$ and $M_2 / \mathfrak{m} M_2 = QL(\overline{M})$. Since $QL(\overline{M})$ is quasilinear, we have $B(M_2 \times M_2) \subset \mathfrak{m}$. Conversely, if such a decomposition $M = M_1 \perp M_2$ is given, with M_1 strictly regular and $B(M_2 \times M_2) \subset \mathfrak{m}$, then $\overline{M}_2 = QL(\overline{M})$. Since $QL(\overline{M})$ is anisotropic, every primitive vector x in M_2 has a value $q(x) \in \mathfrak{o}^*$. $\square$

Lemma 2.16. *Suppose again that M is a reduced nondegenerate quadratic $\mathfrak{o}$-module. Then M is maximal among all finitely generated (thus free) $\mathfrak{o}$-modules $N \subset K \otimes_{\mathfrak{o}} M$ with $q(N) \subset \mathfrak{o}$.*

Proof. We work in the quadratic K-module $E := K \otimes_{\mathfrak{o}} M$. According to Lemma 2.15, there is a decomposition $M = M_1 \perp M_2$ with M_1 strictly regular and $B(M_2 \times M_2) \subset \mathfrak{m}$. Let $x \in E$ and $q(M + \mathfrak{o} x) \subset \mathfrak{o}$. We have to show that $x \in M$.

We write $x = x_1 + x_2$ with $x_1 \in K \otimes M_1$, $x_2 \in K \otimes M_2$. For every $y \in M_1$, we have $B(x_1, y) = B(x, y) = q(x + y) - q(x) - q(y) \in \mathfrak{o}$. Since the bilinear form $B = B_q$ is nondegenerate on M_1, it follows that $x_1 \in M_1$. Therefore, $M + \mathfrak{o} x = M + \mathfrak{o} x_2$ and so $q(x_2) \in \mathfrak{o}$. Now, $x_2 = az$ for a primitive vector $z \in M_2$ and $a \in K$. Since $q(z) \in \mathfrak{o}^*$, we get $a \in \mathfrak{o}$, and so $x_2 \in M$. We conclude that $x \in M$. $\square$

Lemma 2.17 (Extension of Lemma 1.66). *Let M be a reduced nondegenerate quadratic $\mathfrak{o}$-module and let e be a primitive isotropic vector in M. Then e can be completed to a hyperbolic vector pair e, f in M.*

Proof. We work again in $E = K \otimes M$. The ideal $B(e, M)$ of $\mathfrak{o}$ is finitely generated. Therefore we have $B(e, M) = a\mathfrak{o}$ for an element a of $\mathfrak{o}$. If a were equal to zero, then the vector $\overline{e} \in \overline{M}$ would lie in the quasilinear part $QL(\overline{M})$ of $\overline{M}$. However, $\overline{e} \neq 0$ and $\overline{q}(\overline{e}) = 0$, so that $\overline{e} \notin QL(\overline{M})$ due to the anisotropy of $QL(\overline{M})$. Therefore, $a \neq 0$. Since $B(e, M) = a\mathfrak{o}$ and $q(M) \subset \mathfrak{o}$, we have $q(M + \mathfrak{o}(a^{-1}e)) \subset \mathfrak{o}$. Thus, by Lemma 2.16, $a^{-1}e \in M$. Since e is primitive, we conclude that $a^{-1} \in \mathfrak{o}$, i.e. $a \in \mathfrak{o}^*$. Hence, $B(e, M) = \mathfrak{o}$. As in the proof of Lemma 1.66, we choose $z \in M$ with $B(e, z) = 1$ and complete e to a hyperbolic pair with the vector $f := z - q(z)e$. $\square$

Theorem 2.18. *Let M be strictly regular, N a reduced nondegenerate quadratic $\mathfrak{o}$-module and $K \otimes M < K \otimes N$. Then $M < N$. Furthermore, $N \cong M \perp P$ where P is a reduced nondegenerate quadratic $\mathfrak{o}$-module.*

---

[5] Recall that $QL(\overline{M})$ denotes the quasilinear part of $\overline{M}$; see the beginning of §1.6.

Proof. If Q is a strictly regular quadratic submodule of N, then $N = Q \perp P$ with $P := Q^\perp = \{x \in N \mid B(x, Q) = 0\}$, since Q is nondegenerate with respect to the bilinear form $B = B_q$ (Lemma 1.53). We have $\overline{N} = \overline{Q} \perp \overline{P}$ and conclude that $\overline{P}$ is nondegenerate, i.e. P is reduced nondegenerate.

Because of this preliminary remark, it suffices to show that $M < N$. Suppose first of all that M is hyperbolic, i.e. $M \cong r \times H$ for some $r > 0$, where H denotes the quadratic module $\begin{bmatrix} 0 & 1 \\ 1 & 0 \end{bmatrix}$ over $\mathfrak{o}$. We proceed by induction on r.

Since $K \otimes M < K \otimes N$, $K \otimes N$ contains isotropic vectors. We can then choose a *primitive* isotropic vector e in N. This vector can be completed to a hyperbolic vector pair in N, by Lemma 2.17. Hence, $H < N$ and so $N \cong H \perp N'$ where N' is a reduced nondegenerate quadratic $\mathfrak{o}$-module. From $r \times (K \otimes H) < K \otimes N$, we get $K \otimes N \cong r \times (K \otimes H) \perp G$ for some space G over K. On the other hand, $K \otimes N \cong (K \otimes H) \perp (K \otimes N')$. By the Cancellation Theorem over K (Theorem 1.67), we get $(r-1) \times (K \otimes H) \perp G \cong K \otimes N'$. Therefore, $(r-1) \times (K \otimes H) < K \otimes N'$. The induction hypothesis gives us $(r-1) \times H < N'$ and so $r \times H < N$.

In the general case, $K \otimes M < K \otimes N$ implies that $K \otimes (M \perp (-M)) < K \otimes (N \perp (-M))$. Since $M \perp (-M)$ is hyperbolic, it follows from above that $M \perp (-M) < N \perp (-M)$, and so, $N \perp (-M) \cong M \perp (-M) \perp P$ by the remark at the beginning of the proof, where P is another reduced nondegenerate quadratic $\mathfrak{o}$-module. By the Cancellation Theorem over $\mathfrak{o}$ (Theorem 1.67), we may conclude that $N \cong M \perp P$, and so $M < N$.   □

Theorem 2.19 ([32, Lemma 2.8]). *Suppose that M and N are reduced nondegenerate quadratic $\mathfrak{o}$-modules with $K \otimes M < K \otimes N$. The spaces $\overline{M} = M/\mathfrak{m}M$ and $\overline{N} = N/\mathfrak{m}N$ over k have the following properties:*
(a) $\overline{M} < \overline{N}$.
(b) More precisely: there exist quadratic subspaces S and T of $\overline{N}$ with $\overline{M} < S$, $\overline{N} = S \perp T$, T strictly regular, $\dim S \leq \dim QL(\overline{M}) + \dim QL(\overline{N}) + \dim M$.
(c) If $\overline{N}$ is anisotropic, then $M < N$.

Remark. We will not need properties (b) and (c) later on. They can however be obtained in the following proof at no extra cost.

Proof of Theorem 2.19. (i) By Lemma 2.15, we can choose an orthogonal decomposition $M = M_1 \perp M_2$ with M_1 strictly regular and $B(M_2 \times M_2) \subset \mathfrak{m}$. We have $K \otimes M_1 < K \otimes N$. By Theorem 2.18, we have $N \cong M_1 \perp N_2$, where N_2 is a reduced nondegenerate quadratic $\mathfrak{o}$-module. Then $K \otimes M_1 \perp K \otimes M_2 < K \otimes M_1 \perp K \otimes N_2$ implies $K \otimes M_2 < K \otimes N_2$ by the Cancellation Theorem over K (Theorem 1.67) as usual. If we could prove the claims of the theorem for M_2 and N_2 instead of M and N, then they would follow immediately for M and N as well. Hence we assume now, without loss of generality, that $B(M \times M) \subset \mathfrak{m}$.

(ii) If $M = \{0\}$, nothing has to be done. So suppose that $M \neq \{0\}$. Surely k has characteristic 2. We suppose, without loss of generality that $F := K \otimes M$ is a subspace of $E := K \otimes N$ (instead of only: F is isomorphic to a subspace of E). Since $\overline{M}$ is anisotropic, we have $M = \{x \in F \mid q(x) \in \mathfrak{o}\}$ (as already ascertained before). Therefore, $N_1 := N \cap F \subset M$. Moreover, N_1 is a direct summand of the $\mathfrak{o}$-module

N, since N/N_1 is torsion free and finitely generated, and thus free. Hence N_1 itself is also free.

Let $m := \dim M$. By the Elementary Divisor Theorem for valuation rings, there exist bases $x_1, \ldots, x_m$ and $y_1, \ldots, y_m$ of the free $\mathfrak{o}$-modules M and N_1 with $y_i = c_i x_i$, $c_i \in \mathfrak{o}$ $(1 \leq i \leq m)$. We suppose without loss of generality that $c_i = 1$ for $1 \leq i \leq s$ and $c_i \in \mathfrak{m}$ for $s < i \leq m$. {It is allowed that $s = 0$ or $s = m$.}

If $\overline{N}$ is anisotropic, then $q(x) \in \mathfrak{o}^*$ for every primitive vector x of N, and thus for every vector in N_1 which is primitive in N_1. Hence $s = m$, i.e. $N_1 = M$ and $M < N$. This establishes property (c).

(iii) Let V be the image of N_1 in $\overline{N} = N/\mathfrak{m}N$. Since N_1 is a direct summand of N, we may make the identification $V = \overline{N_1}$. Since $B(M \times M) \subset \mathfrak{m}$, we have $B(N_1 \times N_1) \subset \mathfrak{m}$, thus $\overline{B}(V \times V) = 0$. Hence V is quasilinear. We have

$$\overline{M} = k\overline{x}_1 \oplus \cdots \oplus k\overline{x}_m \cong [a_1, \ldots, a_m], \text{ with } a_i := \overline{q}(\overline{x}_i) \in k^*. \quad (2.4)$$

$$V = k\overline{y}_1 \oplus \cdots \oplus k\overline{y}_m \cong [a_1, \ldots, a_s] \perp (m-s) \times [0]. \quad (2.5)$$

Let $R := QL(\overline{N})$ be the quasilinear part of $\overline{N}$ and $V_0 := V \cap R$. Since R is anisotropic, V_0 is also anisotropic. From (2.5) it follows that $V_0 < [a_1, \ldots, a_s]$, and so

$$[a_1, \ldots, a_s] \cong V_0 \perp [b_1, \ldots, b_{s-t}] \quad (2.6)$$

for elements $b_i \in k^*$ and $t := \dim V_0$. {If $t = s$, the right-hand side of (2.6) should be read as V_0.} From (2.5) and (2.6) we get a decomposition $V = V_0 \perp U$, with

$$U \cong [b_1, \ldots, b_{s-t}] \perp (m-s) \times [0]. \quad (2.7)$$

Since $V \cap R = V_0$, we have $U \cap R = \{0\}$. We choose a submodule W of $\overline{N}$ such that

$$\overline{N} = (R + U) \oplus W = R \oplus U \oplus W.$$

Since R is the quasilinear part of $\overline{N}$, $P := U \oplus W$ is strictly regular and $\overline{N} = R \perp P$. Let $u_1, \ldots, u_{m-t}$ be a basis of U, associated with the representation (2.7). P is a symplectic vector space with respect to the bilinear form $\overline{B} = B_{\overline{q}}$. Hence we can complete $u_1, \ldots, u_{m-t}$ to a basis $u_1, \ldots, u_{m-t}, z_1, \ldots, z_{m-t}$ of a subspace P_1 of P with $\overline{B}(u_i, z_j) = \delta_{ij}$ $(1 \leq i, j \leq m - t)$. As a quadratic space, P_1 is of the form

$$P_1 = (ku_1 + kz_1) \perp \cdots \perp (ku_{m-t} + kz_{m-t})$$

and $P = P_1 \perp T$ for a strictly regular space T.

For $s - t < i \leq m - t$, we have $ku_i + kz_i \cong \begin{bmatrix} 0 & 1 \\ 1 & 0 \end{bmatrix}$. Therefore, $[a_{t+i}] < ku_i + kz_i$ for this i. Furthermore we have $V_0 < R$ and $[b_i] < ku_i + kz_i$ for $1 \leq i \leq s - t$, since $b_i = \overline{q}(u_i)$. Putting everything together, (2.4) and (2.6) yield

$$\overline{M} \cong V_0 \perp [b_1, \ldots, b_{s-t}] \perp [a_{s+1}, \ldots, a_m] < R \perp P_1,$$

and

$$R \perp P = R \perp P_1 \perp T = \overline{N}.$$

Let $S := R \perp P_1$, then $\overline{M} < S$ and $\overline{N} = S \perp T$. Since $\overline{M} = QL(\overline{M})$ and $R = QL(\overline{N})$, we have furthermore

$$\begin{aligned}
\dim S &= \dim QL(\overline{N}) + 2(m - t) \\
&= \dim QL(\overline{N}) + \dim QL(\overline{M}) + \dim \overline{M} - 2t \\
&\leq \dim QL(\overline{N}) + \dim QL(\overline{M}) + \dim M. \qquad \square
\end{aligned}$$

As a special case of Theorem 2.19, we obtain

Corollary 2.20. *Let M and M' be two reduced nondegenerate quadratic o-modules with $K \otimes M \cong K \otimes M'$. Then $M/\mathfrak{m}M \cong M'/\mathfrak{m}M'$.*

Now let $\lambda : K \to L \cup \infty$ again be a place and o its associated valuation ring. As before, we denote by $\overline{\lambda}$ the field embedding $k \hookrightarrow L$ determined by λ. By Corollary 2.20 it makes sense to make the following definition:

Definition 2.21. Let E be a space over K which has FR with respect to λ. Let M be a reduced nondegenerate quadratic o-module with $E \cong K \otimes M$. We call the quadratic L-module $L \otimes_\lambda M = L \otimes_{\overline{\lambda}} (M/\mathfrak{m}M)$ (which is uniquely determined by E up to isometry) the *specialization of E with respect to λ*, and denote it by $\lambda_*(E)$.

Note. If E has good reduction with respect to λ, then $\lambda_*(E)$ has the old meaning.

Caution! If char $L = 2$, then $\lambda_*(E)$ can be a degenerate quadratic L-module, even if E is strictly regular. However, this does not happen when the field embedding $\overline{\lambda}$ is separable; see §2.2.

Remark 2.22. If M is a reduced nondegenerate quadratic o-module, then the quadratic K-module $K \otimes M$ is definitely nondegenerate, for we know that $QL(K \otimes M) = K \otimes QL(M)$, and if $QL(K \otimes M)$ were isotropic, then $QL(M)$, and thus $QL(M/\mathfrak{m}M)$, would also be isotropic. Therefore our assumption in Definition 2.21, that the quadratic K-module E is nondegenerate, is a natural one and does not cause a loss of generality.

We illustrate the concept of fair reduction with a few examples.

Example 2.23. Let o be a valuation ring with char $K = 0$, char $k = 2$, and let $\lambda : K \to k \cup \infty$ be the canonical place of o. We want to lift an arbitrary space S over k to a space over K by means of λ. We choose a decomposition

$$S \cong \begin{bmatrix} \alpha_1 & 1 \\ 1 & \beta_1 \end{bmatrix} \perp \cdots \perp \begin{bmatrix} \alpha_m & 1 \\ 1 & \beta_m \end{bmatrix} \perp [\gamma_1, \ldots, \gamma_r]$$

with $\alpha_i, \beta_i, \gamma_j \in k$. {Since the quasilinear part of S is anisotropic, the elements $\gamma_1, \ldots, \gamma_r$ are linearly independent over k^2.} Next we choose pre-images a_i, b_i, c_j of the elements $\alpha_i, \beta_i, \gamma_j$ in o. Suppose that $r = 2s$ is even. We choose elements $t_1, \ldots, t_s \in \mathfrak{m}$. Then the K-space

$$E := \begin{bmatrix} a_1 & 1 \\ 1 & b_1 \end{bmatrix} \perp \cdots \perp \begin{bmatrix} a_m & 1 \\ 1 & b_m \end{bmatrix} \perp \begin{bmatrix} c_1 & t_1 \\ t_1 & c_2 \end{bmatrix} \perp \cdots \perp \begin{bmatrix} c_{2s-1} & t_s \\ t_s & c_{2s} \end{bmatrix}$$

clearly has FR with respect to λ and $\lambda_*(E) \cong S$. If $r = 2s+1$ is odd, we again choose elements $t_1, \ldots, t_s \in \mathfrak{m}$. This time the K-space

$$E := \begin{bmatrix} a_1 & 1 \\ 1 & b_1 \end{bmatrix} \perp \cdots \perp \begin{bmatrix} a_m & 1 \\ 1 & b_m \end{bmatrix} \perp \begin{bmatrix} c_1 & t_1 \\ t_1 & c_2 \end{bmatrix} \perp \cdots \perp \begin{bmatrix} c_{2s-1} & t_s \\ t_s & c_{2s} \end{bmatrix} \perp [c_{2s+1}]$$

has FR with respect to λ and $\lambda_*(E) \cong S$. In the special case that all $t_i = 0$, we obtain a K-space

$$F := \begin{bmatrix} a_1 & 1 \\ 1 & b_1 \end{bmatrix} \perp \cdots \perp \begin{bmatrix} a_m & 1 \\ 1 & b_m \end{bmatrix} \perp [c_1, \ldots, c_r]$$

for every r. Since char $K = 0$, we can interpret F (and more generally E) as a bilinear space,

$$F \cong \begin{pmatrix} 2a_1 & 1 \\ 1 & 2b_1 \end{pmatrix} \perp \cdots \perp \begin{pmatrix} 2a_m & 1 \\ 1 & 2b_m \end{pmatrix} \perp \langle 2c_1, \ldots, 2c_r \rangle.$$

The elements $a_1, \ldots, a_m$ can always be chosen to be $\neq 0$ and we obtain the diagonalization

$$F \cong \langle 2a_1, d_1, \ldots, 2a_m, d_m, 2c_1, \ldots, 2c_r \rangle$$

with

$$d_i := \frac{4a_i b_i - 1}{2a_i} \quad (1 \leq i \leq r).$$

Example 2.24. Let k be an imperfect field of characteristic 2, $\mathfrak{o}$ the power series ring $k[[t]]$ in one variable t, and so $K = k((t))$. Choose $c \in k \setminus k^2$ and let E be the space $\begin{bmatrix} 1 & 1 \\ 1 & ct^{-2} \end{bmatrix}$ over K.

Claim. E has FR, but does not have GR with respect to $\mathfrak{o}$.

Proof. We have $E \cong K \otimes M$, where M is the quadratic $\mathfrak{o}$-module $\begin{bmatrix} 1 & 1 \\ t & c \end{bmatrix}$. Its reduction $\overline{M} = M/\mathfrak{m}M$ is the space $\begin{bmatrix} 1 & 0 \\ 0 & c \end{bmatrix} = [1, c]$ over k, which is anisotropic. Therefore M is reduced nondegenerate and E has FR.

Suppose for the sake of contradiction that $E \cong K \otimes N$ where N is a quadratic space over $\mathfrak{o}$. Since E does not have a quasilinear part, the same is true for N. Hence, $N \cong \begin{bmatrix} \alpha & 1 \\ 1 & \beta \end{bmatrix}$ with $\alpha, \beta \in \mathfrak{o}$. We then have $\begin{bmatrix} 1 & 1 \\ 1 & ct^{-2} \end{bmatrix} \cong \begin{bmatrix} \alpha & 1 \\ 1 & \beta \end{bmatrix}$ over K. An inspection of the Arf-invariants shows that $ct^{-2} = \alpha\beta + x^2 + x$ for some $x \in k((t))$. We must have $x \neq 0$. We write $x = \sum_{i \geq d} a_i t^i$ for some $d \in \mathbb{Z}$ and $a_d \neq 0$. Since $\alpha\beta \in \mathfrak{o}$, we have $d = -1$ and $c = a_{-1}^2$. This is a contradiction since c is not a square in k. We conclude that E does not have GR. $\square$

Example 2.25. Again let $\mathfrak{o} = k[[t]]$ and k a field of characteristic 2. In §1.7 we determined that the space $E := \begin{bmatrix} 1 & 1 \\ 1 & t^{-1} \end{bmatrix}$ does not have GR with respect to $\mathfrak{o}$. Could it be that E has at least FR with respect to $\mathfrak{o}$?

Let us assume that this is so. Then there exists a reduced nondegenerate quadratic $\mathfrak{o}$-module M with $E \cong K \otimes M$. We choose a decomposition $M = M_1 \perp M_2$ with

M_1 strictly regular and $B(M_2 \times M_2) \subset \mathfrak{m}$. We already know that M is degenerate, so $\dim M_2 > 0$. Since $\dim M = 2$ and $\dim M_1$ is even, we must have $M_1 = 0$. Hence $\overline{M}$ is quasilinear. But M is not quasilinear, since E is not quasilinear. Therefore we have a representation

$$M \cong \begin{bmatrix} a & t^n c \\ t^n c & b \end{bmatrix}$$

with $n \in \mathbb{N}$ and units $a, b, c \in \mathfrak{o}^*$. The space $\overline{M} \cong [\overline{a}, \overline{b}]$ is anisotropic and so the element $\overline{a}\overline{b}$ is not a square in k. Over K we have

$$\begin{bmatrix} 1 & 1 \\ 1 & t^{-1} \end{bmatrix} \cong K \otimes M \cong \langle a \rangle \otimes \begin{bmatrix} 1 & t^n ca^{-1} \\ t^n ca^{-1} & ba^{-1} \end{bmatrix} \cong \langle a \rangle \otimes \begin{bmatrix} 1 & 1 \\ 1 & abc^{-2}t^{-2n} \end{bmatrix}.$$

Comparing Arf-invariants shows

$$abc^{-2}t^{-2n} = t^{-1} + x^2 + x \tag{2.8}$$

with $x \in K$. Let $v := K \to \mathbb{Z} \cup \infty$ be the valuation associated to $\mathfrak{o}$. From (2.8) we get $v(x) < 0$ and thus $v(x) = -n$. Therefore $x = t^{-n} \sum_{i=0}^{\infty} x_i t^i$ with $x_i \in k$, $x_0 \neq 0$. Comparing the coefficients of t^{-2n} on the left- and right-hand sides of (2.8) gives

$$\overline{a}\overline{b}\overline{c}^{-2} = x_0^2.$$

So $\overline{a}\overline{b}$ is a square in k after all, a contradiction. Therefore E does not have FR.

We continue with the general theory. So $\lambda : K \to L \cup \infty$ is again an arbitrary place. We immediately obtain a consequence of Theorem 2.19 which is not contained in the results of §1.8 in the case of good reduction.

Corollary 2.26. *Let E and F be spaces over K which have FR with respect to λ and with $F < E$. Then $\lambda_* F < \lambda_* E$.*

This corollary engenders a substitution principle for quadratic forms, modelled on Theorem 1.29.

Theorem 2.27. *Let $(g_{kl}(t))_{1 \leq k, l \leq m}$ and $(f_{ij}(t))_{1 \leq i, j \leq n}$ be symmetric matrices whose coefficients are polynomials in variables $t = (t_1, \ldots, t_r)$ over a field k. Let $L \supset k$ be a field extension of k and $c = (c_1, \ldots, c_r)$ an r-tuple with coefficients in L. Suppose that the quadratic forms $[g_{kl}(t)]$ and $[f_{ij}(t)]$ over $k(t)$ satisfy $[g_{kl}(t)] < [f_{ij}(t)]$. Suppose also that the quadratic forms $[g_{kl}(c)]$ and $[f_{ij}(c)]$ over L are nondegenerate. Then $[g_{kl}(c)] < [f_{ij}(c)]$ (over L).*

We return to an arbitrary place $\lambda : K \to L \cup \infty$.

Theorem 2.28 (Extension of Theorem 1.108(b)). *Let G be a space and F a strictly regular space over K. Suppose that F has GR with respect to λ. Finally, let $E := F \perp G$.*

Claim: E has FR with respect to λ if and only if G has FR with respect to λ. In this case we have $\lambda_* E \cong \lambda_* F \perp \lambda_* G$.

Proof. We have $F = K \otimes M$ where M is a strictly regular quadratic $\mathfrak{o}$-module. If G has FR with respect to λ, then $G = K \otimes P$ where P is a reduced nondegenerate quadratic $\mathfrak{o}$-module. The quadratic $\mathfrak{o}$-module $N := M \perp P$ is then also reduced nondegenerate and

$$L \otimes_\lambda N = (L \otimes_\lambda M) \perp (L \otimes_\lambda P) = F \perp G = E.$$

Hence E has FR with respect to λ and $\lambda_* E \cong \lambda_* F \perp \lambda_* G$.

Suppose now that E has FR with respect to λ. We have $E \cong K \otimes N$ where N is a reduced nondegenerate quadratic $\mathfrak{o}$-module. By Theorem 2.18, $K \otimes M < K \otimes N$ implies that $N \cong M \perp P$ where P is a reduced nondegenerate quadratic $\mathfrak{o}$-module. Hence

$$F \perp (K \otimes P) \cong E \cong F \perp G.$$

Now $G \cong K \otimes P$ by the Cancellation Theorem over K (Theorem 1.67). Therefore G has FR with respect to λ. $\square$

To the currently developed theory of fair reduction, we can associate a variation of the weak specialization theory of §1.7, which we will briefly present in the following.

Theorem 2.29 (Extension of Theorem 1.82). *Let $\mathfrak{o}$ be quadratically henselian. Let (M, q) be a reduced nondegenerate and anisotropic quadratic $\mathfrak{o}$-module. Furthermore, let e be a primitive vector in M. Then $q(e) \in \mathfrak{o}^*$.*

Proof. We choose a decomposition $M = N \perp M'$ with N strictly regular and $B(M' \times M') \subset \mathfrak{m}$. Traversing the proof of Theorem 1.82 and replacing $M^\perp$ by M' everywhere will give the proof, since all arguments will faithfully remain valid. $\square$

As before, $\mathfrak{o}^h$ denotes the henselization of the valuation ring $\mathfrak{o}$.

Lemma 2.30. *Let M be a quadratic $\mathfrak{o}$-module. Then $M^h := \mathfrak{o}^h \otimes_\mathfrak{o} M$ is reduced nondegenerate if and only if M is reduced nondegenerate.*

Proof. This is evident since $\mathfrak{o}^h/\mathfrak{m}^h = k$, so that $M^h/\mathfrak{m}^h M^h$ is canonically isomorphic to the quadratic k-module $M/\mathfrak{m}M$. $\square$

Now the road is clear to extend, in an appropriate way, the main result of §1.7 (Theorem 1.96), using fair reduction. As before, let $\lambda : K \to L \cup \infty$ be a place with associated valuation ring $\mathfrak{o}$. Let S be a system of representatives of $Q(K)/Q(\mathfrak{o})$ in $\mathfrak{o}$, as introduced in §1.7.

Definition 2.31. Let $E = (E, q)$ be a quadratic K-module (always free of finite rank). We say that E is *weakly obedient with respect to λ* (or: with respect to $\mathfrak{o}$) when E has a decomposition

$$E = \bigperp_{s \in S} E_s \tag{2.9}$$

such that the quadratic K-module $(E_s, s^{-1}(q|E_s))$ has FR with respect to λ for every $s \in S$. Every decomposition of the form (2.9) is called a *weakly λ-modular* (or: *weakly $\mathfrak{o}$-modular*) *decomposition* of E.

Remark. Let E be weakly obedient with respect to λ. Then

$$E \cong \underset{s \in S}{\perp} \langle s \rangle \otimes (K \otimes M_s)$$

with reduced nondegenerate quadratic o-modules M_s. For every $s \in S$ we may choose a decomposition $M_s = N_s \perp M'_s$ with strictly regular N_s and $B_s(M'_s \times M'_s) \subset \mathfrak{m}$, where B_s is the bilinear form associated to M_s. Just as indicated in §1.7, we see that

$$G := \underset{s \in S}{\perp} \langle s \rangle \otimes (K \otimes M'_s)$$

is anisotropic. Furthermore,

$$F := \underset{s \in S}{\perp} \langle s \rangle \otimes (K \otimes N_s)$$

is strictly regular. We have $E \cong F \perp G$. Therefore E is definitely not degenerate and is thus a space over K.

Theorem 2.32 (Extension of Theorem 1.96). *Let E be a space over K which is weakly obedient with respect to λ and let*

$$E = \underset{s \in S}{\perp} E_s = \underset{s \in S}{\perp} F_s$$

be two weakly λ-modular decompositions of E. Then[6] $\lambda_*(E_1) \sim \lambda_*(F_1)$.

Proof. The arguments in the proof of Theorem 1.96 remain valid in the current more general situation. One should use Theorem 2.29 and Lemma 2.30 above. □

Now the following definition makes sense:

Definition 2.33. Let E be a quadratic space, weakly obedient with respect to λ and let $E = \underset{s \in S}{\perp} E_s$ be a weakly λ-modular decomposition of E. Then we call $\lambda_*(E_1)$ a *weak specialization* of E with respect to λ. We denote the Witt class of $\lambda_*(E_1)$ by $\lambda_W(E)$. In other words, $\lambda_W(E) := \{\lambda_*(E_1)\} \in \widehat{W}q(L)$. By Theorem 2.32, $\lambda_W(E)$ is uniquely determined by E and λ.

Remark. Our proof of Theorem 2.32 is independent of the main result Theorem 2.19 of the specialization theory developed above. We could also have deduced the important Corollary 2.20 of Theorem 2.19 (specialization by FR is well-defined) from Theorem 2.32, analogous to the proof of Theorem 1.105 in the quadratic case. In other words, we could have established the weak specialization theory first and then develop from this the basic idea of specialization by FR (Definition 2.21 above), just as before in §1.7 and §1.8.

In the theorems of §1.8 about the weak specialization of tensor products (Theorem 1.114(*ii*), Theorem 1.115) we may not simply replace the word GR by FR and

---

[6] See Definition 1.76 for the definition of Witt equivalence $\sim$.

the word "obedient" by "weakly obedient". This already shows us that in our theory, good reduction does not become superfluous in any way after the introduction of fair reduction.

More important even is the observation that in the theory of generic splitting in §2.1 and §2.2 good reduction appears centre stage, and not fair reduction: for example, let φ be a regular quadratic form over a field k and let $L \supset k$ be a field extension. Let $(K_r \mid 0 \leq r \leq h)$ be a generic splitting tower of φ. If $\lambda : K_r \to L \cup \infty$ is a place over k for some $r \in \{1, \ldots, h\}$, then the kernel form φ_r of $\varphi \otimes K_r$ automatically has *good* reduction with respect to λ.

2.4 Unified Theory of Generic Splitting

Just as in §2.1 and §2.2 we will not use geometric language and talk of quadratic forms instead of quadratic spaces in this section. The definitions concerning quadratic spaces over fields, encountered in §2.3, will be faithfully adopted in the language of forms. *In what follows, a "form" will always be understood to be a quadratic form.*

If $\gamma : k \to L \cup \infty$ is a place and φ a *regular* form over k which has GR with respect to γ, then the splitting behaviour of $\gamma_*(\varphi)$ under extensions of the field L is controlled by a given generic splitting tower $(K_r \mid 0 \leq r \leq h)$ of φ, as seen in §1.4 and §2.1 (Theorem 2.5 and generalization of Scholium 1.46). According to §2.2, something similar happens for a place γ and a form φ over k having GR with respect to γ for which $\gamma_*(\varphi)$ is nondegenerate, provided some care is exercised. Now we want to extend these results to the case where φ has just FR with respect to γ.

At the same time we want to unite the results of §2.1 and §2.2 under one roof, starting with the following definition.

Definition 2.34. Let φ be a nondegenerate form over a field K. We call a field extension $K \hookrightarrow L$ φ-*conservative* (or: *conservative for* φ) if the form $\varphi \otimes L$ is again nondegenerate, i.e. if the quasilinear part $QL(\varphi)$ of φ remains anisotropic under a field extension from K to L.

If φ is regular, then $\dim QL(\varphi) \leq 1$ and so every field extension of K is φ-conservative. This is true in particular when char $K \neq 2$. If char $K = 2$, then every separable field extension $K \hookrightarrow L$ is φ-conservative by Theorem 2.7.

Theorem 2.35. *Let* $\lambda : K \to L \cup \infty$ *be a place,* $K' \supset K$ *a field extension and* $\mu : K' \to L \cup \infty$ *an extension of* λ. *Let* φ *be a (nondegenerate) form over* K, *having* FR *(resp.* GR*) with respect to* λ, *and let* $\lambda_*(\varphi)$ *be nondegenerate.*

Claim: the extension $K \hookrightarrow K'$ *is* φ-*conservative. The form* $\varphi \otimes K'$ *has* FR *(resp.* GR*) with respect to* μ *and* $\mu_*(\varphi \otimes K') \cong \lambda_*(\varphi)$.

Proof. In the case of good reduction, this is Theorem 2.8. The proof in the case of fair reduction is similar and goes as follows.

Let $\mathfrak{o} := \mathfrak{o}_\lambda$ and $\mathfrak{o}' := \mathfrak{o}_\mu$. Let $E = (E, q)$ be a space corresponding to φ and M a reduced nondegenerate quadratic $\mathfrak{o}$-module with $E \cong K \otimes_\mathfrak{o} M$. Then $\lambda_*(\varphi)$ corresponds to the space $\lambda_*(E) \cong L \otimes_{\overline{\lambda}} \overline{M}$, where $\overline{\lambda} : \kappa(\mathfrak{o}) \hookrightarrow L$ is the field extension associated to λ. We have the factorization $\overline{\lambda} = \overline{\mu} \circ j$, featuring the inclusion $j : \kappa(\mathfrak{o}) \hookrightarrow \kappa(\mathfrak{o}')$ and the field homomorphism $\overline{\mu} : \kappa(\mathfrak{o}') \hookrightarrow L$, determined by μ. Let $\overline{\varphi}$ be the form over $\kappa(\mathfrak{o})$, associated to $\overline{M}$, i.e. $\lambda_*(\varphi) = \overline{\varphi} \otimes_{\overline{\lambda}} L$. Since $\lambda_*(\varphi)$ is nondegenerate by assumption, $\overline{\lambda} : \kappa(\mathfrak{o}) \hookrightarrow L$ is conservative for $\overline{\varphi}$. Therefore $j : \kappa(\mathfrak{o}) \hookrightarrow \kappa(\mathfrak{o}')$ is conservative for $\overline{\varphi}$ and $\overline{\mu} : \kappa(\mathfrak{o}') \hookrightarrow L$ is conservative for $\overline{\varphi} \otimes \kappa(\mathfrak{o}')$. Now $\overline{\varphi} \otimes \kappa(\mathfrak{o}')$ belongs to the quadratic $\kappa(\mathfrak{o}')$-module $\overline{M}'$ with $M' := \mathfrak{o}' \otimes_{\mathfrak{o}} M$. Therefore $\overline{M}'$ is nondegenerate, in other words M' is reduced nondegenerate. We have $\overline{M}' = \kappa(\mathfrak{o}') \otimes_{\kappa(\mathfrak{o})} \overline{M}$. Furthermore, $K' \otimes E \cong K' \otimes_{\mathfrak{o}'} M'$. Thus $\varphi \otimes K'$ has FR with respect to μ and

$$\mu_*(\varphi \otimes K') \cong (\overline{\varphi} \otimes \kappa(\mathfrak{o}')) \otimes_{\overline{\mu}} L = \overline{\varphi} \otimes_{\overline{\lambda}} L \cong \lambda_*(\varphi).$$

In particular, $\varphi \otimes K'$ is nondegenerate (cf. Remark 2.22). We conclude that $K \hookrightarrow K'$ is φ-conservative.   □

In the following let φ be a nondegenerate form over a field k. Over every field K we denote the form $\begin{bmatrix} 0 & 1 \\ 1 & 0 \end{bmatrix}$ by H.

Theorem 2.36. *Let K and L be extensions of the field k and let $\lambda : K \to L \cup \infty$ be a place over k. Suppose further that L is φ-conservative and let $\varphi \otimes K \cong r \times H \perp \psi$ be the Witt decomposition of φ.*

Claim: K is also a φ-conservative extension of k. The form ψ has GR with respect to λ, and $\varphi \otimes L \cong r \times H \perp \lambda_*(\psi)$. Thus $\mathrm{ind}(\varphi \otimes L) \geq \mathrm{ind}(\varphi \otimes K)$.

Proof. We apply Theorem 2.35 to the trivial place $\gamma : k \hookrightarrow L$ and its extension λ. Now φ has GR with respect to γ and $\gamma_*(\varphi) = \varphi \otimes L$. By Theorem 2.35, K is conservative for φ, in other words $\varphi \otimes K$ is nondegenerate. Again by this theorem, $\varphi \otimes K$ has GR with respect to λ and $\lambda_*(\varphi \otimes K) \cong \varphi \otimes L$. Since $r \times H$ also has GR with respect to λ, it follows that ψ has GR with respect to λ (Theorem 1.108), and we have $\varphi \otimes L \cong r \times H \perp \lambda_*(\psi)$.   □

Corollary 2.37. *Let K and L be specialization equivalent extensions of the field k. Then K is φ-conservative if and only if L is φ-conservative. In this case we have $\mathrm{ind}(\varphi \otimes K) = \mathrm{ind}(\varphi \otimes L)$. Furthermore, if $\lambda : K \to L \cup \infty$ is a place over k, then $\ker(\varphi \otimes K)$ has GR with respect to λ and $\lambda_*(\ker(\varphi \otimes K)) \cong \ker(\varphi \otimes L)$.*

Next we define a *generic splitting tower* $(K_r \mid 0 \leq r \leq h)$ associated to φ just as in §1.4 for char $k \neq 2$ and in §2.1 for regular forms, with *higher kernel forms* φ_r and *higher indices* i_r $(0 \leq r \leq h)$: K_0 is an inessential extension of k, $\varphi_r := \ker(\varphi \otimes K_r)$, $K_{r+1} \sim_k K_r(\varphi_r)$, $i_{r+1} := \mathrm{ind}(\varphi_r \otimes K_{r+1})$, in case $r < h$. The construction stops with step h if $\dim \varphi_h - \dim QL(\varphi_h) \leq 1$.

All this makes sense, since an inductive argument shows that $QL(\varphi_r) = QL(\varphi) \otimes K_r$ is anisotropic for every $r \in \{0, \dots h\}$. Indeed, the field extension $K_r(\varphi_r)/K_r$ is separable for $r < h$. Hence we can conclude by Theorem 2.36 that K_{r+1}/K_r is φ_r-conservative once we already know that $QL(\varphi_r)$ is anisotropic, i.e. that φ_r is nondegenerate. The extension K_h/k is therefore φ-conservative.

Note. If char $k \neq 2$, then $\dim \varphi_h \leq 1$. If char $k = 2$, then $\varphi_h = QL(\varphi_h)$.

We now have to convince ourselves that the field tower $(K_r \mid 0 \le r \le h)$ really accomplishes what we expect from a generic splitting tower. As before in more special situations, it suffices for this purpose to prove a theorem of the following sort.

Theorem 2.38 (Extension of Theorem 2.5 and Theorem 2.10). *Let $\gamma : k \to L \cup \infty$ be a place. Suppose that the form φ has FR with respect to γ and that $\gamma_*(\varphi)$ is nondegenerate. Then $\gamma_*(\varphi)$ is isotropic if and only if γ can be extended to a place $\lambda : k(\varphi) \to L \cup \infty$.*

Proof. We have $\varphi \otimes k(\varphi) \cong H \perp \psi$ for a form ψ over $k(\varphi)$. If there exists a place $\lambda : k(\varphi) \to L \cup \infty$ then, by Theorem 2.35, $\varphi \otimes k(\varphi)$ has FR with respect to λ and $\lambda_*(\varphi \otimes k(\varphi)) \cong \gamma_*(\varphi)$. Then it follows from Theorem 2.28 that ψ has FR with respect to λ and that $\gamma_*(\varphi) \cong H \perp \lambda_*(\psi)$. Hence $\gamma_*(\varphi)$ is isotropic.

Conversely, suppose that the nondegenerate form $\overline{\varphi} := \gamma_*(\varphi)$ is isotropic. Since $H < \overline{\varphi}$ we have $\dim \overline{\varphi} - \dim QL(\overline{\varphi}) \ge 2$. Now, by earlier work, we see that γ can be extended to a place $\tilde{\gamma} : k(\varphi) \to L(\overline{\varphi}) \cup \infty$ (look again at the proof of Theorem 2.3). {Note: here we do not need the assumption that $\overline{\varphi}$ is isotropic, but only that $\dim \overline{\varphi} - \dim QL(\overline{\varphi}) \ge 2$.} Since $\overline{\varphi}$ is isotropic, $L(\overline{\varphi})$ is a purely transcendental extension of L (Lemma 2.1). Thus there exists a place $\rho : L(\overline{\varphi}) \to L \cup \infty$ over L. The place $\lambda := \rho \circ \tilde{\gamma} : k(\varphi) \to L \cup \infty$ is an extension of γ.   □

In what follows, let $(K_r \mid 0 \le r \le h)$ be a generic splitting tower associated to φ with higher kernel forms φ_r and indices i_r.

Theorem 2.39 (Extension of Theorem 1.42 and Theorem 2.12). *Let $\gamma : k \to L \cup \infty$ be a place with respect to which φ has FR (resp. GR). Suppose that the form $\gamma_*(\varphi)$ is nondegenerate. Finally, suppose that $\lambda : K_m \to L \cup \infty$ is a place for some $m \in \{0, \ldots, h\}$, which extends γ and which cannot be extended to K_{m+1} in case $m < h$. Then φ_m has FR (resp. GR) with respect to λ. The form $\gamma_*(\varphi)$ has kernel form $\lambda_*(\varphi_m)$ and Witt index $i_0 + \cdots + i_m$.*

Proof. We give the proof for FR. The case of GR can be treated in an analogous way. We have an isometry

$$\varphi \otimes K_m \cong \varphi_m \perp (i_0 + \cdots + i_m) \times H. \tag{2.10}$$

By Theorem 2.35, $\varphi \otimes K_m$ has FR with respect to λ and $\lambda_*(\varphi \otimes K_m) \cong \gamma_*(\varphi)$. From (2.10) we get, according to Theorem 2.28, that φ_m has FR with respect to λ and that

$$\gamma_*(\varphi) \cong \lambda_*(\varphi_m) \perp (i_0 + \cdots + i_m) \times H. \tag{2.11}$$

In particular, $\lambda_*(\varphi_m)$ is nondegenerate. If $\lambda_*(\varphi_m)$ were isotropic, then we would surely have that $\dim \varphi_m \ge \dim QL(\varphi_m) + 2$, i.e. $m < h$. But then it follows from Theorem 2.38 that λ can be extended to a place $\mu : K_{m+1} \to L \cup \infty$: contradiction! Therefore $\lambda_*(\varphi_m)$ is anisotropic and (2.11) is the Witt decomposition of $\gamma_*(\varphi_m)$.   □

By applying this theorem to the trivial place γ, we obtain

Scholium 2.40 (Extension of Scholium 1.43). *Let $L \supset k$ be a φ-conservative field extension.*

(1) *Let $\lambda : K_m \to L \cup \infty$ be a place over k for some $m \in \{0, \ldots, h\}$, which* cannot *be extended to a place from K_{m+1} to L in case $m < h$. Then φ_m has good reduction with respect to λ and $\lambda_*(\varphi_m)$ is the kernel form of $\varphi \otimes L$.*

(2) *If $\lambda' : K_r \to L \cup \infty$ is a place over k, then $r \leq m$ and λ' can be extended to a place $\mu : K_m \to L \cup \infty$.*

(3) *For a given number t with*

$$0 \leq t \leq \left\lceil \frac{\dim \varphi - \dim QL(\varphi)}{2} \right\rceil = i_0 + \cdots + i_h$$

and $m \in \mathbb{N}_0$ minimal such that $t \leq i_0 + \cdots + i_m$, K_m is a generic φ-conservative field extension of k for the splitting off of t hyperbolic planes of φ (cf. the problem posed at the beginning of §1.4).

Thus the field tower $(K_r \mid 0 \leq r \leq h)$ rightfully merits the name "generic splitting tower for φ" also in the current general situation.

Definition 2.41. As before we call h the *height* of the form φ and write $h = h(\varphi)$. Further, we define the *splitting pattern* $SP(\varphi)$ as the set of all indices $\mathrm{ind}(\varphi \otimes L)$, where L runs through all φ-conservative field extensions of k. By the scholium we have

$$SP(\varphi) = \{i_0 + \cdots + i_r \mid 0 \leq r \leq h\}.$$

Definition 2.42. We call any field extension $E \supset k$ which is specialization equivalent to $k(\varphi)$ over k a *generic zero field of* φ. Further, for $r \in \{0, \ldots, h\}$, we call any field extension $F \supset k$ which is specialization equivalent to K_r over k a *partial generic splitting field of* φ or, more precisely, a *generic splitting field of level r*. This signifies that F is generic for the splitting off of as many hyperbolic planes as the $(r + 1)$-th number $i_0 + \cdots + i_r$ in $SP(\varphi)$ indicates. In case $r = h$, we speak of a *total generic splitting field*.

This terminology generalizes notions from §1.4 and §2.1, but one must exercise some caution; the extensions $E \supset k$ and $F \supset k$ have the mentioned generic properties—as far as we know— *only within the class of φ-conservative field extensions of k.*

Comment on the Theory so far. If the form φ over k is degenerate, but $\dim \varphi - \dim QL(\varphi) \geq 2$, then the function field $k(\varphi)$ can be constructed just as in the nondegenerate case. Thus we can formally define a "generic splitting tower" $(K_r \mid 0 \leq r \leq h)$ as above. We may also inquire about the splitting behaviour of φ under extensions $L \supset k$ of the field k, in other words about the possibilities for the index $\mathrm{ind}(\varphi \otimes L)$ and the kernel form $\ker(\varphi \otimes L)$ (see Definition 1.76). However, can we also extend the theory so far to degenerate forms?

We have a decomposition $\varphi \cong \widehat{\varphi} \perp \delta(\varphi)$ with $\delta(\varphi) \cong t \times [0]$ and $\widehat{\varphi}$ nondegenerate, cf. Definition 1.74 ff. Clearly $k(\varphi)$ is a purely transcendental extension of $k(\widehat{\varphi})$

of transcendence degree t. Therefore $k(\varphi) \sim_k k(\widehat{\varphi})$. Consequently, the above tower $(K_r \mid 0 \le r \le h)$ is a generic splitting tower of $\widehat{\varphi}$. It is now clear that our theorems so far all remain valid for degenerate φ, as long as we complete Definition 2.34 above as follows: for degenerate φ, a field extension $K \hookrightarrow L$ is called conservative for φ if it is conservative for $\widehat{\varphi}$.

This, however, furnishes us with a fairly bland extension of the theory so far to degenerate forms. The obvious, difficult question seems to be: Let φ be a nondegenerate form over a field k and $L \supset k$ a field extension with $\varphi \otimes L$ degenerate. Is it so that the splitting behaviour of $\varphi \otimes L$ under field extensions of L is controlled by a given generic splitting tower $(K_r \mid 0 \le r \le h)$ of φ in a similar way as indicated in Scholium 2.40(1) above? For example, is $\text{SP}(\varphi \otimes L) \subset \text{SP}(\varphi)$?

Our theory does not give any information here. The main problem seems to occur in Theorem 2.35 above. If we do not know there that $\lambda_*(\varphi)$ is nondegenerate, we cannot conclude—as far as I can see—that $\widehat{\varphi \otimes K'}$ has fair reduction with respect to μ. $\qquad\square$

After this digression, we suppose again that φ is a *nondegenerate* form over k and that $(K_r \mid 0 \le r \le h)$ is a generic splitting tower of φ with higher kernel forms φ_r and indices i_r. From Theorem 2.39 above we immediately obtain a literal repetition of Scholium 1.45 in the current, more general situation, if we bear in mind that for every generic splitting tower $(K'_s \mid 0 \le s \le h)$ of φ, all extensions K'_s of k are conservative for φ. Furthermore, we obtain from Theorem 2.39 an extension of Scholium 1.42 in the same way as we obtained Scholium 1.46 from Theorem 1.42:

Scholium 2.43. *Let $\gamma : k \to L \cup \infty$ be a place with respect to which φ has FR (resp. GR). Suppose that $\gamma_*(\varphi)$ is nondegenerate. Then:*

(1) $\text{SP}(\gamma_*(\varphi)) \subset \text{SP}(\varphi)$.

(2) *The higher kernel forms of $\gamma_*(\varphi)$ arise from certain higher kernel forms of φ by means of specialization. More precisely: if $(L_s \mid 0 \le s \le e)$ is a generic splitting tower of $\gamma_*(\varphi)$, then $e \le h$ and, for every s with $0 \le s \le e$, we have*

$$\text{ind}(\gamma_*(\varphi) \otimes L_s) = i_0 + \cdots + i_m$$

with $m \in \{0, \ldots, h\}$. The number m is the biggest integer such that γ can be extended to $\lambda : K_m \to L_s \cup \infty$. The kernel form φ_m of $\varphi \otimes K_m$ has FR (resp. GR) with respect to every extension λ of this kind, and $\lambda_*(\varphi_m)$ is the kernel form of $\gamma_*(\varphi) \otimes L_s$.

(3) *If $\rho : K_r \to L_s \cup \infty$ is a place, which extends $\gamma : k \to L \cup \infty$, then $r \le m$ and ρ can be further extended to a place from K_m to L_s.*

2.5 Regular Generic Splitting Towers and Base Extension

Before turning towards generic splitting towers, we will give two general definitions which will also serve us well in later sections.

As before, the word *form* over a field k will always be understood to mean a nondegenerate quadratic form over k. In the following let φ be a form over k.

Definition 2.44. Let $\dim \varphi$ be even, $\varphi \neq 0$ and $QL(\varphi) = 0$.

(a) The *discriminant algebra* $\Delta(\varphi)$ is defined as follows: If char $k \neq 2$, we let $\Delta(\varphi) :=$ $k[X]/(X^2 - a)$, where a is a representative of the signed determinant $d(\tilde{\varphi}) = ak^{*2}$ of the bilinear form $\tilde{\varphi} = B_\varphi$ associated to φ (cf. §1.2). If char $k = 2$, we let $\Delta(\varphi) := k[X]/(X^2 + X + c)$, where c is a representative of the Arf-invariant $\mathrm{Arf}(\varphi) \in k^+/\wp k$.

(b) We define the *discriminant* of φ to be the isomorphism class of $\Delta(\varphi)$ as k-algebra. We will denote it sloppily by $\Delta(\varphi)$ as well. The discriminant is independent of the choice of a resp. c above.

(c) We say that $\Delta(\varphi)$ *splits* when $\Delta(\varphi)$ is not a field, i.e. when $\Delta(\varphi) \cong k \times k$. We will symbolically write $\Delta(\varphi) = 1$ when φ splits and $\Delta(\varphi) \neq 1$ when φ doesn't split.

Remark 2.45. This notation is not completely groundless, for the isomorphism classes of quadratic separable k-algebras form a group in a natural way with unit element $k \times k$. In this group the equality $\Delta(\varphi \perp \psi) = \Delta(\varphi)\Delta(\psi)$ holds. We will not discuss the group of quadratic separable k-algebras any deeper since we will not make any serious use of it.

Remark 2.46. If K/k is a field extension,[7] then $\Delta(\varphi \otimes K) = \Delta(\varphi) \otimes_k K$.

Remark 2.47. If τ is the norm form of the k-algebra $\Delta(\varphi)$, then $\Delta(\varphi) = \Delta(\tau)$. If further $\dim \varphi = 2$, then $\varphi \cong c\tau$ for some $c \in k^*$. If we write $\tau = \left[\begin{smallmatrix} 1 & 1 \\ 1 & \delta \end{smallmatrix}\right]$, then $\Delta(\varphi) = k[X]/(X^2 + X + \delta)$, also when char $k \neq 2$.

Definition 2.48 (cf. [37, Def.1.1]). Let $\dim \varphi > 1$. We say that "φ *is of outer type*" when $\dim \varphi$ is even, $QL(\varphi) = 0$ and $\Delta(\varphi) \neq 1$. In all other cases (thus in particular when $QL(\varphi) \neq 0$) we say that "φ *is of inner type*".

Remark 2.49. For those readers who are at home in the theory of reductive algebraic groups, we want to remark that Definition 2.48 leans on the concepts of inner/outer type in use there: is φ regular (i.e. $\dim QL(\varphi) \leq 1$) and $\dim \varphi \geq 3$, then the group $SO(\varphi)$ is almost simple. This group is of inner/outer type if and only if this is the case for φ in the sense of Definition 2.48.

We want to construct "regular" generic splitting towers of φ, having particularly convenient properties with respect to extensions of the ground field k, which are nonetheless sufficiently general. Starting with such a regular generic splitting tower of φ we will then construct a regular generic splitting tower of $\varphi \otimes L$ for every φ-conservative field extension L/k.

For every form ψ with $\dim \psi \geq 2 + \dim QL(\psi)$ over a field K we introduced the field extension $K(\psi)$ of K earlier (§1.4; §2.1; §2.2, following Definition 2.9).

---

[7] From now on we will often denote a field extension $F \hookrightarrow E$ by E/F, as has been customary in algebra for a long time.

$K(\psi)$ was defined to be the function field of the affine quadric $\psi(X) = 0$ over K, except when $\psi \cong \begin{bmatrix} 0 & 1 \\ 1 & 0 \end{bmatrix}$, in which case this quadric degenerates in two lines. In this situation we defined $K(\psi) = K(t)$ for some indeterminate t over K. Sometimes it is more natural to use the *projective* quadric $\psi(X) = 0$ over K instead of the affine quadric $\psi(X) = 0$. Thus we arrive at the following:

Definition 2.50. Let ψ be a (nondegenerate quadratic) form over K with $\dim \psi \geq 2 + \dim QL(\psi)$. We define a field extension $K\{\psi\}$ of K as follows: If $\psi \cong \begin{bmatrix} 0 & 1 \\ 1 & 0 \end{bmatrix}$, we let $K\{\psi\} = K$. Otherwise we let $K\{\psi\}$ be the subfield $K(\frac{x_1}{x_i}, \ldots, \frac{x_n}{x_i})$ of $K(\psi) = \mathrm{Quot}\big(K[X_1, \ldots, X_n]/(\psi(X_1, \ldots, X_n))\big) = K(x_1, \ldots, x_n)$ for some $i \in [1, n]$.[8]

In the main case $\psi \not\cong \begin{bmatrix} 0 & 1 \\ 1 & 0 \end{bmatrix}$ this field is obviously independent of the choice of $i \in [1, n]$. We have $K(\psi) = K\{\psi\}(x_i)$ and x_i is transcendental over $K\{\psi\}$. Also in the case $\psi \cong \begin{bmatrix} 0 & 1 \\ 1 & 0 \end{bmatrix}$ we have $K(\psi) = K\{\psi\}(t)$ where t is an indeterminate.

$K(\psi)$ is an inessential extension of $K\{\psi\}$. Thus, all this time we could have used $K\{\psi\}$ instead of $K(\psi)$.

Back to our form φ over k! Let $h = h(\varphi)$ be the height of φ.

Definition 2.51. The *projective standard tower* of φ is the field tower $(K_r \mid 0 \leq r \leq h)$ with $K_0 = k$, $K_{r+1} = K_r\{\varphi_r\}$ $(0 \leq r \leq h - 1)$, where φ_r denotes the kernel form of $\varphi \otimes K_r$. Similarly one obtains the *affine standard tower* $(K'_r \mid 0 \leq r \leq h)$ of φ by replacing the $K_r\{\psi_r\}$ with the function fields $K_r(\varphi_r)$ of the affine quadrics $\varphi_r = 0$.

Remark 2.52. From the definition of generic splitting tower (§2.4, just after Theorem 2.36) it follows immediately that $(K_r \mid 0 \leq r \leq h)$ and $(K'_r \mid 0 \leq r \leq h)$ are both generic splitting towers of φ. For every $r \in \{0, \ldots, h\}$ we have that K'_r is a purely transcendental field extension of K_r of transcendence degree r.

Let us recall that a field extension L/K is called *regular* when it is separable and K is algebraically closed in L. This is synonymous with saying that L and the algebraic closure $\widetilde{K}$ of K in L are linearly disjoint, i.e. that $L \otimes_K \widetilde{K}$ is a field, cf. [45, Chap. 3]. If L/K is regular, then the K-algebra $L \otimes_K E$ does not contain zero divisors for any field extension E/K.

Theorem 2.53. *Let $(K_r \mid 0 \leq r \leq h)$ be the projective standard tower of φ. Then the field extensions K_r/K_{r-1} with $0 < r < h$ are all regular. The same is true for K_h/K_{h-1} in case $h > 0$ and φ is of inner type. Furthermore, the kernel form φ_{h-1} of $\varphi \otimes K_{h-1}$ has dimension ≥ 3 in this situation. If φ is of outer type, then $h \geq 1$ and $K_h = K_{h-1} \otimes_k \Delta(\varphi)$. Furthermore, we then have $\varphi_{h-1} \cong c(\tau \otimes K_{h-1})$, where τ is the norm form of the field extension $\Delta(\varphi)$ of k and $c \in K^*_{h-1}$ is a constant. In particular we have $\dim \varphi_{h-1} = 2$.*

Proof. If ψ is a nondegenerate quadratic form over a field K and if $\dim \psi > 2$, then the extension $K\{\psi\}/K$ is clearly regular. This established the first statement of the theorem. If $h > 0$ and φ is of inner type, then $\varphi \otimes K_{h-1}$ is also of inner type and

---

[8] In [33] this field is denoted by $K(\psi)_0$. $[1, n]$ denotes the set $\{1, 2, \ldots, n\}$.

thus the kernel form φ_{h-1} of $\varphi \otimes K_{h-1}$ is of inner type. This shows that $\dim \varphi_{h-1} \geq 3$ and so $K_h = K_{h-1}\{\varphi_{h-1}\}$ over K_{h-1} is again regular. Suppose now that φ is of outer type. φ does not split, so that $h > 0$. The extension K_{h-1}/K_0 is regular and K_0/k is inessential. Therefore k is algebraically closed in K_{h-1} and $\Delta(\varphi_{h-1}) = \Delta(\varphi \otimes K_{h-1}) = \Delta(\varphi) \otimes_k K_{h-1}$ is a field. If it would be true that $\dim \varphi_{h-1} > 2$, then $K_h = K_{h-1}\{\varphi_{h-1}\}$ over K_{h-1} would be regular. However, $\Delta(\varphi_{h-1}) \otimes_{K_{h-1}} K_h = \Delta(\varphi \otimes K_h)$ is not a field since $\varphi \otimes K_h \sim 0$. Therefore the dimension of φ_{h-1} is 2. By Remarks 2.46 and 2.47 above, we have $\Delta(\varphi_{h-1}) = \Delta(\varphi \otimes K_{h-1}) = \Delta(\varphi) \otimes_k K_{h-1} = \Delta(\tau) \otimes_k K_{h-1} = \Delta(\tau \otimes K_{h-1})$ (and $\tau \otimes K_{h-1}$ is the norm form of K_h/K_{h-1}). We conclude that $\varphi_{h-1} \cong c(\tau \otimes K_{h-1})$ for some element c of K_{h-1}.   □

Theorem 2.54. *Let $r \in [0, h]$ and let E/k be a partially generic splitting field of level r of φ.*
(a) If $r < h$, then k is algebraically closed in E.
(b) If $r = h$ and φ is of inner type, then k is likewise algebraically closed in E.
(c) If $r = h$ and φ is of outer type, then k has algebraic closure $\Delta(\varphi)$ in E.

Proof. By Theorem 2.53 these statements hold when E is the r-field K_r in the projective standard tower $(K_i \mid 0 \leq i \leq h)$ of φ. The theorem now follows in all generality from the following simple lemma.   □

Lemma 2.55. *Let K and L be extensions of the field k which are specialization equivalent, $K \sim_k L$. Then every place $\lambda : K \to L \cup \infty$ over k maps the algebraic closure of k in K isomorphically to the algebraic closure of k in L.*

Proof. Let K° be the algebraic closure of k in K and L° the algebraic closure of k in L. Let $\lambda : K \to L \cup \infty$ and $\mu : L \to K \cup \infty$ be places over k. Since K° and L° are algebraic over k, the places λ and μ are finite on K° resp. L°. Their restrictions to K° and L° are thus field homomorphisms $\lambda' : K^\circ \to L^\circ, \mu' : L^\circ \to K^\circ$, which are both the identity on k. Now $\mu' \circ \lambda'$ is an endomorphism of K°/k and thus automatically an automorphism of K°/k. Likewise, $\lambda' \circ \mu'$ is an automorphism of L°/k. The statement of the lemma is now clear.   □

Now we can precisely formulate a desirable property of generic splitting towers.

Definition 2.56. We call a generic splitting tower $(K_r \mid 0 \leq r \leq h)$ of φ *regular* when the following holds:
(1) For every r with $0 \leq r < h - 1$ the extension K_{r+1}/K_r is regular.
(2) If φ is of inner type and $h > 0$, then K_h/K_{h-1} is also regular. On the other hand, if φ is of outer type, then K_h is regular over the composite $K_{h-1} \cdot \Delta(\varphi) = K_{h-1} \otimes_k \Delta(\varphi)$.

Example. The projective standard tower of φ is regular by Theorem 2.53. The affine standard tower of φ is likewise regular.

Now let $(K_r \mid 0 \leq r \leq h)$ be a regular generic splitting tower of φ and let L/k be a φ-conservative field extension of k. Using $(K_r \mid 0 \leq r \leq h)$ and L we want to construct a generic splitting tower of $\varphi \otimes L$.

Definition 2.57. For every $r \in \{0, 1, \ldots, h\}$ we construct a field composite $K_r \cdot L$ of K_r and L over k as follows: if $r < h$, or if $r = h$ and φ is of inner type, then $K_r \otimes_k L$ is free of zero divisors and we let $K_r \cdot L$ be the quotient field of this ring, i.e. the uniquely determined free composite of K and L over k. If $r = h$ and φ is of outer type, we distinguish the cases where $\varphi \otimes L$ is of outer/inner type.

First, let $\varphi \otimes L$ be of outer type. Then $\Delta(\varphi) \otimes_k L$ is a field. Let $K_h \cdot L$ be the free composite of K_h with $\Delta(\varphi) \otimes_k L$ over $\Delta(\varphi)$, i.e. again the quotient field of $K_h \otimes_{\Delta(\varphi)} (\Delta(\varphi) \otimes_k L) = K_h \otimes_k L$. Finally, let $\varphi \otimes L$ be of inner type. Now $\Delta(\varphi)$ can be embedded over k in L in two ways. We choose one such embedding and set $K_h \cdot L = \mathrm{Quot}(K_h \otimes_{\Delta(\varphi)} L)$.

If necessary, we write more precisely $K_r \cdot_k L$ $(0 \le r \le h)$ for the field composite $K_r \cdot L$. Later on we will call $K_r \cdot L$ sloppily *"the" free composite of K_r and L over k*, also in the case $r = h$, φ of outer type, $\varphi \otimes L$ of inner type. It will never matter which of the two embeddings $\Delta(\varphi) \hookrightarrow L$ we have chosen.

We use φ_r to denote — as before — the kernel form of $\varphi \otimes K_r$ (rth higher kernel form) and i_r to denote the Witt index of $\varphi \otimes K_r$ (rth higher index, $0 \le r \le h$).

Theorem 2.58. *Let J be the set of all $r \in [0, h]$ such that $\varphi_r \otimes K_r \cdot L$ is anisotropic, i.e. such that $\mathrm{ind}(\varphi \otimes K_r) = \mathrm{ind}(\varphi \otimes K_r \cdot L)$. Let $r_0 < r_1 < \cdots < r_e$ be the elements of J.*

Claim:

(a) $(K_{r_i} \cdot L \mid 0 \le i \le e)$ is a regular generic splitting tower of $\varphi \otimes L$.

(b) $K_{r+1} \cdot L / K_r \cdot L$ is a regular inessential extension for every $r \in [0, h] \setminus J$.

For the proof of this technically very important theorem, we need a general lemma about places.

Lemma 2.59. *Let $K \supset k$ and $L \supset k$ be arbitrary extensions of a field k and let $\lambda : K \to L \cup \infty$ be a place over k. Furthermore, let $E \supset k$ be a field extension which is linearly disjoint from K and L over k. Consider the free composites $K \cdot E$ and $L \cdot E$ of K resp. L with E over k.*

Claim: λ has a unique extension $\tilde{\lambda} : K \cdot E \to L \cdot E \cup \infty$ to a place over E.

Proof. $K \cdot E$ is the quotient field of $K \otimes_k E$ and $L \cdot E$ is the quotient field of $L \otimes_k E$. Let $\mathfrak{o} := \mathfrak{o}_\lambda$ and let $\alpha : \mathfrak{o} \otimes_k E \to L \otimes_k E$ be the homomorphism of E-algebras induced by $\lambda|\mathfrak{o} : \mathfrak{o} \to L$. Following the general extension theorem for places [11, §2, Prop.3], we choose a place $\mu : K \cdot E \to \widetilde{(L \cdot E)} \cup \infty$ in the algebraic closure of $L \cdot E$ which extends the homomorphism α. We will now show that μ is the only such extension of α and that $\mu(K \cdot E) \subset (L \cdot E) \cup \infty$.

For this purpose we choose a basis $(\omega_i \mid i \in I)$ of E over k (as k-vector space). Let $z \ne 0$ be an element in $K \cdot E$. We write $z = \frac{x}{y}$, where $x, y \in \mathfrak{o} \otimes E$ are both non-zero. We then have equations

$$x = u \cdot \sum_{i \in I} a_i \otimes \omega_i, \quad y = v \cdot \sum_{i \in I} b_i \otimes \omega_i$$

with $u, v \in K$ and families $(a_i \mid i \in I)$ and $(b_i \mid i \in I)$ in $\mathfrak{o}$, both of them not fully contained in the maximal ideal $\mathfrak{m}$ of $\mathfrak{o}$. Then

$$\mu\Big(\sum_{i \in I} a_i \otimes \omega_i\Big) = \sum_{i \in I} \lambda(a_i)\omega_i \neq 0, \quad \mu\Big(\sum_{i \in I} b_i \otimes \omega_i\Big) = \sum_{i \in I} \lambda(b_i)\omega_i \neq 0,$$

since $(\omega_i \mid i \in I)$ is also a family of elements of $L \cdot E$, linearly independent over L. We thus obtain

$$\mu(z) = \lambda\Big(\frac{u}{v}\Big) \cdot \Big(\sum_{i \in I} \lambda(a_i)\omega_i\Big) \cdot \Big(\sum_{i \in I} \lambda(b_i)\omega_i\Big)^{-1} \in (L \cdot E) \cup \infty. \qquad \square$$

Proof of Theorem 2.58. We assume without loss of generality that φ is anisotropic and use induction on the height h. Remark that $K_0 \sim_k k$ implies that $K_0 \cdot L \sim_L L$ by the lemma. Therefore the extension $K_0 \cdot L/L$ is inessential.

For $h = 0$ nothing more has to be done, so suppose that $h > 0$. First we assume that $\dim \varphi > 2$. We will deal with the (easier) case $\dim \varphi = 2$ at the end.

The form $\varphi_1 = \ker(\varphi \otimes K_1)$ has height $h - 1$ and regular generic splitting tower $(K_r \mid 1 \leq r \leq h)$. We want to apply the induction hypothesis to φ_1, this tower and the field extension $K_1 \cdot L/K_1$.

Clearly the extension $K_1 \cdot L/K_0 \cdot L$ is regular (cf. e.g. [45, p.58]). Furthermore, $K_0 \cdot L/L$ is inessential and L/k φ-conservative. It follows that $K_1 \cdot L/k$ is φ-conservative and then that $K_1 \cdot L/K_1$ is φ_1-conservative. For every r with $1 \leq r \leq h - 1$ we have

$$K_r \cdot_k L = \mathrm{Quot}(K_r \otimes_k L) = \mathrm{Quot}(K_r \otimes_{K_1} (K_1 \otimes_k L)) = K_r \cdot_{K_1} (K_1 \cdot_k L).$$

Also, $K_h \cdot_k L = K_h \cdot_{K_1} (K_1 \cdot_k L)$. This can be seen using the same calculation in case φ is of inner type, and also when φ is of outer type and $\Delta(\varphi)$ cannot be embedded in L. If φ is of outer type and $\Delta(\varphi) \subset L$, we perform the following calculation (bearing Definition 2.57 in mind).

$$K_h \cdot_k L := K_h \cdot_{\Delta(\varphi)} L = \mathrm{Quot}(K_h \otimes_{\Delta(\varphi)} L)$$
$$= \mathrm{Quot}(K_h \otimes_{K_1 \cdot \Delta(\varphi)} (K_1 \cdot \Delta(\varphi) \otimes_{\Delta(\varphi)} L)).$$

We have $K_1 \cdot \Delta(\varphi) = K_1 \otimes_k \Delta(\varphi) = \Delta(\varphi_1)$ and do indeed obtain again

$$K_h \cdot_k L = \mathrm{Quot}(K_h \otimes_{\Delta(\varphi_1)} (K_1 \otimes_k L)) = K_h \cdot_{K_1} (K_1 \cdot_k L).$$

Thus we can apply the induction hypothesis, which tells us among other things that $K_{r+1} \cdot L/K_r \cdot L$ is regular and inessential for every $r \in [1, h] \setminus J$. We now distinguish the cases $r_0 \geq 1$ and $r_0 = 0$.

Suppose that $r_0 \geq 1$. By the induction hypothesis, $(K_{r_i} \cdot L \mid 0 \leq i \leq e)$ is a regular generic splitting tower of $\varphi \otimes K_1 \cdot L$. Let $F_0 := K_0 \cdot L$. The form $\varphi_0 \otimes F_0 = \varphi \otimes F_0$ is isotropic. Therefore the extension $F_0(\varphi_0 \otimes F_0) = K_0(\varphi_0) \cdot_{K_0} F_0$ of F_0 is purely transcendental, and thus inessential. By the lemma, $K_1 \cdot_{K_0} F_0 \sim_{F_0} K_0(\varphi_0) \cdot_{K_0} F_0$. Hence $K_1 \cdot_{K_0} F_0/F_0$ is also inessential. In analogy with the first calculation above, we find that

$$K_1 \cdot_{K_0} F_0 = K_1 \cdot_{K_0} (K_0 \cdot_k L) = K_1 \cdot_k L,$$

so that $K_1 \cdot_k L/K_0 \cdot_k L$ is inessential. This proves statement (b) for the current case.

We saw that $K_0 \cdot L/L$ is inessential. Thus $K_1 \cdot L/L$ is inessential. The generic splitting tower $(K_{r_i} \cdot L \mid 0 \le i \le e)$ of $\varphi \otimes K_1 \cdot L$ is thus also a generic splitting tower of $\varphi \otimes L$.

We come to the case $\dim \varphi > 2$, $r_0 = 0$. Statement (b) is covered by the induction hypothesis. $\varphi \otimes L$ is anisotropic. By the induction hypothesis, $\varphi \otimes K_1 \cdot L$ has regular generic splitting tower $(K_{r_i} \cdot L \mid 1 \le i \le e)$. The extension $K_{r_1} \cdot L/K_0 \cdot L$ is regular. For the proof of statement (a), it remains to be shown that $K_{r_1} \cdot L$ is the generic zero field of $\varphi \otimes K_0 \cdot L$. We already know that $K_{r_1} \cdot L/K_1 \cdot L$ is inessential. Thus it suffices to show that $K_1 \cdot L$ is the generic zero field of $\varphi \otimes (K_0 \cdot L)$.

Let $F_0 := K_0 \cdot L$. We have $\varphi_0 \otimes F_0 = \varphi \otimes F_0$, and thus $F_0(\varphi \otimes F_0) = K_0(\varphi_0) \cdot_{K_0} F_0$. Also $K_1 \sim_{K_0} K_0(\varphi_0)$. By the lemma, this shows that

$$K_1 \cdot_{K_0} F_0 \sim_{F_0} K_0(\varphi_0) \cdot_{K_0} F_0.$$

Furthermore, $K_1 \cdot_{K_0} F_0 = K_1 \cdot_k L$. Thus, $K_1 \cdot_k L \sim_{F_0} F_0(\varphi_0 \otimes F_0)$, what we wanted to show.

Finally, assume that $\dim \varphi = 2$. Now we have $h = 1$. The field $\Delta(\varphi_0) = \Delta(\varphi) \otimes_k K_0$ is a quadratic extension of K_0 and K_1 is a regular extension of $\Delta(\varphi_0)$. We have a place $\lambda : K_1 \to \Delta(\varphi_0) \cup \infty$ over K_0. After composing λ with the non-trivial automorphism of $\Delta(\varphi_0)/K_0$ (if necessary), we obtain a place $\mu : K_1 \to \Delta(\varphi_0) \cup \infty$ over $\Delta(\varphi_0)$. Therefore $K_1/\Delta(\varphi_0)$ is an inessential extension and regular.

Let $F_0 := K_0 \cdot L$, $F_1 := K_1 \cdot L$. We again make a distinction between the cases $\varphi \otimes L$ isotropic, resp. anisotropic.

Suppose first that $\varphi \otimes L$ is isotropic. Now $J = \{1\}$, and $\Delta(\varphi) \otimes_k L$ splits. We choose an embedding $\Delta(\varphi) \hookrightarrow L$ and obtain from this an embedding $K_0 \cdot \Delta(\varphi) = \Delta(\varphi_0) \hookrightarrow K_0 \cdot L = F_0$. By Definition 2.57, we now have

$$K_1 \cdot L := K_1 \cdot_{\Delta(\varphi)} L = \mathrm{Quot}(K_1 \otimes_{\Delta(\varphi)} L) = \mathrm{Quot}(K_1 \otimes_{\Delta(\varphi_0)} (\Delta(\varphi_0) \otimes_{\Delta(\varphi)} L)),$$

and furthermore $\Delta(\varphi_0) \otimes_{\Delta(\varphi)} L = K_0 \otimes_k L$. It follows that

$$F_1 = K_1 \cdot L = K_1 \cdot_{\Delta(\varphi_0)} (K_0 \cdot L) = K_1 \cdot_{\Delta(\varphi_0)} F_0.$$

It is now clear that the extension F_1/F_0 is regular. Since $K_1/\Delta(\varphi_0)$ is inessential, it follows furthermore from the lemma that F_1/F_0 is inessential. F_0/L is also inessential. Thus F_1/L is inessential. This establishes statements (a) and (b) in this case.

Finally, let $\dim \varphi = 2$ and assume that $\varphi \otimes L$ is anisotropic. Now $J = \{0, 1\}$. Statement (b) is vacuous. We already know that F_0/L is inessential. Furthermore, $K_1/\Delta(\varphi_0)$ is regular and inessential. Therefore

$$F_1 = K_1 \cdot_{\Delta(\varphi_0)} (\Delta(\varphi_0) \cdot_{K_0} F_0)$$

is regular and inessential over $\Delta(\varphi_0) \cdot_{K_0} F_0 = \Delta(\varphi_0 \otimes F_0)$. Furthermore, $\Delta(\varphi_0 \otimes F_0) = F_0\{\varphi_0 \otimes F_0\}$. Hence F_1 is the generic zero field of the form $\varphi_0 \otimes F_0$. Therefore (F_0, F_1) is a regular generic splitting tower of $\varphi \otimes L$. $\square$

Corollary 2.60. *In the situation of Theorem 2.58, $\varphi \otimes L$ has height e and generic splitting pattern*

$$\mathrm{SP}(\varphi \otimes L) = \{i_0 + \cdots + i_{r_j} \mid 0 \le j \le e\}.$$

For every $j \in [0, e]$ is $\varphi_{r_j} \otimes (K_{r_j} \cdot L)$ furthermore the j-th higher kernel form of $\varphi \otimes L$ with respect to the generic splitting tower constructed here.

2.6 Generic Splitting Towers of a Specialized Form

In the following, let $\gamma : k \to L \cup \infty$ be a place and φ a quadratic form over k having FR with respect to γ. Assume that the form $\overline{\varphi} := \gamma_*(\varphi)$ over L is nondegenerate. Finally, let $(K_r \mid 0 \le r \le h)$ be a generic splitting tower of φ.

We have seen in §2.4 that the splitting behaviour of $\overline{\varphi}$ with respect to $\overline{\varphi}$-conservative extensions of the field L is controlled by the tower $(K_r \mid 0 \le r \le h)$, cf. Theorem 2.39 and Scholium 2.43. Hence we may hope that it is actually possible to construct a generic splitting tower of $\overline{\varphi}$ in a natural way from $(K_r \mid 0 \le r \le h)$. We will devote ourselves to the task of constructing such a tower.

In the special case where the place γ is trivial and the tower $(K_r \mid 0 \le r \le h)$ is regular, we already dealt with this task in the previous section (Theorem 2.58). In general we cannot expect to find a solution as nice as the solution in §2.5.

Let us continue by fixing some notation. Let i_r be the higher indices and φ_r the higher kernel forms of φ with respect to the given splitting tower $(K_r \mid 0 \le r \le h)$ of φ. Thus we have $\mathrm{SP}(\varphi) = \{i_0 + \cdots + i_r \mid 0 \le r \le h\}$. Further, let J be the subset of $\{0, 1, \ldots, h\}$ with $\mathrm{SP}(\overline{\varphi}) = \{i_0 + \cdots + i_r \mid r \in J\}$. The set J contains $e := h(\overline{\varphi})$ elements. Let us list them as

$$0 \le t(0) < t(1) < \cdots < t(e) = h.$$

We formally set $t(-1) = -1$.

Let $(L_s \mid 0 \le s \le e)$ be a generic splitting tower of $\overline{\varphi}$. Following Scholium 2.43 we choose places $\mu_s : K_{t(s)} \to L_s \cup \infty$ $(0 \le s \le e)$ such that μ_0 extends the place γ and such that every μ_s with $1 \le s \le e$ extends the place μ_{s-1}. For every $r \in \{0, \ldots, h\}$ we have an $s \in \{0, \ldots, e\}$ with $t(s-1) < r \le t(s)$. Let $\lambda_r : K_r \to L_s \cup \infty$ be the restriction of the place μ_s to K_r. {In particular, $\mu_s = \lambda_{t(s)}$.} Let $\psi_s = \ker(\overline{\varphi} \otimes L_s)$ be the sth higher kernel form of $\overline{\varphi}$. By §2.4 $\varphi_{t(s)}$ has FR with respect to μ_s and $(\mu_s)_*(\varphi_{t(s)}) \cong \psi_s$. Furthermore, for $t(s-1) < r < t(s)$, φ_r also has FR with respect to λ_r and

$$(\lambda_r)_*(\varphi_r) \cong \psi_s \perp (i_{r+1} + \cdots + i_{t(s)}) \times H.$$

Given a place $\alpha : E \to F \cup \infty$ and a subfield K of E, we denote the image of K under α in general sloppily with $\alpha(K)$, in other words,

$$\alpha(K) := \alpha(\mathfrak{o}_\alpha \cap K).$$

In the important case that the place γ is surjective, i.e. $\gamma(k) = L$, we would like to choose the places μ_s in such a way that $(\mu_s(K_{t(s)}) \mid 0 \leq s \leq e)$ is a generic splitting tower of $\bar{\varphi}$. For not necessarily surjective γ, one could try to achieve that $(L \cdot \mu_s(K_{t(s)}) \mid 0 \leq s \leq e)$ is a generic splitting tower of $\bar{\varphi}$, where $L \cdot \mu_s(K_{t(s)})$ denotes the subfield of L_s generated by L and $\mu_s(K_{t(s)})$ in L_s.[9] In any case, for arbitrary choice of μ_s—as above—the following theorem holds.

Theorem 2.61. *For $t(s-1) < r \leq t(s)$ the subfield $\mu_s(K_r) \cdot L$ of L_s, generated by $\mu_s(K_r)$ and L, is a generic splitting field of $\bar{\varphi}$ of level s.*

Proof. Let $F_r := \lambda_r(K_r) \cdot L$. By restricting the range, we obtain from λ_r a place from K_r to F_r which extends γ. Hence

$$\mathrm{ind}(\bar{\varphi} \otimes F_r) \geq \mathrm{ind}(\varphi \otimes K_r) = i_0 + \cdots + i_r.$$

Since $\mathrm{ind}(\bar{\varphi} \otimes F_r)$ is a number of $\mathrm{SP}(\bar{\varphi})$, we get

$$\mathrm{ind}(\bar{\varphi} \otimes F_r) \geq i_0 + \cdots + i_{t(s)}.$$

Thus there is a place from L_s to F_r over L. Since $F_r \subset L_s$, we get $F_r \sim_L L_s$. Thus F_r is a generic splitting field of $\bar{\varphi}$ of level s. □

This theorem does *not* imply that $(K_{t(s)} \cdot L \mid 0 \leq s \leq e)$ is a generic splitting tower of $\bar{\varphi}$. However, in the following we will produce more special situations for which this *is* the case.

Theorem 2.62. *There exists a regular[10] generic splitting tower $(L_s \mid 0 \leq s \leq e)$ of $\bar{\varphi}$ and a place $\lambda : K_h \to L_e \cup \infty$ having the following properties:*

(i) *λ extends γ.*
(ii) *For every $s \in \{0, 1, \ldots, e\}$ and every r with $t(s-1) < r \leq t(s)$ we have $L_s = \lambda(K_r) \cdot L$.*
(iii) *$L_0 = L$.*

Proof. We inductively construct fields $L_s \supset L$ and places $\mu_s : K_{t(s)} \to L_s \cup \infty$ as follows.

$\underline{s = 0}$: We let $L_0 := L$. We choose for μ_0 any place from $K_{t(0)}$ to L which extends γ. This is possible by Scholium 2.43. Obviously we have $\mu_0(K_r) \cdot L = L$ for every r with $0 \leq r \leq t(0)$.

$\underline{s \to s+1}$: By the induction hypothesis, $(L_j \mid 0 \leq j \leq s)$ is the beginning of a generic splitting tower of $\bar{\varphi}$. In particular, L_s is a generic splitting field of $\bar{\varphi}$ of level s. Furthermore there is a place $\mu_s : K_{t(s)} \to L_s \cup \infty$. According to §2.4 the form $\varphi_{t(s)}$ has FR with respect to μ_s and $(\mu_s)_*(\varphi_{t(s)}) \cong \psi_s$.

---

[9] Now we are *not* using the notation established in Definition 2.57.
[10] cf. Definition 2.56.

Now we extend μ_s to a place $\tilde{\mu}_s$ from $K_{t(s)}(\varphi_{t(s)})$ to $L_s(\psi_s)$ (cf. the proof of Theorem 2.3). Let $u := t(s) + 1$ and let $\alpha : K_u \to K_{t(s)}(\varphi_{t(s)}) \cup \infty$ be a place over $K_{t(s)}$, which does indeed exist since K_u is a generic zero field of $\varphi_{t(s)}$. Then we have a place $\tilde{\mu}_s \circ \alpha$ from K_u to $L_s(\psi_s)$ and define L_{s+1} to be the composite of the fields L_s and $(\tilde{\mu}_s \circ \alpha)(K_u)$ in $L_s(\psi_s)$,

$$L_{s+1} := L \cdot (\tilde{\mu}_s \alpha)(K_u).$$

Let $\nu_s : K_u \to L_{s+1} \cup \infty$ be the place obtained from $\tilde{\mu}_s \circ \alpha$ by restricting the range. We have

$$\operatorname{ind}(\overline{\varphi} \otimes L_{s+1}) \geq \operatorname{ind}(\varphi \otimes K_u) = i_0 + \cdots + i_u > i_0 + \cdots + i_{t(s)},$$

and thus $\operatorname{ind}(\overline{\varphi} \otimes L_{s+1}) \geq i_0 + \cdots + i_{t(s+1)}$. This implies

$$\operatorname{ind}(\psi_s \otimes L_{s+1}) \geq i_u + \cdots + i_{t(s+1)} > 0.$$

Thus $\psi_s \otimes L_{s+1}$ is isotropic. Since $L_{s+1} \subset L_s(\psi_s)$, we conclude that L_{s+1} is a generic zero field of ψ_s. By Scholium 2.43(3) we now have that ν_s can be extended to a place from $K_{t(s+1)}$ to L_{s+1}. We choose $\mu_{s+1} : K_{t(s+1)} \to L_{s+1} \cup \infty$ to be such a place. Since $L_{s+1} = \nu_s(K_u) \cdot L$, we have $L_{s+1} = \mu_{s+1}(K_r) \cdot L$ for every r with $u \leq r \leq t(s)$.

Eventually our construction yields a generic splitting tower

$$(L_s \mid 0 \leq s \leq e)$$

of $\overline{\varphi}$ and a place $\lambda := \mu_e$ from K_h to L_e which satisfy properties (i) and (ii) of the theorem.

Let us check that the tower $(L_s \mid 0 \leq s \leq e)$ is regular! For $0 \leq s \leq e - 1$ our construction gives $L_s \subset L_{s+1} \subset L_s(\psi_s)$. If $\dim \psi_s > 2$, then $L_s(\psi_s)$ is regular over L_s, and thus L_{s+1} is regular over L_s. If $\dim \psi_s = 2$, then we must have $s = e - 1$ and $t(s) = h - 1$. Next, we modify the last step of the construction above in the sense that we replace $K_{h-1}(\varphi_{h-1})$ by the algebraic closure E of K_{h-1} in this field and also $K_{e-1}(\psi_{e-1})$ by the algebraic closure F of K_{e-1} in $K_{e-1}(\psi_{e-1})$.

Let $\mu := \mu_{h-1}$. We have $\varphi_{h-1} \cong \langle a \rangle \otimes \begin{bmatrix} 1 & 1 \\ 1 & b \end{bmatrix}$ with $a \in \mathfrak{o}_\mu^*$, $b \in \mathfrak{o}_\mu^*$, and $\psi_{e-1} \cong \langle \overline{a} \rangle \otimes \begin{bmatrix} 1 & 1 \\ 1 & \overline{b} \end{bmatrix}$ with $\overline{a} := \mu(a)$, $\overline{b} := \mu(b)$. We obtain $E = K_{h-1}(\xi)$ and $F = L_{e-1}(\eta)$ with generators ξ, η which satisfy the minimal equations $\xi^2 + \xi + b = 0$, $\eta^2 + \eta + \overline{b} = 0$. The extension K_h/E is inessential. This can be verified as in the proof of Theorem 2.58 towards the end. (There for the case $\dim \varphi = 2$.)

The place $\mu = \mu_{e-1}$ from K_{h-1} to L_{e-1} has exactly one extension $\tilde{\mu}_{e-1} : E \to F \cup \infty$ with $\tilde{\mu}_{e-1}(\xi) = \eta$ (cf. Lemma 2.2). If we prepend a place from K_h to E over E to $\tilde{\mu}_{e-1}$, we obtain a place $\mu_e : K_h \to F \cup \infty$ which extends μ_{e-1}. The field $L_e := \mu_e(K_h) \cdot L$ differs from L_{e-1} since it splits the form $\overline{\varphi}$ totally. Hence we must have $L_e = F$. Our generic splitting tower $(L_s \mid 0 \leq s \leq e)$ is regular.   □

What we have obtained by now can perhaps best be understood in the important special case where the place $\gamma : k \to L \cup \infty$ is surjective (i.e. $\gamma(k) = L$). For the given generic splitting tower $(K_r \mid 0 \leq r \leq h)$ of φ we have constructed a place $\lambda : K_h \to L_e \cup \infty$ which extends γ and for which the tower $(\lambda(K_r) \mid 0 \leq r \leq h)$ is an almost regular generic splitting tower of $\overline{\varphi}$ "with repetitions", i.e. a generic splitting tower in which the storeys are listed possibly more than once: $\lambda(K_r) = \lambda(K_{t(s)})$ for $t(s-1) < r \leq t(s)$. The place λ furnishes a connection between the tower $(K_r \mid 0 \leq r \leq h)$ and the generic splitting tower $(L_s \mid 0 \leq s \leq e)$ of $\overline{\varphi}$, $L_s = \lambda(K_{t(s)})$ which reflects in an obvious way how the splitting behaviour of $\overline{\varphi}$ over L is a coarsening of the splitting behaviour of φ over K.

In many situations it is more natural or, for some given problem, more useful (see e.g. [37]) to associate to the tower $(K_r \mid 0 \leq r \leq h)$ a different generic splitting tower of $\overline{\varphi}$. Such a situation (with γ trivial) was depicted in §2.5. Next we give a further construction which applies to an arbitrary place γ with respect to which φ has FR and with $\overline{\varphi} = \gamma_*(\varphi)$ nondegenerate, as above.

Specifically, let $(K_r \mid 0 \leq r \leq h)$ be the projective standard tower of φ (see Definition 2.51). We define a field tower $(L_r \mid 0 \leq r \leq h)$ and a sequence of places $(\lambda_r : K_r \to L_r \cup \infty \mid 0 \leq r \leq h)$ with $L_r \supset L$ and $L_r = \lambda_r(K_r) \cdot L$ for all r, $0 \leq r \leq h$, as follows.[11]

We start with $L_0 := L$, $\lambda_0 := \gamma$. If $\lambda_r : K_r \to L_r \cup \infty$ is already defined and $r < h$, then φ_r has FR with respect to λ_r by §2.4. Let $\overline{\varphi}_r := (\lambda_r)_*(\varphi_r)$. We set $L_{r+1} := L_r\{\overline{\varphi}_r\}$. Let $\tilde{\lambda}_r : K_r(\varphi_r) \to L_r(\overline{\varphi}_r) \cup \infty$ be an extension of the place λ_r, obtained using the procedure in the proof of Theorem 2.3.

Next, let $\overline{\varphi}_r \not\cong H$. The place $\tilde{\lambda}_r$ maps $K_{r+1} = K_r\{\varphi_r\}$ into $L_{r+1} \cup \infty$. From the construction in §2.1 it is clear that $L_{r+1} = \tilde{\lambda}_r(K_{r+1}) \cdot L$. We define $\lambda_{r+1} : K_{r+1} \to L_{r+1} \cup \infty$ by restricting the place $\tilde{\lambda}_r$. We have $L_{r+1} = \lambda_{r+1}(K_{r+1}) \cdot L$.

Finally, let $\overline{\varphi}_r \cong H$ (and so $r = h - 1$). Now we have $L_{r+1} = L_r$. From the construction in §2.1, we have $\tilde{\lambda}_r(K_r(\varphi_r)) \cdot L = L_r$. Again we define λ_{r+1}, i.e. λ_h, to be the restriction of $\tilde{\lambda}_r$ to a place from K_{r+1} to $L_{r+1} = L_r$. Then, clearly, $\lambda_{r+1}(K_{r+1}) \cdot L = L_{r+1}$.

Definition 2.63. We call the tower $(L_r \mid 0 \leq r \leq h)$ the *transfer* of the projective standard tower by the place γ and we call $\lambda := \lambda_h : K_h \to L_h \cup \infty$ a *transferring place*.

Remark. By Remark 2.4 the places $\lambda_r : K_r \to L_r \cup \infty$ for $0 \leq r \leq h - 1$ in the construction above are uniquely determined by γ. If $\overline{\varphi}_{h-1} \not\cong H$ this is also the case for $\lambda_h = \lambda$. If $\overline{\varphi}_{h-1} \cong H$, then there are at most two possibilities for λ_h, since K_h is a quadratic extension of K_{h-1} in this situation. The fields $L_r = \lambda_r(K_r) \cdot L$ are all uniquely determined by φ and γ.

Theorem 2.64. *Let $(L_r \mid 0 \leq r \leq h)$ be the transfer of the projective standard tower $(K_r \mid 0 \leq r \leq h)$ of φ by the place $\gamma : k \to L \cup \infty$ and let $\lambda : K_h \to L_h \cup \infty$ be a transferring place. Let $\overline{\varphi}_r := (\lambda_r)_*(\varphi_r)$ be the specialization of φ_r with respect to the restriction $\lambda_r : K_r \to L_r \cup \infty$ of λ $(0 \leq r \leq h)$.*

---

[11] The equation $L_r = \lambda_r(K_r) \cdot L$ only signifies that L_r is generated as a field by both subfields $\lambda_r(K_r)$ and L.

Claim: $(L_r \mid r \in J)$ *is a generic splitting tower of* $\overline{\varphi}$ *and for every* $r \in J$ *we have* $\overline{\varphi}_r = \ker(\overline{\varphi} \otimes L_r)$. *For* $r \notin J$, L_{r+1}/L_r *is purely transcendental. In particular, the extension* $L_{t(0)}/L$ *is purely transcendental.*

Proof. Let $r \in \{0, \dots, h\}$ be given. From the Witt decomposition

$$\varphi \otimes K_r \cong (i_0 + \cdots + i_r) \times H \perp \varphi_r$$

we obtain

$$\overline{\varphi} \otimes L_r \cong (i_0 + \cdots + i_r) \times H \perp \overline{\varphi}_r,$$

since $(\lambda_r)_*(\varphi \otimes K_r) \cong \overline{\varphi} \otimes L_r$.

If $r \notin J$, then $\mathrm{ind}(\overline{\varphi} \otimes L_r) > i_0 + \cdots + i_r$. Now $\overline{\varphi}_r$ is isotropic and hence $L_{r+1} = L_r\{\overline{\varphi}_r\}$ is purely transcendental over L_r.

Let a number s be given with $0 \leq s \leq e$. Then $L_{t(s)+1} = L_{t(s)}\{(\overline{\varphi}_{t(s)})\}$ and $L_{t(s+1)}/L_{t(s)+1}$ is purely transcendental. Hence $L_{t(s+1)}/L_{t(s)}$ is a generic zero field of $\overline{\varphi}_{t(s)}$. Furthermore, $L_{t(0)}/L$ is purely transcendental. Thus

$$\mathrm{ind}(\overline{\varphi} \otimes L_{t(0)}) = \mathrm{ind}(\overline{\varphi}) = i_0 + \cdots + i_{t(0)}.$$

This shows that $\overline{\varphi}_{t(0)}$ is the kernel form of $\overline{\varphi} \otimes L_{t(0)}$. Suppose now that for some $s \in \{1, \dots, e\}$ we already showed that $\overline{\varphi}_{t(k)}$ is the kernel form of $\overline{\varphi} \otimes L_{t(k)}$ for $0 \leq k < s$. Then $(L_{t(k)} \mid 0 \leq k \leq s)$ is the beginning of a generic splitting tower of $\overline{\varphi}$. Hence $\overline{\varphi} \otimes L_{t(s)}$ has index $i_0 + \cdots + i_{t(s)}$, and it follows that $\overline{\varphi}_{t(s)}$ is the kernel form of $\overline{\varphi} \otimes L_{t(s)}$. Thus, this holds for all $s \in \{0, \dots, e\}$. Now it is clear that $(L_{t(s)} \mid 0 \leq s \leq e)$ is a generic splitting tower of $\overline{\varphi}$.   □

Chapter 3
Some Applications

3.1 Subforms which have Bad Reduction

The theory of weak specialization, developed in §1.3, §1.7 and the end of §2.3, has until now played only an auxiliary role, which we could have done without when dealing with quadratic forms (due to Theorem 2.19). For the first time we now come to independent applications of weak specialization.

Let $\lambda : K \to L \cup \infty$ be a place and φ a bilinear or quadratic form over K having good reduction with respect to λ. If φ contains subforms which have bad reduction, there are consequences for the form $\lambda_*(\varphi)$ which we will discuss now. We will look at bilinear forms first.

Theorem 3.1. *Let φ be a nondegenerate bilinear form over K, having good reduction with respect to λ. Suppose that there exists a nondegenerate subform $\langle b_1, \ldots, b_m \rangle$ of φ with $\lambda(b_i c^2) = 0$ or ∞ for every $i \in \{1, \ldots, m\}$ and every $c \in K^*$. Then $\lambda_*(\varphi)$ has Witt index $\geq \left\{ \frac{m}{2} \right\}$. (As usual, $\left\{ \frac{m}{2} \right\}$ denotes the smallest integer $\geq \frac{m}{2}$.)*

Proof. Let $\psi := \langle b_1, \ldots, b_m \rangle$. We have a decomposition $\varphi \cong \psi \perp \eta$, to which we apply the additive map $\lambda_W : W(K) \to W(L)$. We get $\lambda_W(\psi) = 0$.[1] Hence $\{\lambda_*(\varphi)\} = \lambda_W(\varphi) = \lambda_W(\eta)$. Let ρ be the anisotropic form in the Witt class $\lambda_W(\eta)$. Then $\lambda_*(\varphi) \sim \rho$ and $\dim \rho \leq \dim \eta = \dim \varphi - m$. Therefore $\mathrm{ind}\, \lambda_*(\varphi) \geq \left\{ \frac{m}{2} \right\}$. $\square$

Corollary 3.2. *Let φ be a nondegenerate bilinear form over K, having good reduction with respect to λ. Suppose that $\lambda_*(\varphi)$ is anisotropic. Then every subform of φ is nondegenerate and has good reduction with respect to λ.*

Proof. Clearly φ is now also anisotropic. Therefore every subform of φ is anisotropic and so definitely nondegenerate. By Theorem 3.1, φ does not contain any one-dimensional subforms $\langle b \rangle$ with $\lambda(bc^2) = 0$ or ∞ for every $c \in K^*$. Hence every subform of φ has good reduction. $\square$

---

[1] We use the shorter notation $\lambda_W(\psi)$ instead of $\lambda_W(\{\psi\})$, cf. §1.7.

M. Knebusch, *Specialization of Quadratic and Symmetric Bilinear Forms*,
Algebra and Applications 11, DOI 10.1007/978-1-84882-242-9_3,
© Springer-Verlag London Limited 2010

Remark. If char $L = 2$, then $\lambda_*(\varphi)$ is in general only determined by λ and φ up to stable isometry. However, if a representative of $\lambda_*(\varphi)$ is anisotropic, then $\lambda_*(\varphi)$ is uniquely determined up to genuine isometry, since $\lambda_*(\varphi)$ is up to isometry the only anisotropic form in the Witt class $\lambda_W(\varphi)$ in this case.

Next we can derive the following statement, which is similar to the Substitution Principle in §1.3 (Theorem 1.29).

Theorem 3.3 ([32, Prop. 3.3]). *Let $(f_{ij}(t))$ be an $n \times n$-matrix of polynomials $f_{ij}(t) \in k[t]$ over an arbitrary field k, where $t = (t_1, \ldots, t_r)$ is a set of indeterminates over k. Let $g_1(t), \ldots, g_m(t)$ be further polynomials in $k[t]$. Finally, let $c = (c_1, \ldots, c_r)$ be an r-tuple of coordinates in a field extension L of k, such that $\det(f_{ij}(c)) \neq 0$, and such that c is a non-singular zero of every polynomial $g_p(t)$ $(1 \leq p \leq m)$, i.e. $g_p(c) = 0$, $\frac{\partial g_p}{\partial t_q}(c) \neq 0$ for an element $q \in \{1, \ldots, n\}$, which depends on p. If $\langle g_1(t), \ldots, g_m(t) \rangle$ is a subform of the bilinear form $(f_{ij}(t))$ over $k(t)$, then the bilinear form $(f_{ij}(c))$ over L has Witt index $\geq \left\{\frac{m}{2}\right\}$.*

Proof ([32, p.291]). Going from $k[t]$ to $L[t]$, we may suppose without loss of generality that $L = k$.

For $r = 1$, Theorem 3.3 follows immediately from Theorem 3.1, applied to the place $\lambda : k(t_1) \to k \cup \infty$ over k with $\lambda(t_1) = c_1$.

Now let $r > 1$. We first assume that k is an infinite field. Then there exists an r-tuple $(a_1, \ldots, a_r) \in k^r$ with

$$\sum_{q=1}^{r} a_q \frac{\partial g_p}{\partial t_q}(c) \neq 0$$

for $1 \leq p \leq m$. By applying an obvious coordinate transformation, we obtain that $\frac{\partial g_p}{\partial t_1}(c) \neq 0$ for every $p \in \{1, \ldots, m\}$.

Let c' be the $(r-1)$-tuple $(c_2, \ldots, c_r)$. We choose a place $\alpha : k(t) \to k(t_1) \cup \infty$ over $k(t_1)$ with $\alpha(t_i) = c_i$ for $2 \leq i \leq r$. Applying the additive map $\alpha_W : W(k(t)) \to W(k)$, just as in the proof of Theorem 3.1, shows that the space $(f_{ij}(t, c'))$ over $k(t_1)$ is Witt equivalent to a space

$$\langle g_1(t_1, c'), \ldots, g_m(t_1, c'), h_1(t_1), \ldots, h_s(t_1) \rangle$$

with polynomials $h_p(t_1)$, $1 \leq p \leq s$, and $m + s \leq n$. {Possibly $s = 0$.} Let $\lambda : k(t_1) \to k \cup \infty$ be the place over k with $\lambda(t_1) = c_1$. We have $\lambda_W(\langle g_p(t_1, c') \rangle) = 0$ for $1 \leq p \leq m$. Then, after applying λ_W, we see that the form $(f_{ij}(c))$ over k is Witt equivalent to a form of dimension $\leq s$. Since $s \leq n - m$, $(f_{ij}(c))$ certainly has Witt index $\geq \left\{\frac{m}{2}\right\}$.

Finally, let us consider the case where the field k is finite. Let u be an indeterminate over k. By what we proved above, the form $(f_{ij}(c))$ over $k(u)$ has Witt index $\geq \left\{\frac{m}{2}\right\}$. Upon specializing with an arbitrary place $\beta : k(u) \to k \cup \infty$ over k, we see that $(f_{ij}(c))$ over k has Witt index $\geq \left\{\frac{m}{2}\right\}$. $\square$

Remark. If char $k \neq 2$, then the part of the proof dealing with $r > 1$, k infinite, can be shortened by first applying the Substitution Principle, Theorem 1.29, followed by Theorem 3.1. If char $k = 2$ however, when applying Theorem 1.29, we need that the form $(f_{ij}(t_1, c'))$ over $k(t_1)$ is anisotropic, which is not necessarily the case.

If we now want to obtain theorems for quadratic forms, similar to Theorems 3.1 and 3.3, at best with fair instead of good reduction, then we cannot use weak specialization as easily as above. For instance, let φ, ψ be strictly regular forms over a field K with $\psi < \varphi$, and let $\lambda : K \to L \cup \infty$ be a place. We then have an orthogonal decomposition $\varphi \cong \psi \perp \eta$. Suppose now that φ has FR (= fair reduction) with respect to λ and that ψ is weakly obedient (see Definition 2.31), but far from having FR. We would like to apply the "operator" λ_W to the relation $\varphi \cong \psi \perp \eta$ in order to learn something about $\lambda_*(\varphi) = \lambda_W(\varphi)$. However, this is only possible when η is also weakly obedient with respect to λ, something we cannot assume just like that. To circumnavigate this cliff, we will not simply apply the theory about λ_W of §1.7 and §2.3, but rather orientate ourselves towards the argumentation which led to this theory. For quadratic forms we will obtain theorems similar, but not exactly analogous, to Theorem 3.1 and 3.3. Broadly speaking we will make more assumptions, but will also be able to draw stronger conclusions from these.

Thus, let $\lambda : K \to L \cup \infty$ be a place, $\mathfrak{o} = \mathfrak{o}_\lambda$ the associated valuation ring and $\mathfrak{m} = \mathfrak{m}_\lambda$ its maximal ideal. As before (§1.7) we choose a system of representatives S of $Q(K)/Q(\mathfrak{o})$ in K^* with $1 \in S$.

In two of the following three theorems, we will make statements for quadratic and bilinear forms in parallel. Throughout we will give the proofs for the more complicated quadratic case. In the bilinear case, the proofs go in the same manner.

Theorem 3.4. *Let $E = (E, q)$ be a quadratic (resp. bilinear) space over K which has FR (resp. GR) with respect to λ. Further, let F be a linear subspace of E with $\lambda(q(x)) = 0$ or ∞ (resp. $\lambda(B(x, x)) = 0$ or ∞) for every $x \in F$. Then $\lambda_*(E)$ has Witt index $\geq \dim F$.*

Proof in the quadratic case. We may assume that L coincides with the residue class field $k = \mathfrak{o}/\mathfrak{m}$ of $\mathfrak{o}$ and that λ is the canonical place from K to k associated to $\mathfrak{o}$. Let $m := \dim F$. In E we choose a reduced nondegenerate quadratic $\mathfrak{o}$-module M with $E = KM$. Let N be the $\mathfrak{o}$-module $M \cap F$. Then $M = N \oplus N'$ where N' is a further $\mathfrak{o}$-module.

The image $\overline{N}$ of N in $M/\mathfrak{m}M$ is a k-vector space of dimension m. By our assumption about F, we have $q(x) \in \mathfrak{m}$ for every $x \in N$, and so $\overline{q}(\overline{N}) = 0$ where $\overline{q}$ is the quadratic form on $\overline{M} := M/\mathfrak{m}M$ induced by q. The quadratic space $\overline{M} = (\overline{M}, \overline{q}) = \lambda_*(E)$ is nondegenerate by assumption. Since $QL(\overline{M})$ is anisotropic, we have $\overline{N} \cap QL(\overline{M}) = \{0\}$. We choose a decomposition $\overline{M} = QL(\overline{M}) \oplus G = QL(\overline{M}) \perp G$ with $\overline{N} \subset G$. The quadratic space G is strictly regular and contains the totally isotropic subspace $\overline{N}$ of dimension m. Therefore $\operatorname{ind} G \geq m$ (cf. Lemma 1.54), and so $\operatorname{ind} \overline{M} \geq m$. $\qquad\square$

Theorem 3.5. *Let φ be a quadratic form over K, having FR with respect to $\lambda : K \to L \cup \infty$ and containing a subform ψ of dimension m, which is weakly obedient with respect to λ and such that $\lambda_W(\psi) \sim 0$. Then $\lambda_*(\varphi)$ has Witt index $\geq \left\{ \frac{m}{2} \right\}$.*

Proof. By going from K to the henselization K^h with respect to $\mathfrak{o}$ and accordingly from λ to the place $\lambda^h : K^h \to L \cup \infty$, we may assume without loss of generality that $\mathfrak{o}$ itself is henselian. We choose a weakly λ-modular decomposition of ψ,

$$\psi \cong \underset{s \in S}{\perp} s\psi_s \tag{3.1}$$

where the ψ_s are forms which all have FR with respect to λ.

For every $s \in S$ with $s \neq 1$ we have a Witt decomposition $\psi_s \cong \psi_s^\circ \perp r_s \times H$ with $r_s \geq 0$, ψ_s° anisotropic. Thus

$$\psi \cong (\psi_1 \perp (\sum_{s \neq 1} r_s) \times H) \perp \underset{s \neq 1}{\perp} s\psi_s^\circ.$$

Upon replacing (3.1) by this weakly λ-modular decomposition of ψ, we may assume at the outset that the forms ψ_s with $s \neq 1$ in (3.1) are all anisotropic. {Of course $\psi_s = 0$ for almost all s.}

The assumption $\lambda_W(\psi) = 0$ signifies that $\lambda_*(\psi_1) \sim 0$. In particular, $\lambda_*(\psi_1)$ does not have a quasilinear part $\neq 0$. Therefore ψ_1 is strictly regular and actually has GR with respect to λ. We have a decomposition $\varphi \cong \psi_1 \perp \eta$. By Theorem 2.28, η also has FR with respect to λ and $\lambda_*(\varphi) \cong \lambda_*(\psi_1) \perp \lambda_*(\eta)$. Since $\lambda_*(\psi_1)$ is hyperbolic, we have

$$\operatorname{ind} \lambda_*(\varphi) = \frac{1}{2} \dim \psi_1 + \operatorname{ind} \lambda_*(\eta).$$

Furthermore, $\rho := \underset{s \neq 1}{\perp} s\psi_s$ is a subform of η. If we can verify the statement of the theorem for η and ρ instead of for φ and ψ, then the statement will follow for φ and ψ as well. Hence we assume from now on without loss of generality that $\psi_1 = 0$.

Let $E = (E, q)$ be a space, isometric to φ and $F = (F, q|F)$ a subspace of E, isometric to ψ. By an old argument from §1.7, we have that for no vector $x \in F$ the value $q(x)$ can lie in $\mathfrak{o}^*$ because, $\mathfrak{o}$ being henselian, the form ψ_s can only represent elements $c^2 \varepsilon$ with $c \in K$, $\varepsilon \in \mathfrak{o}^*$ for every $s \neq 1$, cf. Theorem 1.82. By Theorem 3.4, we then have $\operatorname{ind} \varphi \geq \dim \psi$. So, obviously $\operatorname{ind} \varphi \geq \left\{ \frac{1}{2} \dim \psi \right\}$. $\qquad \square$

Theorem 3.6. *Let φ be a quadratic form over K which has FR with respect to λ. Suppose that the specialization $\lambda_*(\varphi)$ is anisotropic. Then every subform ψ of φ that is weakly obedient with respect to λ also has FR. {Note that therefore $\lambda_*(\psi)$ is a subform of $\lambda_*(\varphi)$ by Theorem 2.19.}*

Proof. Again we may assume that λ is the canonical place associated to $\mathfrak{o}$ and that $\mathfrak{o}$ is henselian. Just as in the previous proof, we see that ψ has a weakly λ-modular decomposition

$$\psi \cong \underset{s \in S}{\perp} s\psi_s,$$

where the forms $\lambda_*(\psi_s)$ for $s \neq 1$ are all anisotropic. Now $\underset{s \neq 1}{\perp} s\psi_s$ is a subform of ψ to which we can apply Theorem 3.4. Since $\lambda_*(\varphi)$ is anisotropic, this form has to be the zero form. Thus $\psi \cong \psi_1$. Hence ψ has FR with respect to λ.   □

Theorem 3.7 (cf. [32, Prop.3.5]). *Let $(f_{ij}(t))$ be a symmetric $n \times n$-matrix of polynomials $f_{ij}(t) \in k[t]$ over an arbitrary field k, where $t = (t_1, \ldots, t_r)$ is a set of indeterminates. Let $g_1(t), \ldots, g_m(t)$ be further polynomials in $k[t]$. Finally, let $c = (c_1, \ldots, c_r)$ be an r-tuple of coordinates in a field extension L of k such that all $g_p(c) = 0$ $(1 \leq p \leq m)$ and such that the $m \times r$-matrix $\left(\frac{\partial g_p}{\partial t_q}(c) \right)$ has rank m.*

   (i) *If the quadratic form $[f_{ij}(c)]$ over L is nondegenerate, and if the quasi-linear form $[g_1(t), \ldots, g_m(t)]$ over $k(t)$ is a subform of the quadratic form $[f_{ij}(t)]$, then the form $[f_{ij}(c)]$ over L has Witt index $\geq m$.*
   (ii) *If the bilinear form $(f_{ij}(c))$ over L is nondegenerate and if the bilinear form $\langle g_1(t), \ldots, g_m(t) \rangle$ over $k(t)$ is a subform of $(f_{ij}(t))$, then the form $(f_{ij}(c))$ over L has Witt index $\geq m$.*

Proof of (i) ([32, p. 292 ff.]). By going from $k[t]$ to $L[t]$, we may assume without loss of generality that $k = L$. Next we replace the indeterminates t_i by $t_i - c_i$, thus assuming without loss of generality that $c = 0$. Finally we subject the variables t_i to a linear transformation with coefficients in k and obtain

$$\frac{\partial g_p}{\partial t_q}(0) = \delta_{pq} \quad (1 \leq p \leq m, \ 1 \leq q \leq r). \tag{3.2}$$

Now we introduce the field $K = k(t, s)$ with indeterminate s over $k(t)$ as well as the elements $u_i = \frac{t_i}{s}$ $(1 \leq i \leq r)$ and the field $k(u) = k(u_1, \ldots, u_r)$. We have $K = k(u, s)$. Let $\lambda : K \to k(u) \cup \infty$ be the place over $k(u)$ with $\lambda(s) = 0$. Its valuation is discrete of rank 1. It associates to every polynomial $f(u, s) \in k[u, s]$ the order $\mathrm{ord}_s f(u, s) \in \mathbb{N}_0$. This is the biggest number $k \in \mathbb{N}_0$ such that s^k divides the polynomial $f(u, s)$ in $k[u, s]$ or in $k(u)[s]$. Let (F, q) be a quadratic space, isometric to $[g_1(t), \ldots, g_m(t)]$. We will show:

$$\lambda(q(x)) = 0 \text{ or } \infty \text{ for every } x \in F. \tag{3.3}$$

By Theorem 3.4, this will then imply that the form $[f_{ij}(0)]$ over $k(u)$ has Witt index $\geq m$. By the Substitution Principle, Theorem 2.27, the form $[f_{ij}(0)]$ will also have Witt index $\geq m$ over k. {Note that $[f_{ij}(0)]$ obviously has the same Witt index over k and $k(u)$.}

We come to the proof of statement (3.3). Let $x \in F$ and $x \neq 0$. Then $q(x) = ZN^{-1}$, where Z, N are polynomials in $k[u, s]$ of the following form:

$$N = h(u, s)^2, \quad Z = \sum_{i=1}^{m} a_i(u, s)^2 g_i(u_1 s, \ldots, u_r s).$$

Here $h(u, s)$ and the $a_i(u, s)$ are polynomials in $k[u, s]$ with $h(u, s) \neq 0$ and not all $a_i(u, s) = 0$.

Now we interpret all these polynomials as polynomials in s with coefficients in $k[u]$ and consider the lowest terms of N and S. We have an $l \in \mathbb{N}_0$ with

$$a_i(u, s) = b_i(u)s^l \quad + \quad \text{higher terms}$$

and at least one coefficient $b_i(u) \neq 0$ $(1 \leq i \leq m)$. By (3.2), $g_i(u_1 s, \ldots, u_r s)$ has $u_i s$ as lowest term. Hence Z has $c(u)s^{2l+1}$ as lowest term, with

$$c(u) := \sum_{i=1}^{m} b_i(u)^2 u_i,$$

provided that $c(u) \neq 0$. This is the case however, for one can easily see that the quadratic form $[u_1, \ldots, u_m]$ over $k(u)$ is anisotropic. (Note that $m \leq r$.) The anisotropy is also clear by Example 1.99 in §1.7.

Therefore $\mathrm{ord}_s Z = 2l + 1$. On the other hand, $\mathrm{ord}_s N = 2\,\mathrm{ord}_s h$ is even. We conclude that (3.3) holds. $\square$

3.2 Some Forms of Height 1

In this section k can be any field. However, we are mostly interested in fields of characteristic 2. For any nondegenerate quadratic form φ of height $h \geq 1$ (i.e. φ not split) over k, the last but one kernel form φ_{h-1} has height 1. Thus, in order to gain an understanding of quadratic forms, it is of prime importance to have an overview of all anisotropic forms of height 1.

If $\mathrm{char}\,k \neq 2$, it is well-known (cf. the end of §1.4) and it will be shown in §3.6 and §3.9 that an anisotropic form φ over k has height 1 iff there exists an element $a \in k^*$ and a Pfister form τ over k with $\varphi \cong a\tau$, $\dim \tau \geq 2$, in case $\dim \varphi$ is even, $\varphi \cong a\tau'$, $\dim \tau \geq 4$, in case $\dim \varphi$ is odd. Here τ' denotes the pure part of τ, cf. the end of §1.4. Moreover, in both cases τ is uniquely determined by φ up to isomorphism.

Pfister forms can also be defined in case k has characteristic 2 (see below). We will see that diverse forms of height 1 can be obtained from quadratic Pfister forms in this case as well. Nonetheless, the question remains open what other anisotropic quadratic forms of height 1 there are. This is currently perhaps the most important unsolved problem about quadratic forms in characteristic 2.

Definition 3.8.

(a) Let $d \in \mathbb{N}_0$. A *bilinear d-fold Pfister form* over k is a form

$$\tau = \langle 1, a_1 \rangle \otimes \cdots \otimes \langle 1, a_d \rangle$$

with coefficients $a_i \in k^*$. {We let $\tau = \langle 1 \rangle$, in case $d = 0$.}

(b) Let $d \in \mathbb{N}$. A *quadratic d-fold Pfister form* over k is a form

$$\rho \otimes \begin{bmatrix} 1 & 1 \\ 1 & b \end{bmatrix},$$

where ρ is a bilinear Pfister form of degree $d - 1$ and $b \in k$, $1 - 4b \neq 0$.[2]

If τ is a bilinear d-fold Pfister form with $d \geq 1$, then we have an orthogonal decomposition $\tau \cong \langle 1 \rangle \perp \tau'$. It is well-known [5, p. 257] that also in characteristic 2, the form τ' is uniquely determined by τ up to isomorphism.[3] We call τ' the *pure part of the Pfister form* τ. It seems that for a quadratic Pfister form ρ in characteristic 2, it does not make any sense to speak of a pure part of ρ (however, see §3.12, Definition 3.107).

Next we will discuss some elementary properties of Pfister forms.

Definition 3.9. If φ is a quadratic or bilinear form over k, then $D(\varphi)$ denotes the set of all elements $a \in k^*$ which are values of the form φ, $a = \varphi(c_1, \ldots, c_n)$ for an n-tuple $(c_1, \ldots, c_n) \in k^n$.[4] We denote by $N(\varphi)$ the set of all $a \in k^*$ with $a\varphi \cong \varphi$. We call the $a \in D(\varphi)$ *the elements represented by* φ and the $a \in N(\varphi)$ the *similarity norms of* φ.

Note. $D(\varphi)$ is the set of all $a \in k^*$ with $\langle a \rangle < \varphi$ in the bilinear case and $[a] < \varphi$ in the quadratic case. Obviously, $N(\varphi)$ is a subgroup of k^* which contains all squares of elements of k^*. Furthermore, $D(\varphi)$ is a union of cosets of $N(\varphi)$ in k^*.

We call $N(\varphi)$ the *norm group* of φ. Analogously, we define for a quadratic or bilinear module E over k the norm group $N(E)$ and the set $D(E)$. For instance, if $E = (E, q)$ is a quadratic module, then $D(E)$ is the set of all $a \in k^*$ for which there exists an $x \in E$ with $q(x) = a$, and $N(E)$ is the set of all $a \in k^*$ such that there exists a similarity $\sigma : E \xrightarrow{\sim} E$ with norm a, i.e. a linear automorphisms σ of E with $q(\sigma(x)) = aq(x)$.

Theorem 3.10. *If τ is a bilinear or quadratic Pfister form over k, then $D(\tau) = N(\tau)$.*

For this—in characteristic $\neq 2$—well-known theorem, due to Pfister, Witt has given a particularly short proof which belongs to the standard repertoire of all modern books on the algebraic theory of quadratic forms (e.g. [55, p. 69 ff.], [52, p. 27]). The proof can be extended to characteristic 2 without effort. We recall it here for the convenience of the reader.

Proof of Theorem 3.10. We use induction on the degree d of τ. In the bilinear case the start of the induction, $d = 0$, is trivial. If $d = 1$ in the quadratic case, then τ is the hyperbolic plane $\begin{bmatrix} 0 & 1 \\ 1 & 0 \end{bmatrix}$ or the norm form of a quadratic separable field extension of k. In both cases the statement is clear.

Next we deduce d from $d - 1$. We write $\tau = \rho \perp a\rho$, where ρ is a (bilinear or quadratic) Pfister form of degree $d - 1$. Since $1 \in D(\tau)$, we clearly have

---

[2] The form [1] is thus *not* considered to be a quadratic Pfister form, not even if char $k \neq 2$.

[3] If τ is anisotropic, this follows easily from §3.6, Theorem 3.50 below.

[4] If φ is bilinear, and $c \in k^n$, we define $\varphi(c) := \varphi(c, c)$.

$N(\tau) \subset D(\tau)$. It is also clear that $N(\rho) \subset N(\tau)$. Now, let $c \in D(\tau)$ be given. We write $c = \rho(x) + a\rho(y)$ with vectors $x, y \in k^m$, $m = 2^{d-1}$. If $\rho(x) = 0$ or $\rho(y) = 0$, then c is a similarity norm of τ since a is one and since $D(\rho) = N(\rho) \subset N(\tau)$ by our induction hypothesis. Suppose now that $\rho(x)$ and $\rho(y)$ are both nonzero. Then we have:

$$\tau = \rho \perp a\rho \cong \rho(x)\rho \perp a\rho(y)\rho \cong \langle \rho(x), a\rho(y) \rangle \otimes \rho$$
$$\cong \langle \rho(x) + a\rho(y), \; a\rho(x)\rho(y)(\rho(x) + a\rho(y)) \rangle \otimes \rho$$
$$= (\rho(x) + a\rho(y))\langle 1, a\rho(x)\rho(y) \rangle \otimes \rho$$
$$\cong (\rho(x) + a\rho(y))\langle 1, a \rangle \otimes \rho = (\rho(x) + a\rho(y))\tau.$$

Here we several times used the fact that $\rho(x)$ and $\rho(y)$ are similarity norms of ρ. □

Theorem 3.11.

(i) *Let τ be an isotropic bilinear Pfister form over k. Then there exists a bilinear Pfister form σ over k with $\tau \cong \langle 1, -1 \rangle \otimes \sigma$. In particular, τ is metabolic.*

(ii) *Every isotropic quadratic Pfister form over k is hyperbolic.*

Proof. (i): We use induction on the degree d of τ. For $d = 1$ the statement is clear. So suppose $d > 1$. We choose a factorization $\tau \cong \langle 1, a \rangle \otimes \rho$ where ρ is a Pfister form of degree $d - 1$. If ρ is isotropic then, by our induction hypothesis, $\rho \cong \langle 1, -1 \rangle \otimes \gamma$ for some Pfister form γ. Then $\tau \cong \langle 1, -1 \rangle \otimes \sigma$ with $\sigma := \langle 1, a \rangle \otimes \gamma$. Now suppose that ρ is anisotropic. We have vectors $x, y \in k^m$, $m = 2^{d-1}$, with $\rho(x) + a\rho(y) = 0$, but x and y not both zero. Since ρ is anisotropic, we must have $\rho(x) \neq 0$ and $\rho(y) \neq 0$. Theorem 3.10 then implies $-a = \rho(x)\rho(y)^{-1} \in N(\rho)$. Therefore, $\tau \cong \langle 1, a \rangle \otimes \rho \cong \langle 1, -1 \rangle \otimes \rho$.

(ii): The argumentation is completely analogous.   □

Corollary 3.12. *If τ is an anisotropic quadratic Pfister form over k, then τ has height 1.*

Proof. $\tau \otimes k(\tau)$ is an isotropic Pfister form over $k(\tau)$. Thus, $\tau \otimes k(\tau) \sim 0$ by Theorem 3.11.   □

If char $k \neq 2$, τ is a quadratic Pfister form over k and φ is a subform of τ with $\dim \varphi = \dim \tau - 1$, then we have: If φ is isotropic, then φ splits. Indeed, we have $\tau \cong \varphi \perp [a]$ for some $a \in k^*$. If φ is isotropic, then τ is isotropic, thus $\tau \sim 0$ by Theorem 3.11 and hence $\varphi \sim [-a]$. As above, this implies: if φ is anisotropic, then φ has height 1. In characteristic 2, we cannot come to the same conclusion since φ is not an orthogonal summand of τ. As a replacement we have the following theorem:

Theorem 3.13. *Let char $k = 2$. Let ρ be a bilinear Pfister form of degree $d \geq 1$, let $\beta = \begin{bmatrix} 1 & 1 \\ 1 & b \end{bmatrix}$ be a binary quadratic form and let $c \in D(\beta)$. Define $\varphi := \rho' \otimes \beta \perp [c]$, then:*

(i) *If φ is isotropic, then $\varphi \cong (2^d - 1) \times H \perp [c]$.*

(ii) *If φ is anisotropic, then φ has height 1.*

(iii) *$k(\varphi)$ is specialization equivalent with $k(\rho \otimes \beta)$ over k.*

Proof. Let φ be isotropic. Since φ is a subform of $\rho \otimes \beta$, the form $\rho \otimes \beta$ is isotropic. Hence, by Theorem 3.11,

$$\rho \otimes \beta = (\langle 1 \rangle \perp \rho') \otimes \beta \sim 0.$$

It follows that $\rho' \otimes \beta \sim \beta$. {Note that $\beta = -\beta$.} Thus $\varphi \sim \beta \perp [c] \sim [c]$, since $\beta \perp [c]$ is isotropic and $\dim \beta = 2$. This proves (i).

Next, suppose that φ is anisotropic. Then $\varphi \otimes k(\varphi)$ is isotropic and so $\varphi \otimes k(\varphi)$ splits by what we already proved. Therefore $h(\varphi) = 1$.

Finally, let $K \supset k$ be a field extension with $(\rho \otimes \beta)_K$ isotropic.[5] Then $(\rho \otimes \beta)_K \sim 0$. Hence, $(\rho' \otimes \beta)_K \sim \beta_K$ and $\varphi_K \sim (\beta \perp [c])_K \sim [c]$. It is now clear that $\rho \otimes \beta$ and φ become isotropic over the same field extensions of k. Thus $k(\rho \otimes \beta)$ is a generic zero field of φ. □

The class of examples of forms of height 1, given by this theorem, can be expanded considerably, as we will show now. But first some generalities. We need the following theorem from the elementary theory of quadratic forms.

Theorem 3.14. *Let $E = (E, q)$ be a strictly regular space over k and let $F_1 = (F_1, q|F_1)$, $F_2 = (F_2, q|F_2)$ be quadratic submodules of E. Further, let $\alpha : F_1 \xrightarrow{\sim} F_2$ be an isometry. Then there exists an automorphism σ of E, i.e. an isometry $\sigma : E \xrightarrow{\sim} E$, which extends α.*

For a proof, we refer to [10, §4, Th.1]. If F_1 (thus also F_2) is strictly regular, then Theorem 3.14 can also be formulated as a cancellation theorem and is then a special case of Theorem 1.67. In characteristic $\neq 2$, this is just the Witt Cancellation Theorem [58, Satz 4]. We dub Theorem 3.14 the "*Witt Extension Theorem*" (although Theorem 3.14 was not formulated in this generality by Witt).

Let $E = (E, q)$ again be a strictly regular space. Let F be a quadratic submodule of E. Then

$$F^\perp := \{x \in E \mid B_q(x, E) = 0\}$$

is also a subspace of E. Since B_q is nondegenerate, we have $(F^\perp)^\perp = F$. We call $F^\perp$ the *polar of F in E* and write $F^\perp = \text{Pol}_E(F)$. If F_1 is a submodule of E (w.r.t. q), isometric to F, then $\text{Pol}_E(F_1)$ is also isometric to $\text{Pol}_E(F)$ by Theorem 3.14. This fact supplies us with very convenient terminology in the language of forms (instead of spaces) as follows.

Definition 3.15. Let φ be a quadratic form over k and τ a strictly regular form over k. Suppose that $\varphi < \tau$. We define a (possibly degenerate) form ψ over k as follows: let $(E, q) = E$ be a space, isometric to τ and $(F, q|F) = F$ a submodule of E, isometric to φ. Then we define ψ to be given by the submodule $\text{Pol}_E(F)$ of E. ψ is up to isometry uniquely determined by φ and τ. We write $\psi = \text{Pol}_\tau(\varphi)$ and call ψ "*the*" polar of φ in τ.

---

[5] From now on, we sometimes write ψ_K instead of $\psi \otimes K$ for a form ψ over K, as announced in §2.1.

We return to the geometric language. We suppose now that k has characteristic 2. On a strictly regular space (E, q) over k we can then use symplectic geometry, since $B_q(x, x) = 0$ for every $x \in E$.

Lemma 3.16. *Let F be a quadratic submodule of a strictly regular space E and let $P = QL(F)$ be the quasilinear part of F. Then we also have $QL(\mathrm{Pol}_E(F)) = P$. More precisely: let G be a complement of P in F, i.e. $F = P \oplus G = P \perp G$. Then there exists an orthogonal decomposition $E = M \perp G \perp R$ with $P \subset M$, $\dim M = 2 \dim P$, $F = P \perp G$. For every such decomposition, we have $\mathrm{Pol}_E(F) = P \perp R$.*

Proof. The space G is strictly regular. Let $T = \mathrm{Pol}_E(G)$, i.e. $E = T \perp G$. T is also strictly regular and $P \subset T$. Let $x_1, \ldots, x_r$ be a basis of the vector space P. Since P is quasilinear, we have $B(x_i, x_j) = 0$ for all $i, j \in \{1, \ldots, r\}$. By elementary symplectic geometry, T contains vectors $y_1, \ldots, y_r$ with $B(x_i, y_j) = \delta_{ij}$ for all $i, j \in \{1, \ldots, r\}$. Let M be the vector space spanned by $x_1, \ldots, x_r, y_1, \ldots, y_r$. The bilinear form $B|M \times M$ is clearly nondegenerate. Hence the quadratic submodule M of T is nondegenerate, i.e. is a quadratic space, and we have $T = M \perp R$ with a further quadratic space R. We have $E = M \perp R \perp G$.

Now let $E = M \perp R \perp G$ be a given decomposition, with $P \subset M$, $\dim M = 2 \dim P$, $F = P \perp G$. Then

$$\mathrm{Pol}_E(F) = \mathrm{Pol}_E(P) \cap \mathrm{Pol}_E(G) = \mathrm{Pol}_E(P) \cap (M \perp R) = \mathrm{Pol}_M(P) \perp R.$$

By elementary symplectic geometry, we get $\mathrm{Pol}_M(P) = P$, and so $\mathrm{Pol}_E(F) = P \perp R$. □

In the language of forms, the lemma allows us to say the following.

Scholium 3.17. *Let τ be a strictly regular quadratic form and φ an arbitrary quadratic form with $\varphi < \tau$. Then there is an orthogonal decomposition $\tau = \mu \perp \gamma \perp \rho$ with*

$$QL(\varphi) < \mu, \quad \dim \mu = 2 \dim QL(\varphi), \quad \varphi \cong QL(\varphi) \perp \gamma, \quad \mathrm{Pol}_\tau(\varphi) \cong QL(\varphi) \perp \rho.$$

If $QL(\varphi) \cong [a_1, \ldots, a_r]$, then $\mu \cong \begin{bmatrix} a_1 & 1 \\ 1 & b_1 \end{bmatrix} \perp \ldots \perp \begin{bmatrix} a_r & 1 \\ 1 & b_r \end{bmatrix}$ with further elements $b_1, \ldots, b_r \in k$. We have $\mathrm{Pol}_\tau(\mathrm{Pol}_\tau(\varphi)) = \varphi$. If φ is nondegenerate, then $\mathrm{Pol}_\tau(\varphi)$ is nondegenerate, since φ and $\mathrm{Pol}_\tau(\varphi)$ have the same quasilinear part.

Later on, we will be particularly interested in the case where φ itself is quasilinear. Then $\gamma = 0$. We thus have a decomposition $\tau = \mu \perp \rho$ with $\varphi < \mu$, $\dim \mu = 2 \dim \varphi$, and so $\mathrm{Pol}_\tau(\varphi) = \varphi \perp \rho$.

Remark. The scholium remains valid in characteristic $\neq 2$. Then $QL(\varphi)$ is simply the radical of φ, hence $a_1 = \cdots = a_r = 0$. Then we can also establish that $b_1 = \cdots = b_r = 0$. If φ is nondegenerate (i.e. $QL(\varphi) = 0$), then $\mathrm{Pol}_\tau(\varphi)$ is nondegenerate and $\tau = \varphi \perp \mathrm{Pol}_\tau(\varphi)$. In characteristic $\neq 2$ the notion of polar is not as interesting as in characteristic 2.

Theorem 3.18. *Again let* $\operatorname{char} k = 2$. *Let* τ *be a quadratic Pfister form over* k *of degree* $d \geq 2$ *and let* χ *be an* anisotropic quasilinear *form over* k *with* $\dim \chi < 2^{d-1}$. *Suppose that* $\chi < a\tau$ *for some* $a \in k^*$. *Then, for the polar* $\varphi = \operatorname{Pol}_{a\tau}(\chi)$, *we have the following:*

(i) φ *is nondegenerate,* $QL(\varphi) = \chi$.

(ii) *If* τ *is isotropic, then* φ *splits.*

(iii) *If* φ *is anisotropic, then* φ *has height 1.*

(iv) *The fields* $k(\varphi)$ *and* $k(\tau)$ *are specialization equivalent over* k.

(v) *If* $\dim \chi < \frac{1}{4} \dim \tau$, *then for every form* γ *with* $\varphi \cong \chi \perp \gamma$, $k(\gamma)$ *is also specialization equivalent to* $k(\varphi)$ *over* k.

Proof. 1) We assume without loss of generality that $a = 1$, thus $\chi < \tau$. We already know that (i) holds. We choose a decomposition $\varphi = \chi \perp \gamma$. Then $\operatorname{Pol}_\tau(\varphi) = \chi$ and so, in particular, $QL(\varphi) = \chi$. Upon applying the scholium to φ, we see that $\rho = 0$ there. We thus have a decomposition $\tau = \mu \perp \gamma$ with $\chi < \mu$ and $2 \dim \chi = \dim \mu$. Since $2 \dim \chi < \dim \tau$, we certainly have $\gamma \neq 0$.

2) Now let τ be isotropic. By Theorem 3.11, we have $\tau \sim 0$. It follows that $\gamma \sim \mu$ (note that $\mu = -\mu$) and then $\varphi \sim \chi \perp \mu$. We have

$$\chi \cong [a_1, \ldots, a_r], \quad \mu \cong \begin{bmatrix} a_1 & 1 \\ 1 & b_1 \end{bmatrix} \perp \ldots \perp \begin{bmatrix} a_r & 1 \\ 1 & b_r \end{bmatrix}$$

with elements $a_i, b_i \in k$. Now,

$$[a_i] \perp \begin{bmatrix} a_i & 1 \\ 1 & b_i \end{bmatrix} \cong [a_i] \perp \begin{bmatrix} 0 & 1 \\ 1 & b_i \end{bmatrix} \cong [a_i] \perp H.$$

It follows that $\chi \perp \mu \cong \chi \perp r \times H$, also $\varphi \sim \chi$. This proves (ii).

3) If φ is anisotropic, then $\varphi \otimes k(\varphi)$ is isotropic. Nevertheless, $\chi \otimes k(\varphi)$ remains anisotropic, since $k(\varphi)$ is separable over k (Theorem 2.7). If we apply what we already proved to $\chi \otimes k(\varphi)$ and $\tau \otimes k(\varphi)$, we see that $\varphi \otimes k(\varphi)$ splits. Hence $h(\varphi) = 1$.

4) If φ is isotropic, then the field extensions $k(\varphi)$ and $k(\tau)$ of k are both purely transcendental. Now suppose that φ is anisotropic. $\varphi \otimes k(\tau)$ is isotropic (even split). Of course, $\tau \otimes k(\varphi)$ is also isotropic. Hence $k(\varphi) \sim_k k(\tau)$ in this case also.

5) Finally, suppose that $\dim \chi < \frac{1}{4} \dim \tau = 2^{d-2}$. Now $\dim \mu < 2^{d-1}$, thus $\dim \gamma > 2^{d-1}$. We have $\tau \otimes k(\tau) \sim 0$, hence $\gamma \otimes k(\tau) \sim \mu \otimes k(\tau)$. Since $\dim \mu < \dim \gamma$, it follows that $\gamma \otimes k(\tau)$ is isotropic. Clearly $\tau \otimes k(\gamma)$ is also isotropic. We conclude that $k(\gamma) \sim_k k(\tau)$. $\qquad\qquad\square$

We want to understand the connection between the forms φ and $a\tau$ in Theorem 3.18 better. For this purpose we give another general definition for quadratic forms over a field k of characteristic 2.

Definition 3.19. Let φ be an arbitrary form and τ a strictly regular form over k. We say that τ is a *strictly regular hull of* φ, and write $\varphi <_s \tau$, when the following conditions hold:

(a) $\varphi < \tau$.

(b) Let $(E, q) = E$ be a space, isometric to τ and let $(F, q|F) = F$ be a quadratic sub-
   module of E, isometric to φ. Then there *does not exist* a strictly regular subspace
   $(E', q|E') = E'$ of E with $F \subsetneq E' \subsetneq E$.

By the Witt Extension Theorem (Theorem 3.14), the relation $\varphi <_s \tau$ only depends
on the isometry classes of φ and τ, and is independent of the choice of the spaces E
and F.

Lemma 3.20. *Let φ be an arbitrary quadratic form and τ a strictly regular quadratic
form with $\varphi < \tau$. Then τ is a strictly regular hull of φ if and only if $\mathrm{Pol}_\tau(\varphi)$ is quasilinear.*

Proof. Let $F \subset E$ be quadratic modules associated to φ and τ, as indicated in Def-
inition 3.19. If τ is not a strictly regular hull of φ, then E contains a strictly regu-
lar subspace E' with $F \subset E'$ and $E' \neq E$. We have an orthogonal decomposition
$E = E' \perp E''$. Clearly, $E'' \subset \mathrm{Pol}_E(F)$. The space E'' is $\neq 0$ and strictly regular.
Hence $\mathrm{Pol}_E(F)$ is not quasilinear.
   Now suppose that $\varphi <_s \tau$. We choose decompositions $F = P \perp G$, $E = M \perp G \perp R$ which have the properties exhibited in Lemma 3.16. We have $\mathrm{Pol}_E(F) = P \perp R$. Since $M \perp G$ is strictly regular and contains the space F, we must have
$M \perp G = E$, i.e. $R = 0$. It follows that $\mathrm{Pol}_E(F) = P$ is quasilinear. $\square$

Remark. The definition of a strictly regular hull is also meaningful in characteristic
$\neq 2$ and Lemma 3.20 still holds (a quasilinear form then looks like $[0, \dots, 0]$), but
again this term is far less interesting now. If φ and τ are both nondegenerate, then
$\varphi <_s \tau$ holds only if $\varphi \cong \tau$.

Definition 3.21. Let φ be a quadratic form over k. We call φ a *close Pfister neighbour*[6] if φ is nondegenerate and not quasilinear and if there exist a Pfister form τ
and an $a \in k^*$ with $\varphi <_s a\tau$. More precisely, we then call φ a *close neighbour of τ.*

If φ is a close neighbour of τ and if φ is anisotropic, then Theorem 3.18 tells us
that φ has height 1 and that $k(\tau)$ is a generic zero field of φ.
   The relation "strictly regular hull" is compatible with taking tensor products, as
we will show now.

Lemma 3.22. *Let φ be an arbitrary quadratic form and τ a strictly regular quadratic
form over k with $\varphi < \tau$. Let σ be a bilinear nondegenerate form over k. Then
$\sigma \otimes \varphi < \sigma \otimes \tau$ (this is clear) and $\mathrm{Pol}_{\sigma \otimes \tau}(\sigma \otimes \varphi) \cong \sigma \otimes \mathrm{Pol}_\tau(\varphi)$.*

Proof. Let $(E, q) = E$ be a quadratic space, isometric to τ and $(F, q|F) = F$ a
submodule, isometric to φ. Furthermore, let S be a bilinear space, isometric to σ.
By Lemma 3.16, we have orthogonal decompositions $F = P \perp G$, $E = Q \perp G \perp R$
with $P = QL(F) \subset Q$ and $\dim Q = 2 \dim P$. From those we obtain analogous
decompositions

---

[6] In §3.8 we will introduce the more general notion of Pfister neighbour.

$$S \otimes F = (S \otimes P) \perp (S \otimes G), \ S \otimes E = (S \otimes Q) \perp (S \otimes G) \perp (S \otimes R).$$

Clearly, $S \otimes P$ is the quasilinear part of $S \otimes F$ and $\dim(S \otimes Q) = 2 \dim(S \otimes P)$. By Lemma 3.16, we get

$$\mathrm{Pol}_E(F) = P \perp R \qquad \text{and} \qquad \mathrm{Pol}_{S \otimes E}(S \otimes F) = (S \otimes P) \perp (S \otimes R).$$

Therefore, $S \otimes \mathrm{Pol}_E(F) = \mathrm{Pol}_{S \otimes E}(S \otimes F)$.   □

Theorem 3.23. *If $\varphi <_s \tau$ and if σ is a nondegenerate bilinear form, then $\sigma \otimes \varphi <_s \sigma \otimes \tau$.*

Proof. By Lemma 3.20, $\mathrm{Pol}_\tau(\varphi)$ is quasilinear. Using Lemma 3.22, it follows that $\mathrm{Pol}_{\sigma \otimes \tau}(\sigma \otimes \varphi) = \sigma \otimes \mathrm{Pol}_\tau(\varphi)$ is quasilinear. Again by Lemma 3.20, we have $\sigma \otimes \varphi <_s \sigma \otimes \tau$.   □

Corollary 3.24. *If φ is a close neighbour of a quadratic Pfister form τ, and if σ is a bilinear Pfister form, then $\sigma \otimes \varphi$ is a close neighbour of the Pfister form $\sigma \otimes \tau$.*

Thus it is shown that for a large class of forms of height 1, if φ is any such form and σ is any bilinear Pfister form, then the tensor product $\sigma \otimes \varphi$ again has height 1, in so far as it is anisotropic. Such a phenomenon does not happen in characteristic $\neq 2$. If φ is an anisotropic form of height 1 and not of the form $a\tau$ for some Pfister form τ, then (as mentioned above) we know in characteristic $\neq 2$ that φ has the form $a\tau'$, where τ' is the pure part of some Pfister form. If σ is a further Pfister form $\neq \langle 1 \rangle$, then $\sigma \otimes \varphi$ *does not have* height 1, since neither $\dim(\sigma \otimes \varphi)$ nor $1 + \dim(\sigma \otimes \varphi)$ is a 2-power in this case.

The following question suggests itself: Let φ be a close neighbour of some Pfister form τ. Is τ uniquely determined by φ (up to isometry)?

We will show that this is indeed so. We will also derive other theorems about Pfister forms. However, for this purpose, we need more tools from the general theory of quadratic forms, namely the so-called Subform Theorem and a "Norm Theorem". Both theorems are well known in characteristic $\neq 2$ and still do hold in characteristic 2. We will derive them next, see §3.3 and §3.5.

3.3 The Subform Theorem

The so-called Subform Theorem of Cassels and Pfister (e.g. [55, Th.3.7]) is one of the most important tools in the algebraic theory of quadratic forms over fields of characteristic $\neq 2$. In 1969/70, Pfister's Egyptian student Mahmud Amer proved the Subform Theorem in characteristic 2 in his dissertation in a manner which does not leave anything to be desired. Unaware of this fact, R. Baeza proved a slightly more special version of the Subform Theorem in characteristic 2 in 1974 [4]. Unfortunately Amer's proof was never published. In Pfister's book [52] one can find a sketch of Amer's proof which leaves a gap, however. Since the Subform Theorem will play an important role in what follows, and will furnish us with a description of all strictly regular forms of height 1 in particular, we will now present Amer's proof in its completeness. Apart from this, the paper [4] on this topic is also worth reading in our opinion.

Theorem 3.25 ("Cassels's Lemma"). *Let k be a field of arbitrary characteristic, x an indeterminate over k and $p \in k[x]$ a polynomial. Let φ be a quadratic form in n variables over k which represents the polynomial p over the field $k(x)$. Then φ already represents the polynomial p over the ring $k[x]$, i.e. there exist polynomials $f_1, \ldots, f_n \in k[x]$ with $\varphi(f_1, \ldots, f_n) = p$.*

Proof. 1) We may assume that $p \neq 0$. We may further assume that φ is nondegenerate upon replacing φ by $\widehat{\varphi}$ (cf. Definition 1.74). If φ is isotropic, then $\begin{bmatrix} 0 & 1 \\ 1 & 0 \end{bmatrix}$ is a subform of φ (cf. Lemma 1.66). Clearly $\begin{bmatrix} 0 & 1 \\ 1 & 0 \end{bmatrix}$ already represents every polynomial over $k[x]$. Thus we assume from now on that φ is anisotropic.

2) We choose a representation

$$p = \varphi\left(\frac{f_1}{f_0}, \frac{f_2}{f_0}, \ldots, \frac{f_n}{f_0}\right) \tag{3.4}$$

with $f_i \in k[x]$ $(0 \leq i \leq n)$, $f_0 \neq 0$ and deg f_0 minimal. ("deg" denotes the degree of a polynomial.) We assume that deg $f_0 > 0$. This should lead to a contradiction.

Consider the form $\psi := \langle -p \rangle \perp \varphi$ in $n + 1$ variables over $k(x)$. For every polynomial f_i with $1 \leq i \leq n$, let

$$f_i = f_0 g_i + r_i,$$

the division with remainder of f_i by f_0, thus with $g_i, r_i \in k[x]$, deg $r_i <$ deg f_0.

It can happen that $r_i = 0$, i.e. that deg $r_i = -\infty$. However, for at least one $i \in \{1, \ldots, n\}$ we have $r_i \neq 0$. We introduce the vectors $f := (f_0, f_1, \ldots, f_n)$ and $g := (g_0, g_1, \ldots, g_n)$ with $g_0 := 1$. By (3.4) we have $\psi(f) = 0$. By the minimality of deg f_0 we certainly have $\psi(g) \neq 0$ however. In particular, the vectors f and g are linearly independent over $k(x)$.

In what follows we will denote the "inner product" $B_\psi(f, g)$ simply by fg and we will use such a notation (as in Amer) also for forms different from ψ. Let

$$u := \psi(g)f - (fg)g.$$

We have

$$\psi(u) = \psi(g)^2 \psi(f) + (fg)^2 \psi(g) - (fg)^2 \psi(g) = 0,$$

since $\psi(f) = 0$. From the linear independence of f and g it follows that $u \neq 0$.

The form ψ is also isotropic over $k(x)$. Hence the first coordinate u_0 of u is likewise different from zero. We have

$$u_0 = \psi(g)f_0 - fg = \frac{1}{f_0}[\psi(g)f_0^2 - (fg)f_0] = \frac{1}{f_0}\psi(f - f_0 g).$$

Every coordinate of $f - f_0 g = (0, r_1, \ldots, r_n)$ is of lower degree than f_0. Hence $\deg \psi(f - f_0 g) \le 2 \deg f_0 - 2$ and so $\deg u_0 \le \deg f_0 - 2$. Since $\psi(u) = 0$ we thus obtained a representation of p of the form (3.4) in which the denominator polynomial u_0 is of lower degree than f_0: contradiction! $\square$

Addendum (Amer). *In the situation of Theorem 3.25, let φ be anisotropic and $p \ne 0$. Furthermore, let M be a subset of $\{1, \ldots, n\}$ and $p = \varphi(h_1, \ldots, h_n)$ a representation with $h_i \in k(x)$ for $1 \le i \le n$ and even $h_i \in k[x]$ for $i \in M$. Then there exist polynomials $g_1, \ldots, g_n \in k[x]$ with $p = \varphi(g_1, \ldots, g_n)$ and $g_i = h_i$ for $i \in M$.*

Proof. We choose a representation

$$p = \varphi \left(\frac{f_1}{f_0}, \ldots, \frac{f_n}{f_0} \right) \tag{3.5}$$

with $f_i \in k[x]$ $(0 \le i \le n)$, $f_0 \ne 0$, $\frac{f_i}{f_0} = h_i$ for $i \in M$ and $\deg f_0$ minimal. By way of contradiction, suppose again that $\deg f_0 > 0$. Using these polynomials f_i, we go through the proof above again, starting from step 2. This time we have $r_i = 0$ and $g_i = h_i$ for every $i \in M$. For $i \in M$, the vector u above has ith coordinate

$$u_i = \psi(g) f_i - (fg) g_i = \psi(g) f_0 h_i - (fg) h_i$$
$$= [\psi(g) f_0 - (fg)] h_i = u_0 h_i.$$

Thus $\frac{u_i}{u_0} = h_i$ for every $i \in M$. By the above considerations we get $\deg u_0 < \deg f_0$: contradiction! $\square$

Remark. If $\operatorname{char} k \ne 2$, then the Addendum is dispensable since it can easily be obtained from Theorem 3.25 as follows: as usual we interpret φ as a quadratic form over the k-vector space $V := k^n$ with standard basis $e_1, \ldots, e_n$. Suppose without loss of generality that $M = \{1, \ldots, m\}$ for an $m \le n$ and $W := \sum_{i=1}^{m} k e_i$. The form $\varphi_1 := \varphi | W$ is anisotropic and thus strictly regular. We have an orthogonal decomposition $\varphi \cong \varphi_1 \perp \varphi_2$. Now φ_2 represents $f_1 := p - \varphi_1(h_1, \ldots, h_m)$ over $k(x)$ and thus over $k[x]$ by Theorem 3.25. This yields the representation of p by φ, as given in the Addendum.

Theorem 3.26 (Subform Theorem). *Let ψ be an anisotropic quadratic form over k and let φ be a nondegenerate quadratic form over k in m variables. Let $x_1, \ldots, x_m$ be indeterminates over k. Suppose that the form ψ represents the element $\varphi(x_1, \ldots, x_m)$ over the field $k(x_1, \ldots, x_m)$. Then φ is a subform of ψ.*

The proof involves several lemmata. We assume all the time that ψ is an n-dimensional anisotropic quadratic form over k and that k *has characteristic* 2. For $\operatorname{char} k \ne 2$ one can find the proof of the Subform Theorem in many books, in particular in [52], [55].

Lemma 3.27. *Let $a, b, c \in k$, $b \ne 0$ and let x be an indeterminate over k. Suppose that ψ represents the element $ax^2 + bx + c$ over $k(x)$. Then $\begin{bmatrix} a & b \\ b & c \end{bmatrix} < \psi$.*

Proof. By Theorem 3.25, there exists a vector $f \in k[x]^n$ with $\psi(f) = ax^2 + bx + c$. Since ψ is anisotropic, every coordinate of f has to be a polynomial of degree ≤ 1. Hence $f = xu + v$, for certain vectors $u, v \in k^n$. Again using the notation $uv := B_\psi(u, v)$, we then have the equation

$$\psi(u)x^2 + (uv)x + \psi(v) = ax^2 + bx + c.$$

Therefore, $\psi(u) = a$, $\psi(v) = c$ and $uv = b$. Hence the value matrix of $\psi|ku + kv$ is $\begin{bmatrix} a & b \\ b & c \end{bmatrix}$. $\qquad\qquad\square$

Lemma 3.28. *Let $E = (E, q)$ be an anisotropic quadratic module over k and let u, v be vectors in E with $q(u) = q(v) \neq 0$. Then there exists an isometry $\sigma : E \xrightarrow{\sim} E$ with $\sigma(u) = v$ and $\sigma(v) = u$.*

Proof. Suppose without loss of generality that $u \neq v$. Then $u + v \neq 0$, so that $q(u + v) = 2q(u) + uv = uv \neq 0$. {As before, we write uv instead of $B_q(u, v)$.} The quadratic submodule $ku + kv$ of E is therefore strictly regular and it follows that

$$E = (ku + kv) \perp F$$

with $F := (ku + kv)^\perp$ (cf. Lemma 1.53). The linear map $\sigma : E \to E$ with $\sigma(u) = v$, $\sigma(v) = u$, $\sigma(w) = w$ for all $w \in F$ is the one we are looking for. $\qquad\square$

Remark. The map σ is nothing else than the reflection in the hyperplane $(u + v)^\perp$, perpendicular to the vector $u + v \neq 0$ (cf. [10, §6, No. 4]),

$$\sigma(z) = z - \frac{(u + v)z}{q(u + v)}(u + v), \qquad \text{for all } z \in E.$$

Lemma 3.29. *Let $\psi = \varphi \perp \chi$ be an orthogonal decomposition with φ strictly regular. Let γ be a further quadratic form over k with $\varphi \perp \gamma < \psi$. Then $\gamma < \chi$.*

This lemma is a special case of a much more general theorem, which deserves attention in its own right. We will prove this theorem at the end of this section.

Lemma 3.30. *Let $\psi = \varphi \perp \chi$ with $\dim \varphi = 2$ and φ strictly regular. Let x_1, x_2 be indeterminates over k and d an element $\neq 0$ of k. Suppose that ψ represents the polynomial $\varphi(x_1, x_2) + d$ over $k(x_1, x_2)$. Then $d \in D(\chi)$.*[7]

Proof. Let $\varphi = \begin{bmatrix} a & b \\ b & c \end{bmatrix}$ and $\dim \psi = n$. We have $b \neq 0$. Therefore, evaluating ψ at the first two standard vectors $e_1 = (1, 0, \ldots, 0)$, $e_2 = (0, 1, 0, \ldots, 0)$ of the vector space k^n yields the values $\psi(e_1) = a$, $\psi(e_2) = c$, $e_1 e_2 = b$. Let us now work in the vector space $V = K^n$ over the field $K := k(x_2)$ which we also equip with the form ψ. (More precisely, we are dealing with the form $\psi \otimes K$ here, but we simply write ψ.) By our assumption, ψ represents the element

$$\varphi(x_1, x_2) + d = ax_1^2 + (bx_2)x_1 + (cx_2^2 + d)$$

---

[7] One recalls the notation in Definition 3.9.

over $K(x_1)$. Lemma 3.27 tells us now that there exist vectors e_1', h' in V with $\psi(e_1') = a$, $\psi(h') = cx_2^2 + d$, $e_1'h' = bx_2$. By Lemma 3.28 there is an isometry $\sigma : V \xrightarrow{\sim} V$ with respect to ψ with $\sigma(e_1') = e_1$. Letting $h := \sigma(h') \in V$, we then have

$$\psi(e_1) = a, \quad \psi(h) = cx_2^2 + d, \quad e_1h = bx_2.$$

If $h_1, \ldots, h_n$ are the coordinates of h, i.e. $h = \sum_{i=1}^{n} h_ie_i$, then the last equation gives $e_1h = h_2b = x_2b$, and so $h_2 = x_2$. {Bear in mind that $e_ie_j = 0$ for $i \leq 2$, $j > 2$, and $uu = 0$ for every $u \in V$.} By the Addendum there exists a vector $g = (g_1, \ldots, g_n) \in k[x_2]^n$ with $g_2 = h_2 = x_2$ and $\psi(g) = cx_2^2 + d$. Since ψ is anisotropic, all polynomials g_i have degree ≤ 1. Thus, $g = x_2u + v$ with vectors $u = (u_1, \ldots, u_n)$, $v = (v_1, \ldots, v_n)$ in k^n and $u_2 = 1$, $v_2 = 0$. We have

$$cx_2^2 + d = \psi(u)x_2^2 + (uv)x_2 + \psi(v),$$

so that $\psi(u) = c$, $uv = 0$ and $\psi(v) = d$. Furthermore, $e_1u = u_2b = b$, $e_1v = v_2b = 0$. Therefore $\psi|ke_1 + ku \cong \begin{bmatrix} a & b \\ b & c \end{bmatrix}$ and $v \in (ke_1 + ku)^\perp$. This shows that

$$\begin{bmatrix} a & b \\ b & c \end{bmatrix} \perp [d] \quad < \quad \psi = \begin{bmatrix} a & b \\ b & c \end{bmatrix} \perp \chi.$$

By Lemma 3.29 we get $[d] < \chi$. $\square$

Lemma 3.31. *Let $\psi = \varphi \perp \chi$ with φ strictly regular, $\dim \varphi = 2r$ and let $x_1, \ldots, x_{2r}$ be indeterminates over k. Further, let d be an element $\neq 0$ of k such that ψ represents the element $\varphi(x_1, \ldots, x_{2r}) + d$ over $k(x_1, \ldots, x_{2r})$. Then $d \in D(\chi)$.*

Proof. After a linear coordinate transformation of the variables $x_1, \ldots, x_{2r}$ we may assume that

$$\varphi = \begin{bmatrix} a_1 & b_1 \\ b_1 & c_1 \end{bmatrix} \perp \ldots \perp \begin{bmatrix} a_r & b_r \\ b_r & c_r \end{bmatrix}$$

with elements $a_i, b_i, c_i \in k$, $b_i \neq 0$. The form ψ represents the element $\sum_{i=1}^{r}(a_ix_{2i-1}^2 + b_ix_{2i-1}x_{2i}+c_ix_{2i}^2)+d$ over $k(x_1, \ldots, x_{2r})$. The statement now follows from Lemma 3.30 by induction on r. $\square$

Lemma 3.32. *Let $\varphi = [d_1, \ldots, d_m]$ be an anisotropic quasilinear form over k. Assume that there is a form ψ which represents the element $\varphi(x_1, \ldots, x_m) = d_1x_1^2 + \cdots + d_mx_m^2$ over the field $k(x_1, \ldots, x_m)$ with indeterminates $x_1, \ldots, x_m$ over k. Then $\varphi < \psi$.*

Proof. By induction on m. Let $m = 1$. Then ψ represents the element d_1 over $k(x_1)$. Using the Substitution Principle, Theorem 2.27, it follows that ψ represents the element d_1 over k as well.

Suppose now that $m > 1$. By the Substitution Principle we immediately get that $d_m \in D(\psi)$. We choose a vector $u \in k^n$ with $\psi(u) = d_m$. Let $d := d_1x_1^2 + \cdots +$

$d_{m-1}x_{m-1}^2$. By Theorem 3.25, ψ represents the element $d + d_m x_m^2$ over $K[x_m]$ with $K := k(x_1, \ldots, x_{m-1})$. Since ψ is anisotropic, we have vectors $w', u' \in K^n$ with $\psi(x_m u' + w') = d_m x_m^2 + d$. This implies

$$\psi(u') = d_m, \ \psi(w') = d, \ u'w' = 0.$$

By Lemma 3.28, the vector space K^n has an isometry $\sigma : K^n \overset{\sim}{\longrightarrow} K^n$ with respect to ψ with $\sigma(u') = u$. Letting $w := \sigma(w') \in K^n$, we then have

$$\psi(u) = d_m, \ \psi(w) = d, \ uw = 0.$$

Thus, the form $\psi_1 := \psi|(ku)^\perp$ represents the element d over K, and so $[d_1, \ldots, d_{m-1}] < \psi_1$, by our induction hypothesis. Hence there is a system of linearly independent vectors $v_1, \ldots, v_{m-1}$ in $(ku)^\perp$ with $\psi(v_i) = d_i$ $(1 \le i \le m - 1)$ and $v_i v_j = 0$ for $i \ne j$. The vector u cannot be a linear combination of $v_1, \ldots, v_{m-1}$, for if there was a relation $u = a_1 v_1 + \cdots + a_{m-1} v_{m-1}$ with coefficients $a_i \in k$, applying ψ to this relation would give $d_m = a_1^2 d_1 + \cdots + a_{m-1}^2 d_{m-1}$, contradicting the anisotropy of φ. Thus $W := kv_1 + \cdots + kv_{m-1} + ku$ is an m-dimensional subspace of k^n and $\psi|W \cong [d_1, \ldots, d_m]$.   □

Theorem 3.26 now follows in a straightforward way from Lemmas 3.31 and 3.32. We conclude this section with the generalization of Lemma 3.29, announced above.

Theorem 3.33. *Let E and F be free quadratic modules over a valuation ring $\mathfrak{o}$ and let G be a strictly regular quadratic space over $\mathfrak{o}$. If $G \perp F < G \perp E$, then $F < E$.*

Proof. We work in the quadratic module $U := G \perp E$ with associated quadratic form q. By our assumption, U contains a submodule V, which is a direct summand of U (as an $\mathfrak{o}$-module), so that $V = (V, q|V)$ has an orthogonal decomposition $V = G' \perp F'$ with $G \cong G'$, $F \cong F'$. Since G' is strictly regular, we have a further orthogonal decomposition $U = G' \perp E'$. This gives $E \cong E'$ by the Cancellation Theorem, Theorem 1.67. Hence, there is an isometry $\sigma : U \overset{\sim}{\longrightarrow} U$ with $\sigma(G') = G$ and $\sigma(E') = E$. Since F' is orthogonal to G', $F'' := \sigma(F')$ has to be orthogonal to G. Therefore, $F'' \subset E$. The $\mathfrak{o}$-module $\sigma(V) = G \perp F''$ is a direct summand of $U = G \perp E$, and so the $\mathfrak{o}$-module $E/F'' \cong U/\sigma(V)$ is free. Thus F'' is a direct summand of E, and so $F < E$.   □

Remark. The statement of Theorem 3.33 remains valid for a local ring $\mathfrak{o}$ because the Cancellation Theorem, Theorem 1.67 holds more generally over local rings.

3.4 Milnor's Exact Sequence

Let k be an arbitrary field. We want to derive an exact sequence, going back to J. Milnor [48], which establishes a close relationship between the Witt ring $W(k(t))$

of the rational function field $k(t)$ in one variable t and the Witt rings of field extensions of k, generated by one element. We need this sequence in order to prove the announced Norm Theorem in §3.5. It should be emphasized that we will deal with nondegenerate *bilinear forms* now, although we are mostly interested in quadratic forms where applications are concerned. Milnor assumes in [48, §5] that char $k \neq 2$, but this restriction is not necessary here.

Let $\mathcal{P}$ be the set of monic irreducible polynomials $p(t) \in k[t]$. For every $p \in \mathcal{P}$, let k_p denote the residue class field $k[t]/(p)$ and $\lambda_p : k(t) \to k_p \cup \infty$ the canonical place with valuation ring $\mathfrak{o}_p := k[t]_{(p(t))}$. {Note: except for the ring $\mathfrak{o}_\infty := k[t^{-1}]_{(t^{-1})}$, these rings are exactly all nontrivial valuation rings of $k(t)$ which contain the field k. They are all discrete.}

For every polynomial $p \in \mathcal{P}$ and every nondegenerate (symmetric) bilinear form φ over $k(t)$, let $\partial_p(\varphi)$ denote the weak specialization of the form $\langle p \rangle \varphi = \langle p \rangle \otimes \varphi$,

$$\partial_p(\varphi) :\, = (\lambda_p)_W(\langle p \rangle \varphi) \in W(k_p),$$

(cf. §1.3). The Witt class $\partial_p(\varphi)$ can be described in an elementary way for one-dimensional forms as follows: Let $f(t) \in k[t]$ be a polynomial $\neq 0$. Consider the decomposition $f(t) = p(t)^m g(t)$, with $m \in \mathbb{N}_0$ and $g(t)$ not divisible by $p(t)$. If m is even, then $\partial_p(\langle f \rangle) = 0$. If m is odd, then $\partial_p(\langle f \rangle) = \langle \overline{g} \rangle$, where $\overline{g}$ denotes the residue class of g in k_p. {Actually, we should write $[\langle \overline{g} \rangle]$ instead of $\langle \overline{g} \rangle$, but we avoid such intricacies, now and in the future. From now on we will usually not make a notational distinction between a form and its associated Witt class.}

Thus we obtain an additive map

$$\partial_p : W(k(t)) \longrightarrow W(k_p).$$

Furthermore we have a natural ring homomorphism

$$i : W(k) \longrightarrow W(k(t)), \quad \varphi \longmapsto \varphi \otimes k(t).$$

$W(k(t))$ is a $W(k)$-algebra by means of this ring homomorphism. Analogously, we interpret every ring $W(k_p)$, $p \in \mathcal{P}$, as a $W(k)$-algebra. In what follows, only the structure of $W(k(t))$ and the $W(k_p)$ as modules over $W(k)$ will be important however. The maps ∂_p are $W(k)$-linear. We combine them in a $W(k)$-linear map

$$\partial = (\partial_p \mid p \in \mathcal{P}) : W(k(t)) \longrightarrow \bigoplus_{p \in \mathcal{P}} W(k_p).$$

Theorem 3.34 (Milnor [48]). *The sequence of $W(k)$-modules*

$$0 \longrightarrow W(k) \xrightarrow{\; i \;} W(k(t)) \xrightarrow{\; \partial \;} \bigoplus_{p \in \mathcal{P}} W(k_p) \longrightarrow 0$$

is split exact. If $\lambda : k(t) \longrightarrow k \cup \infty$ is an arbitrarily chosen place over k, then $\lambda_W : W(k(t)) \longrightarrow W(k)$ is a retraction for i, in other words $\lambda_W \circ i = \mathrm{id}_{W(k)}$.

We will carry out the *proof* in several steps, broadly following the style of Milnor's procedure [48, p. 335 ff]. In what follows, a "form" is always a nondegenerate symmetric bilinear form, which will be identified with its Witt class.

(1) If φ is a form over k, then $\varphi \otimes k(t)$ has good reduction with respect to every place $\lambda : k(t) \to k \cup \infty$ over k and

$$\lambda_W(\varphi \otimes k(t)) = \lambda_*(\varphi \otimes k(t)) = \varphi.$$

This proves the last statement of the theorem. In particular, i is injective. In the following, we think of $W(k)$ as a subring of $W(k(t))$ via the map i.

(2) We consider an ascending filtration $(L_d \mid d \in \mathbb{N}_0)$ of the $W(k)$-algebra $W(k(t))$. For every $d \in \mathbb{N}_0$, let L_d be the subring of $W(k(t))$, generated by one-dimensional forms $\langle f \rangle$ for polynomials $f \in k[t]$ with $f \neq 0$, $\deg f \leq d$. Clearly we then have

$$W(k) = L_0 \subset L_1 \subset L_2 \subset \ldots,$$

and $W(k(t))$ is the union of all L_d. The forms $\langle f_1 \ldots f_s \rangle$ for products of finitely many polynomials $f_1, \ldots, f_s$ with $\deg f_i \leq d$ $(1 \leq i \leq s)$ generate L_d additively. Further, let $(U_d \mid d \in \mathbb{N}_0)$ be the filtration of $\bigoplus_{p \in \mathcal{P}} W(k_p)$ with

$$U_0 := 0, \qquad U_d := \bigoplus_{\substack{p \in \mathcal{P} \\ \deg p \leq d}} W(k_p) \qquad (d > 0).$$

Clearly $\partial(L_d) \subset U_d$ for every $d \in \mathbb{N}_0$ and in particular, $\partial(L_0) = 0$. Therefore ∂ induces for every $d \geq 1$ a $W(k)$-linear homomorphism

$$\bar{\partial}_d : L_d/L_{d-1} \longrightarrow U_d/U_{d-1}.$$

For every $d \in \mathbb{N}$ we will verify that $\bar{\partial}_d$ is bijective. As an immediate consequence we then obtain that for every $d \in \mathbb{N}$, ∂ induces an isomorphism from L_d/L_0 to U_d and thus an isomorphism from $W(k(t))/W(k)$ to $\bigoplus_p W(k_p)$, which concludes the proof.

We let $\mathcal{P}_d$ denote the set of all $p \in \mathcal{P}$ of degree d and we identify U_d/U_{d-1} with the direct sum of the $W(k)$-modules $W(k_p)$, $p \in \mathcal{P}_d$. The component of index $p \in \mathcal{P}_d$ of the map

$$\bar{\partial}_d : L_d/L_{d-1} \longrightarrow \bigoplus_{p \in \mathcal{P}_d} W(k_p)$$

is the $W(k)$-linear map

$$\bar{\partial}_p : L_d/L_{d-1} \longrightarrow W(k_p),$$

induced by $\partial_p : W(k(t)) \to W(k_p)$.

(3) **Proof of the surjectivity of $\bar{\partial}_d$ $(d \geq 1)$.** It suffices to prove that, for given $p \in \mathcal{P}_d$ and $\eta \in W(k_p)$, there exists a $\xi \in L_d$ with $\partial_p(\xi) = \eta$, $\partial_p(\xi) = 0$ for $q \in \mathcal{P}_d$, $q \neq p$. Furthermore, we may assume that η is the Witt class of a one-dimensional form, i.e.

$\eta = \langle \overline{f} \rangle$ for an $f \in k[t]$ with $\deg f < d$, $f \neq 0$, where $\overline{f}$ obviously denotes the image of f in $k_p = k[t]/p(t)$. Letting $\xi := \langle pf \rangle$ clearly accomplishes our goal.

(4) For the proof of the injectivity of $\overline{\partial}_d$ we employ the following lemma (statement and proof are exactly as in Milnor [48, Lemma 5.5]).

Lemma 3.35. *Let $d \geq 1$. The additive group L_d is modulo L_{d-1} generated by elements $\langle pg_1 \ldots g_s \rangle$ with $s \in \mathbb{N}_0$, $p \in \mathcal{P}_d$ and every g_i a polynomial of degree $< d$. Moreover, if f is the polynomial of degree $< d$ with $f \equiv g_1 \ldots g_s$ mod p, then*

$$\langle pf \rangle \equiv \langle pg_1 \ldots g_s \rangle \text{ mod } L_{d-1}.$$

Proof of Lemma 3.35. We recall that the identity

$$\langle a + b \rangle = \langle a \rangle + \langle b \rangle - \langle ab(a + b) \rangle$$

holds in the Witt ring of any field, cf. §1.3. Let $\langle f_1 \ldots f_r \, g_1 \ldots g_s \rangle$ be a generator of L_d with pairwise different $f_i \in \mathcal{P}_d$ and polynomials $g_j \neq 0$ of degree $< d$. If $r \geq 2$, then $f_1 = f_2 + h$ with $h \neq 0$, $\deg h < d$, and we obtain the identity

$$\langle f_1 \rangle = \langle f_2 \rangle + \langle h \rangle - \langle f_1 f_2 h \rangle.$$

Multiplying with $\langle f_2 \ldots f_r \, g_1 \ldots g_s \rangle$ and omitting quadratic factors gives

$$\langle f_1 \ldots f_r \, g_1 \ldots g_s \rangle = \langle f_3 \ldots g_s \rangle + \langle h f_2 \ldots g_s \rangle - \langle f_1 h f_3 \ldots g_s \rangle.$$

Every term on the right-hand side contains at most $r - 1$ factors of degree d. By induction on r we now obtain that L_d modulo L_{d-1} is additively generated by forms $\langle fg_1 \ldots g_s \rangle$, where f is monic of degree d, and every g_j is a nonzero polynomial of degree $< d$. Furthermore, we may assume that f is irreducible.

Consider such a generator $\langle pg_1 \ldots g_s \rangle$ with $p \in \mathcal{P}_d$, $\deg g_i < d$. Assume that $s \geq 2$. Then $g_1 g_2 = pk + h$ with polynomials k, h of degree $< d$ and $h \neq 0$. Hence,

$$\langle g_1 g_2 \rangle = \langle h \rangle + \langle pk \rangle - \langle pkhg_1 g_2 \rangle.$$

Multiplying with $\langle pg_3 \ldots g_s \rangle$ gives

$$\langle pg_1 g_2 \ldots g_s \rangle = \langle phg_3 \ldots g_s \rangle + \langle kg_3 \ldots g_s \rangle - \langle khg_1 \ldots g_s \rangle,$$

and so

$$\langle pg_1 \ldots g_s \rangle \equiv \langle phg_3 \ldots g_s \rangle \text{ mod } L_{d-1}.$$

Furthermore,

$$g_1 \ldots g_s \equiv hg_3 \ldots g_s \text{ mod } p.$$

The proof of the lemma can now be concluded by means of a simple induction on s.

$\square$

(5) **Proof of the injectivity of $\overline{\partial}_d$** $(d \geq 1)$. Assume for the sake of contradiction that $\overline{\partial}_d$ is not injective. Then there exists a $\xi \in L_d \setminus L_{d-1}$ with $\partial_p(\xi) = 0$ for every

$p \in \mathcal{P}_d$. By the lemma, we have a congruence

$$\xi \equiv \sum_{i=1}^{s} \langle p_i \rangle \varphi_i \bmod L_{d-1}$$

with finitely many $p_i \in \mathcal{P}_d$ and forms $\varphi_i = \langle f_{i1}, \ldots, f_{i,r_i} \rangle$ for nonzero polynomials f_{ij} of degree $< d$. So, for every $i \in \{1, \ldots, s\}$,

$$\partial_{p_i}(\langle p_i \rangle \varphi_i) = \partial_{p_i}(\xi) = 0.$$

Furthermore, $\partial_q(\langle p_i \rangle \varphi_i) = 0$ for $q \in \mathcal{P}_d$, $q \neq p_i$. Hence every summand $\langle p_i \rangle \varphi_i$ of ξ modulo L_{d-1} is already in the kernel of $\bar{\partial}_d$. Thus we certainly have a $p \in \mathcal{P}_d$ and polynomials $f_1, \ldots, f_r \in k[t] \setminus \{0\}$ of degree $< d$ such that

$$\xi := \langle pf_1, \ldots, pf_r \rangle \notin L_{d-1},$$

but $\partial_p(\xi) = 0$, and so $\partial_q(\xi) = 0$ for every $q \in \mathcal{P}_d$. This should now lead to a contradiction.

We assume that the sequence $f_1, \ldots, f_r$ is chosen to be of minimal length r. We have $\partial_p(\xi) = \langle \bar{f}_1, \ldots, \bar{f}_r \rangle$. Thus, the form $\langle \bar{f}_1, \ldots, \bar{f}_r \rangle$ over k_p is metabolic and so in particular isotropic. Hence, after permuting the $\bar{f}_i$ if necessary, there certainly exists an index $s \in \{2, \ldots, r\}$ and polynomials $u_1, \ldots, u_s \in k[t] \setminus \{0\}$ of degree $< d$, such that

$$\bar{u}_1^2 \bar{f}_1 + \cdots + \bar{u}_s^2 \bar{f}_s = 0$$

in k_p. By the lemma we have

$$\langle pu_i^2 f_i \rangle \equiv \langle ph_i \rangle \bmod L_{d-1},$$

where $h_i \in k[t]$ is the polynomial of degree $< d$ with $u_i^2 f_i \equiv h_i \bmod p$. {Note: every $h_i \neq 0$.} Certainly $h_1 + h_2 \neq 0$, for otherwise we could write ξ modulo L_{d-1} in a shorter way, contradicting the minimality of r. Applying the relation

$$\langle h_1, h_2 \rangle = \langle h_1 + h_2, \, h_1 h_2 (h_1 + h_2) \rangle,$$

while keeping in mind that by the lemma

$$\langle ph_1 h_2 (h_1 + h_2) \rangle \equiv \langle pg_1 \rangle \bmod L_{d-1}$$

for a polynomial $g_1 \neq 0$ of degree $< d$, we obtain

$$\xi \equiv \langle p(h_1 + h_2), \, pg_1, ph_3, \ldots ph_s, \, pf_{s+1}, \ldots pf_r \rangle \bmod L_{d-1}.$$

Continuing in this way, we finally obtain a congruence

$$\xi \equiv \langle p(h_1 + \cdots + h_{s-1}), \, ph_s, pg_1, \ldots, pg_{s-2}, \, pf_{s+1}, \ldots, pf_r \rangle \bmod L_{d-1}.$$

The polynomial $h_1 + \cdots + h_s$ is of degree $< d$ and is divisible by p. Hence, $h_1 + \cdots + h_s = 0$. It follows that

$$\xi \equiv \langle pg_1, \ldots, pg_{s-2}, \ pf_{s+1}, \ldots, pf_r \rangle \bmod L_{d-1},$$

which contradicts the minimality of r. This establishes the injectivity of $\overline{\partial}_d$ and concludes the proof of the theorem. $\qquad\square$

3.5 A Norm Theorem

Let φ be a (symmetric) bilinear or quadratic form over a field k. We are interested in the *norm group* $N(\varphi)$ of φ, which is the group of similarity norms of φ. It consists of all $a \in k^*$ with $\varphi \cong a\varphi$ and was already introduced in §3.2.

If φ is a quadratic form, then we have a decomposition

$$\varphi = \widehat{\varphi} \perp r \times [0]$$

with $\widehat{\varphi}$ nondegenerate and $r \in \mathbb{N}_0$ (cf. §1.6). Clearly, $N(\varphi) = N(\widehat{\varphi})$. Therefore we may assume without loss of generality that φ is nondegenerate. By an analogous argument we may also assume this in the bilinear case.

Thus we always assume that φ is nondegenerate in what follows. In the quadratic case we have a Witt decomposition

$$\varphi \cong \varphi_0 \perp r \times H$$

with φ_0 anisotropic, $r \in \mathbb{N}_0$. Clearly, $N(\varphi) = N(\varphi_0)$. Thus we may always assume that φ is anisotropic in the quadratic case. In the bilinear case and when char $k = 2$ we only have a stable isometry

$$\varphi \approx \varphi_0 \perp r \times \widetilde{H},$$

with φ_0 anisotropic, $r \in \mathbb{N}_0$ and φ_0 is up to isometry uniquely determined by φ. Now $N(\varphi)$ is a subgroup of $N(\varphi_0)$ and is possibly different from $N(\varphi_0)$. In what follows we will almost always assume that φ is anisotropic.

Let $K := k(t)$ be the rational function field over k for a set of indeterminates $t = (t_1, \ldots, t_r)$. We want to compare the norm groups of φ and $\varphi \otimes K$. Clearly, $N(\varphi)$ is a subgroup of $N(\varphi \otimes k(t))$. Furthermore, $N(\varphi \otimes k(t))$ contains the group K^{*2} of all squares f^2 with $f \in K$, $f \neq 0$. Thus $N(\varphi \otimes K)$ is obtained from the group $N(\varphi)K^{*2}$ by adjoining certain square-free polynomials $f(t) \in k[t]$, i.e. polynomials $f(t) \neq 0$ which are products of pairwise different irreducible polynomials. Our task will now be to determine these polynomials $f(t)$.

We endow the monomials $t_1^{\alpha_1} \ldots t_r^{\alpha_r}$ $(\alpha_i \in \mathbb{N}_0)$ with the lexicographic ordering with respect to the sequence of variables $t_1 \geq t_2 \geq \cdots \geq t_r$. Thus, if $(\alpha_1, \ldots, \alpha_r) \in \mathbb{N}_0^r$ and $(\beta_1, \ldots, \beta_r) \in \mathbb{N}_0^r$ are two multi-indices, then the monomial $t_1^{\alpha_1} \ldots t_r^{\alpha_r}$ precedes the monomial $t_1^{\beta_1} \ldots t_r^{\beta_r}$ if there exists a $j \in \{1, \ldots, r\}$ with $\alpha_i = \beta_i$ for $i < j$, but $\alpha_j < \beta_j$.

For every polynomial $f \in k[t]$ with $f \neq 0$, let $f^* \in k^*$ denote the leading coefficient of f, i.e. the coefficient of the highest monomial occurring in f. Clearly we have $(fg)^* = f^* g^*$. We call the polynomial f *monic* if $f \neq 0$ and $f^* = 1$.

The following theorem tells us, among other things, that we may restrict ourselves to *monic* square-free polynomials in the task above.

Theorem 3.36. *Let φ be anisotropic. Let $f(t) \in k[t]$ be a similarity norm of $\varphi \otimes K$ and let $t_1^{m_1} \ldots t_r^{m_r}$ be the highest monomial occurring in f. Then the exponents m_i are all even and f^* is a similarity norm of φ.*

Proof. By induction on r. Let $r = 1$. Write $t_1 = t$. Let $m := \deg f$. We use the place $\lambda : k(t) \to k \cup \infty$ over k with $\lambda(t) = \infty$. {Note: the associated valuation ring is $k[t^{-1}]_{(t^{-1})}$.}

We choose an element $a \in D(\varphi)$. Then $af \in D(\varphi \otimes k(t))$. The form $\varphi \otimes k(t)$ has good reduction with respect to λ and $\lambda_*(\varphi \otimes k(t)) = \varphi$ is anisotropic. By Theorem 3.4, there exists a $\xi \in k(t)^*$ with $\lambda(af\xi^2)$ finite and $\neq 0$. This shows that the degree m of f is even. It follows that

$$\varphi \otimes k(t) \cong \langle t^{-m} f \rangle \otimes (\varphi \otimes k(t)). \tag{3.6}$$

Now $\lambda(t^{-m} f) = f^*$. In particular, $\langle t^{-m} f \rangle$ has good reduction with respect to λ. Applying λ_* to (3.6) gives

$$\varphi \cong \langle f^* \rangle \otimes \varphi.$$

This proves the theorem for $r = 1$. Assume that $r > 1$. We write $t = (t_1, t')$ with $t' = (t_2, \ldots, t_r)$ and then have $K = K'(t_1)$ with $K' := k(t_2, \ldots, t_r)$. Applying the result for $r = 1$ to the anisotropic form $\varphi \otimes K'$, we get that m_1 is even and that the highest coefficient $h \in k[t']$ of f, as a polynomial in t_1, is a similarity norm of $\varphi \otimes K'$. By the induction hypothesis, $m_2, \ldots, m_r$ are also even and $h^* = f^*$ is a similarity norm of φ. $\qquad\square$

If $p(t) \in k[t]$ is an *irreducible* polynomial, then $k(p)$ denotes the quotient field of $k[t]/(p)$, i.e. the function field of the affine hyperplane $p(t) = 0$.

Theorem 3.37 (Norm Theorem, [32] for bilinear φ, [7] for quadratic φ). *Assume that the form φ is anisotropic. In the quadratic case, assume moreover that φ is strictly regular. Further, let $p(t) \in k[t] = k[t_1, \ldots, t_r]$ be a monic irreducible polynomial. Then the following are equivalent:*

(i) $p(t) \in N(\varphi \otimes k(t))$.
(ii) $p(t)$ divides a square-free polynomial $f(t) \in k[t]$ with $f(t) \in N(\varphi \otimes k(t))$.
(iii) $\varphi \otimes k(p) \sim 0$.

Under the same assumptions as in the theorem, we immediately obtain the following

Corollary 3.38. *A polynomial $f(t) \in k[t]$, $f(t) \neq 0$ is a similarity norm of $\varphi \otimes k(t)$ if and only if $f^* \in N(\varphi)$ and $\varphi \otimes k(p) \sim 0$ for every monic irreducible polynomial $p \in k[t]$ such that the exponent of p in f is odd.*

The rest of this section will be devoted to the proof of Theorem 3.37. In the main part of the proof (showing that *(iii)* $\Rightarrow$ *(i)*) we will use induction on r. A priori we note that if the theorem holds for a given fixed $r \in \mathbb{N}$, then the corollary also holds for this r.

The implication *(i)* $\Rightarrow$ *(ii)* is trivial. In order to prove *(ii)* $\Rightarrow$ *(iii)*, we use the place $\lambda_p : k(t) \to k(p) \cup \infty$ over k with $\lambda_p(p) = 0$ (cf. §3.4; there we write k_p instead of $k(p)$). By assumption we have

$$\varphi \otimes k(t) \cong \langle f \rangle \otimes (\varphi \otimes k(t)).$$

To this isometry we apply the map $(\lambda_p)_W$, i.e. weak specialization with respect to λ_p, as introduced in §1.3 in the bilinear case and in §1.7 in the quadratic case. We obtain $\varphi \otimes k(p) \sim 0$, as desired.

Now we come to the proof of the implication *(iii)* $\Rightarrow$ *(i)*, which will keep us busy for a longer time. Suppose that $r = 1$ and let φ be an anisotropic *bilinear* form. As before we write $t := t_1$. We use the exact sequence from §3.4,

$$0 \longrightarrow W(k) \overset{i}{\longrightarrow} W(k(t)) \overset{\partial}{\longrightarrow} \bigoplus_{q \in \mathcal{P}} W(k(q)) \longrightarrow 0,$$

with retraction

$$(\lambda_\infty)_* : W(k(t)) \longrightarrow W(k)$$

with respect to the place $\lambda_\infty : k(t) \to k \cup \infty$ over k, which maps t to ∞. Let[8]

$$z := \langle 1, -p \rangle \otimes \varphi_{k(t)} \in W(k(t)).$$

Clearly, $\partial_q(z) = 0$ for every monic irreducible polynomial $q \neq p$. Furthermore, $\partial_p(z)$ is the Witt class of $-\varphi \otimes k(p)$ which is also zero by assumption.

Assume that the polynomial p has degree m. By a well-known theorem of Springer (e.g. [55, Th.5.3]) which also holds for bilinear forms in characteristic 2 by Theorem 2.7[9], $\varphi \otimes E$ is anisotropic for every finite extension E/k of odd degree. Since $\varphi \otimes k(p) \sim 0$ by assumption, m has to be even. Hence $(\lambda_\infty)_W(z) = 0$ and it follows from §3.4 that $z = 0$. Since φ is anisotropic, this gives

$$\varphi_{k(t)} \cong \langle p \rangle \otimes \varphi_{k(t)},$$

as required.

Since we never proved Springer's Theorem for char $k \neq 2$, we will show again that m has to be even, using a different argument.

---

[8] Now we will often write $\varphi_{k(t)}$ instead of $\varphi \otimes k(t)$ etc.

[9] For φ one considers the quadratic form $\varphi \otimes [1]$.

We have seen that $\partial(z) = 0$. Hence, by §3.4, there exists an anisotropic form ψ over k with $z \sim \psi_{k(t)}$. Assume that m is odd. Then $(\lambda_\infty)_W(z) = (\lambda_\infty)_*(\varphi_{k(t)}) = \varphi$. On the other hand, $(\lambda_\infty)_W(z) = (\lambda_\infty)_*(\psi_{k(t)}) = \psi$. Therefore, $\psi = \varphi$. We obtain

$$\langle 1, -p \rangle \otimes \varphi_{k(t)} \sim \varphi_{k(t)},$$

and so $\langle -p \rangle \otimes \varphi_{k(t)} \sim 0$. This is absurd since $\varphi_{k(t)}$ is anisotropic. Thus m has to be even. This proves the theorem for $r = 1$ in the bilinear case.

Let us now consider the case where $r = 1$ and φ is an anisotropic strictly regular quadratic form over k. By assumption we have $\varphi \otimes k(p) \sim 0$. From this we can deduce that p is a similarity norm of $\varphi \otimes k(t)$. In case k has characteristic $\neq 2$, we already know this since φ can then be interpreted as a bilinear form.

So, suppose that char $k = 2$. We want to "lift" φ to a form over a field of characteristic zero in the easiest way possible, cf. [7]. For this purpose we choose a henselian discrete valuation ring A with maximal ideal $\mathfrak{m} = 2A \neq 0$, generated by 2, and residue class field $A/\mathfrak{m} = k$. That this is possible is well-known, cf. [16], [17]. The quotient field $K := \mathrm{Quot}(A)$ has characteristic zero.

We choose elements $a_1, \ldots, a_n$ in A such that

$$p(t) := t^n + \bar{a}_1 t^{n-1} + \cdots + \bar{a}_n,$$

where $\bar{a}_i$ of course denotes the image of a_i in $A/\mathfrak{m} = k$. The polynomial

$$\tilde{p}(t) := t^n + a_1 t^{n-1} + \cdots + a_n$$

is irreducible in $A[t]$ and thus also irreducible in $K[t]$. Let

$$A(\tilde{p}) := A[t]/(\tilde{p}(t)).$$

This is a henselian discrete valuation ring with maximal ideal $2A(\tilde{p})$ and residue class field

$$A(\tilde{p})/2A(\tilde{p}) = k[t]/(p(t)) = k(p).$$

The valuation ring $A(\tilde{p})$ has $K(\tilde{p})$ as its quotient field.

Let us use the geometric language now. Suppose that the form φ corresponds to the quadratic space (E, q) over k. Here we may choose $E = M/\mathfrak{m}M$ for some free A-module M of rank m. We equip M with a strictly regular quadratic form $\tilde{q} : M \to A$ in such a way that (E, q) is the reduction of $(M, \tilde{q})$ modulo $\mathfrak{m}$, which is possible. Let $\tilde{E} := K \otimes_A M$. Our form $\tilde{q}$ extends to a strictly regular quadratic form $\tilde{q}_K : \tilde{E} \to K$. This gives us a strictly regular quadratic space $\tilde{E} := (\tilde{E}, \tilde{q}_K)$ over K which has good reduction with respect to the canonical place $\lambda : K \to k \cup \infty$ associated to A, and whose specialization $\lambda_*(\tilde{E})$ is equal to the given space $E = (E, q)$ over k. By assumption we have $k(p) \otimes E \sim 0$.

We may interpret the canonical place $\mu : \tilde{K}(p) \to k(p) \cup \infty$ associated to the valuation ring $A(\tilde{p})$ as an extension of the place λ. By Theorem 1.116, the space $\tilde{K}(p) \otimes_{\tilde{K}} \tilde{E}$ again has good reduction with respect to μ and

$$\mu_*(K(p) \otimes \widetilde{E}) = \lambda_*(k(p) \otimes \widetilde{E}) = k(p) \otimes E \sim 0.$$

From the quadratic space $M = (M, \widetilde{q})$ over A we obtain a quadratic space $A(\widetilde{p}) \otimes_A M$ over $A(\widetilde{p})$. The statement $\mu_*(k(p) \otimes \widetilde{E}) \sim 0$ tells us that the reduction of $A(\widetilde{p}) \otimes_A M$ to the maximal ideal of $A(\widetilde{p})$ is hyperbolic. Since $A(\widetilde{p})$ is henselian, $A(\widetilde{p}) \otimes_A M$ itself is hyperbolic and then

$$K(\widetilde{p}) \otimes_K \widetilde{E} = K(\widetilde{p}) \otimes_{A(\widetilde{p})} (A(\widetilde{p}) \otimes_A M) \sim 0.$$

Since char $K \neq 2$, it follows from what we proved above (the bilinear case, $r = 1$) that $\widetilde{p}(t)$ is a similarity norm of $K(t) \otimes_K \widetilde{E}$, i.e.

$$K(t) \otimes_K \widetilde{E} \cong \langle \widetilde{p} \rangle \otimes (K(t) \otimes_K \widetilde{E}). \tag{3.7}$$

We want to specialize this isometry with respect to a suitable place from $K(t)$ to $k(t)$. Let $S \subset A[t]$ be the multiplicatively closed set of all polynomials $f \in A[t]$ for which at least one coefficient is not in the maximal ideal $\mathfrak{m} = 2A$ ("unimodular" polynomials). We construct the localization $A(t) := S^{-1}A[t]$ of $A[t]$ at this set. One can easily verify that $A(t)$ is a discrete valuation ring with maximal ideal $2A(t)$, residue class field $A(t)/2A(t) = k(t)$ and quotient field Quot $A(t) = K(t)$. Let $\rho : K(t) \rightarrow k(t) \cup \infty$ be the canonical place associated to the valuation ring $A(t)$. It extends the canonical place $\lambda : K \rightarrow k \cup \infty$ associated to the valuation ring A. Clearly the quadratic space $K(t) \otimes_K \widetilde{E}$ has good reduction with respect to ρ and

$$\rho_*(K(t) \otimes_K \widetilde{E}) = k(t) \otimes_k \lambda_*(\widetilde{E}) = k(t) \otimes_k E.$$

Applying ρ_W to the isometry (3.7) above gives

$$k(t) \otimes_k E \cong \langle p \rangle \otimes (k(t) \otimes_k E).$$

This shows that p is a similarity norm of $\varphi \otimes k(t)$ and we have completely proved the theorem for $r = 1$.

To conclude, we prove the implication $(iii) \Rightarrow (i)$ for $r > 1$, simultaneously for φ bilinear or quadratic. We may suppose that the theorem, and thus also the corollary, is valid for $r - 1$ variables. Also, we already established that the implications $(i) \Rightarrow (ii) \Rightarrow (iii)$ hold for r variables. Next we will prove the implication $(ii) \Rightarrow (i)$ for r variables and only then the implication $(iii) \Rightarrow (i)$.

Thus, let $t = (t_1, t_2, \ldots, t_r) = (t_1, t')$ with $t' = (t_2, \ldots, t_r)$. Let $p \in k[t]$ be monic and irreducible and let $f \in k[t] \setminus \{0\}$. Assume that the exponent of p in f is odd. We have to show that p is a similarity norm of $\varphi_{k(t)}$.

Suppose first that the field k contains infinitely many elements. Let $n := \deg_{t_1} p$ be the degree of p, as a polynomial over $k[t']$. We distinguish the cases $n = 0, n > 0$.

Let $n = 0$. We have a decomposition

$$f(t) = p(t')h(t)$$

in $k[t]$. We choose an element c in k such that $p(t')$ in $k[t']$ does not divide the polynomial $h(c, t')$. This is possible since $|k| = \infty$. {One considers the image of $h(t)$ in $(k[t']/(p))[t_1]$.} By the Substitution Principle (Theorem 1.29 in the bilinear case, Theorem 2.27 in the quadratic case) we have $f(c, t') \in N(\varphi_{k(t')})$. By the implication $(ii) \Rightarrow (i)$ for $r - 1$ variables, it follows that $p = p(t') \in N(\varphi_{k(t')})$ and thus also $p \in N(\varphi_{k(t)})$.

Now let $n > 0$. Let $a(t')$ be the highest coefficient of $p(t)$, as a polynomial in t_1. The polynomial $a(t') \in k[t']$ is monic and $\widetilde{p} := a^{-1}p \in k(t')[t_1]$ is a monic polynomial in the variable t_1. In the ring $k(t')[t_1]$, the exponent of $\widetilde{p}$ in f is odd. Thus, by the established case $r = 1$, $\widetilde{p}$ is a similarity norm of $\varphi_{k(t')}$, and hence also of $\varphi_{k(t)}$. Therefore,

$$\langle p \rangle \otimes \varphi_{k(t)} \cong \langle a \rangle \otimes \varphi_{k(t)}. \tag{3.8}$$

Thus, it suffices to show that a is a similarity norm of $\varphi_{k(t')}$. Let $\pi(t') \in k[t']$ be a monic irreducible polynomial whose exponent in a in the ring $k[t']$ is odd (if such a polynomial π exists at all). We will show that π is a similarity norm of $\varphi_{k(t')}$ and know then that π is indeed a similarity norm of $\varphi_{k(t)}$.

$\pi(t')$ does not divide any of the coefficients of $p(t)$ as a polynomial in t_1. Since $|k| = \infty$, there exists an element c in k with $\pi(t') \nmid p(c, t')$ in the ring $k[t']$. By the Substitution Principle, we obtain from (3.8) that

$$\langle p(c, t') \rangle \otimes \varphi_{k(t')} \cong \langle a(t') \rangle \otimes \varphi_{k(t')},$$

and so

$$a(t')p(c, t') \in N(\varphi_{k(t')}).$$

The exponent of $\pi(t')$ in $a(t')p(c, t')$ is odd. From the implication $(ii) \Rightarrow (i)$ for $r - 1$ variables it follows that $\pi \in N(\varphi_{k(t')})$ and we are finished.

Now suppose that k is a finite field. We consider the field $k(u)$ with u an indeterminate. Upon applying what we proved before to $\varphi_{k(u)}$ and p, f, considered as polynomials over $k(u)$, we see that p is a similarity norm of $\varphi_{k(u)}$. By the Substitution Principle we see that p is a similarity norm of φ, upon specializing u to 0. This completely proves the implication $(ii) \Rightarrow (i)$ for r variables.

Finally, we come to the proof of the implication $(iii) \Rightarrow (i)$ for r variables. As before, let $a(t')$ be the highest coefficient of $p(t)$ as a polynomial in t_1 and let $\widetilde{p} := a^{-1}p$. Then $k(p) = k(t')(\widetilde{p})$ and by our assumption we have $\varphi \otimes k(p) \sim 0$. The implication $(iii) \Rightarrow (i)$ in the case $r = 1$ tells us that $\widetilde{p}$ is a similarity norm of $\varphi_{k(t)}$. Hence, $a(t')p(t) \in N(\varphi_{k(t)})$. By the implication $(ii) \Rightarrow (i)$ for r variables, proved above, $p(t)$ is a similarity norm of $\varphi_{k(t)}$. The proof of Theorem 3.37 is now complete.

3.6 Strongly Multiplicative Forms

Let φ be a (symmetric) bilinear or quadratic form over a field k in n variables and let $t = (t_1, \ldots, t_n)$ be a tuple of n unknowns over k.

Definition 3.39. The form φ is called *strongly multiplicative* when $\varphi(t)$ is a similarity norm of the form $\varphi_{k(t)} = \varphi \otimes k(t)$,[10] in other words when

$$\langle \varphi(t) \rangle \otimes \varphi_{k(t)} \cong \varphi_{k(t)}.$$

Example 3.40.
(1) Clearly every hyperbolic quadratic form $r \times H$ is strongly multiplicative.
(2) If τ is a bilinear or quadratic Pfister form over k, then $\tau_{k(t)}$ is likewise such a form over $k(t)$. Thus Theorem 3.10 tells us that $N(\tau_{k(t)}) = D(\tau_{k(t)})$. In particular we have $\tau(t) \in N(\tau_{k(t)})$. Thus every Pfister form is strongly multiplicative.
(3) If φ is a strongly multiplicative bilinear form, then $\varphi \otimes [1]$ is clearly a strongly multiplicative quasilinear quadratic form. In particular, for every bilinear Pfister form τ, the quadratic form $\tau \otimes [1]$ is strongly multiplicative.

The main aim of this section is to determine all strongly multiplicative bilinear and quadratic forms over k. We start with two very simple theorems whose proofs only require the Substitution Principle.

As before we use $D(\varphi)$ to denote the set of elements of k^* which are represented by φ, i.e. the set of all $a \in k^*$ with $\langle a \rangle < \varphi$ in the bilinear case and $[a] < \varphi$ in the quadratic case. This set consists of cosets of the norm group $N(\varphi)$ in the group k^*.

Theorem 3.41. *For a bilinear or quadratic form φ over k the following statements are equivalent:*
(a) φ is strongly multiplicative.
(b) For every field extension L of k we have $D(\varphi_L) \subset N(\varphi_L)$, and so $D(\varphi_L) = N(\varphi_L)$.

Proof. (a) $\Rightarrow$ (b): We have $\varphi_{k(t)} \cong \langle \varphi(t) \rangle \otimes \varphi_{k(t)}$. This implies $\varphi_{L(t)} \cong \langle \varphi(t) \rangle \otimes \varphi_{L(t)}$ and so, by the Substitution Principle, $\varphi_L \cong \langle \varphi(c) \rangle \otimes \varphi_L$ for every $c \in L^n$ with $\varphi(c) \neq 0$.
(b) $\Rightarrow$ (a): Trivial. $\square$

Theorem 3.42. *If φ is a quadratic form with associated nondegenerate quadratic form $\widehat{\varphi}$, i.e. (cf. Definition 1.74)*

$$\varphi = \widehat{\varphi} \perp \delta(\varphi) = \widehat{\varphi} \perp s \times [0]$$

for some $s \in \mathbb{N}_0$, then φ is strongly multiplicative if and only if $\widehat{\varphi}$ is strongly multiplicative.

Proof. $\varphi(t_1, \ldots, t_n) = \widehat{\varphi}(t_1, \ldots, t_{n-s})$ is a similarity norm of $\varphi_{k(t)}$ if and only if this polynomial is a similarity norm of $\widehat{\varphi}_{k(t)}$. This will be the case exactly when $\widehat{\varphi}(t_1, \ldots, t_{n-s})$ is a similarity norm of $\widehat{\varphi}_{k(t_1, \ldots, t_{n-s})}$, as can be seen by the Substitution Principle for instance. $\square$

In the study of strongly multiplicative quadratic forms we may thus restrict ourselves to nondegenerate forms φ. By an analogous argument we may also do this in the bilinear case. Thus we always assume that φ is nondegenerate in what follows. Let us commence our study with strongly multiplicative *quadratic* forms.

---

[10] In the bilinear case we defined $\varphi(t) := \varphi(t, t)$. Thus $\varphi(t) = (\varphi \otimes [1])(t)$.

Theorem 3.43. *Let φ be a nondegenerate strongly multiplicative quadratic form over k. Then φ is either quasilinear or strictly regular.*

Proof. We recall that the quasilinear part $QL(\varphi)$ of φ is uniquely determined by φ up to isometry. Thus

$$\langle \varphi(t) \rangle \otimes \varphi_{k(t)} \cong \varphi_{k(t)}$$

implies that

$$\langle \varphi(t) \rangle \otimes QL(\varphi)_{k(t)} \cong QL(\varphi)_{k(t)}.$$

Assume that $QL(\varphi) \neq 0$. We choose an element $b \in D(QL(\varphi))$. Then $b\varphi(t) \in D(QL(\varphi)_{k(t)})$. The form $QL(\varphi)$ is anisotropic. Thus $b\varphi$ is a subform of $QL(\varphi)$ by the Subform Theorem of §3.3. Since dim $\varphi \geq$ dim $QL(\varphi)$ it follows that $\varphi = QL(\varphi)$. $\square$

Theorem 3.44. *Let φ be a strictly regular strongly multiplicative quadratic form over k which is not hyperbolic. Assume that* dim $\varphi > 1$.

  (i) φ is an anisotropic Pfister form.

  (ii) If τ is another quadratic Pfister form over k with $\tau < \varphi$, then there exists a bilinear Pfister form σ over k with $\varphi \cong \sigma \otimes \tau$.

Proof. (i) Let φ_0 be the kernel form of φ, i.e. $\varphi \cong \varphi_0 \perp r \times H$ for some $r \geq 0$. We have $\varphi_0 \neq 0$. Now $\langle \varphi(t) \rangle \otimes \varphi_{k(t)} \cong \varphi_{k(t)}$ implies that

$$\langle \varphi(t) \rangle \otimes (\varphi_0)_{k(t)} \cong (\varphi_0)_{k(t)}$$

and then $b\varphi(t) \in D(\varphi_0 \otimes k(t))$ for an arbitrarily chosen $b \in D(\varphi_0)$. By the Subform Theorem of §3.3, $b\varphi$ is a subform of φ_0. Therefore dim $\varphi \leq$ dim φ_0 and so $r = 0$. Thus φ is anisotropic.

  (ii) By Theorem 3.41 we have $N(\varphi) = D(\varphi)$. In particular $1 \in D(\varphi)$. Since φ is strictly regular, there exists a two-dimensional form $\left[\begin{smallmatrix} 1 & 1 \\ 1 & a \end{smallmatrix}\right] < \varphi$. This form is a Pfister form.

  Now let some Pfister form $\tau < \varphi$ be given. We choose a bilinear Pfister form σ of maximal dimension such that $\sigma \otimes \tau < \varphi$. {Note that possibly $\sigma = \langle 1 \rangle$.} The theorem will be proved if we can show that $\varphi \cong \sigma \otimes \tau$.

  By way of contradiction, assume that $\varphi \not\cong \sigma \otimes \tau$. Then we have a decomposition

$$\varphi \cong \sigma \otimes \tau \perp \psi$$

with $\psi \neq 0$. Assume that the Pfister form $\rho := \sigma \otimes \tau$ has dimension m (a power of 2). Let $t = (t_1, \ldots, t_m)$ be a tuple of m unknowns over k. Since φ is strongly multiplicative, we have by Theorem 3.41 that

$$\rho(t)\rho_{k(t)} \perp \rho(t)\psi_{k(t)} \cong \rho_{k(t)} \perp \psi_{k(t)}.$$

Now $\rho(t)\rho_{k(t)} \cong \rho_{k(t)}$. By the Cancellation Theorem (Theorem 1.67) it follows that

$$\rho(t)\psi_{k(t)} \cong \psi_{k(t)}.$$

We choose a $b \in D(\psi)$ and then have $b\rho(t) \in D(\psi_{k(t)})$. By the Subform Theorem we get $b\rho < \psi$ and so

$$(\langle 1, b \rangle \otimes \sigma) \otimes \tau = \langle 1, b \rangle \otimes \rho < \varphi.$$

This contradicts the maximality of $\dim \sigma$. Thus $\sigma \otimes \tau \cong \varphi$. □

In §3.2 we determined that every anisotropic quadratic Pfister form τ has height 1. The same is therefore true for $a\tau$ where $a \in k^*$ is an arbitrary scalar. We are now in the position to prove a converse of this fact.

Theorem 3.45. *Let φ be an anisotropic strictly regular quadratic form over k of height 1. Choose $a \in D(\varphi)$. Then $\varphi \cong a\tau$ for some Pfister form τ.*

Proof. Upon replacing φ by $a\varphi$ we may assume without loss of generality that $a = 1$. As before let $n := \dim \varphi$ and let $t = (t_1, \ldots, t_n)$ be a tuple of n unknowns over k. We have $n \geq 2$.

The polynomial $\varphi(t)$ over k is irreducible. Moreover, after an appropriate linear transformation of the variables $t_1, \ldots, t_n$ over k, we have that $\varphi(t)$ is monic.

Since the height of φ is 1, $\varphi \otimes k(\varphi) \sim 0$. From the Norm Theorem of §3.5 it follows that $\varphi(t)$ is a similarity norm of φ, thus φ is strongly multiplicative. And so, by Theorem 3.44, φ is a Pfister form.[11] □

We now move on to *quasilinear* strongly multiplicative (quadratic) forms. Let char $k = 2$ and let us contemplate arbitrary anisotropic quasilinear forms over k. We will utilize the geometric language to this end.

Thus, let $E = (E, q)$ be an anisotropic quasilinear quadratic space over k. As usual, we denote by $k^{1/2}$ the subfield of the algebraic closure $\tilde{k}$, consisting of all $\lambda \in \tilde{k}$ with $\lambda^2 \in k$. Recall that for every $a \in k$ there is a unique element $\sqrt{a} \in k^{1/2}$ such that $(\sqrt{a})^2 = a$ and also that the map $a \mapsto \sqrt{a}$ is an isomorphism from the field k to $k^{1/2}$.

Clearly

$$f : E \longrightarrow k^{1/2}, \; f(x) := \sqrt{q(x)}$$

is a k-linear map from E to $k^{1/2}$. This map is injective since (E, q) is anisotropic. We equip the k-vector space $k^{1/2}$ with the quadratic form sq : $k^{1/2} \to k$, sq$(x) = x^2$. {sq as in "square".} Then $k^{1/2}$ becomes an—in general infinite-dimensional—quasilinear quadratic k-module and f an isometry from (E, q) to a submodule U of $(k^{1/2}, \text{sq})$. The following stronger statement is now obvious.

Theorem 3.46. *For every anisotropic quasilinear quadratic space (E, q) over k there exists* exactly one *isometry $f : E \xrightarrow{\sim} U$, where U is a finite-dimensional quadratic submodule of $(k^{1/2}, \text{sq})$.*

Now it is easy to classify the strongly multiplicative anisotropic quasilinear forms over k.

---

[11] More precisely we should say "φ is isometric to a Pfister form". In the following we frequently allow inaccuracies like this one, in order to simplify the language.

Theorem 3.47. *Let φ be an anisotropic quasilinear quadratic form over k. The following statements are equivalent:*

(1) *φ is strongly multiplicative.*

(2) *$D(\varphi) = N(\varphi)$.*

(3) *There exists exactly one subfield K of $k^{1/2}$ with $k \subset K$, $[K : k]$ finite, such that the quadratic form $(K, \mathrm{sq}|K)$ is isometric to φ.*

(4) *$\varphi \cong \tau \otimes [1]$ where τ is an anisotropic bilinear Pfister form.*

Proof. The implication (1) $\Rightarrow$ (2) is clear by Theorem 3.42.

(2) $\Rightarrow$ (3): By Theorem 3.46 there exists exactly one finite-dimensional k-subspace U of $k^{1/2}$ such that the form $\mathrm{sq}|U$ is isometric to φ. By assumption (2) we certainly have $1 \in D(\varphi)$. Hence U contains an element z with $z^2 = 1$. We must have $z = 1$ and thus $k \subset U$. Furthermore, for every two elements x, y in U there exists an element z in U such that $z^2 = x^2 \cdot y^2$, and so $z = xy$. Finally, if $x \neq 0$ is any element in U, then $x^{-1} = (x^{-1})^2 \cdot x$ is also in U. We conclude that U is a subfield of $k^{1/2}$.

(3) $\Rightarrow$ (4): We have $K = k(\sqrt{a_1}, \ldots, \sqrt{a_d})$ with finitely many elements $a_1, \ldots, a_d$ in k which are linearly independent over the field $k^{(2)} = \{x^2 | x \in k\}$. The associated quadratic form φ is the orthogonal sum of the one-dimensional forms $[a_{i_1} a_{i_2} \ldots a_{i_r}]$ with $1 \leq i_1 < i_2 < \cdots < i_r \leq d$, i.e.

$$\varphi = \langle 1, a_1 \rangle \otimes \cdots \otimes \langle 1, a_d \rangle \otimes [1].$$

(4) $\Rightarrow$ (1): We already established above (Example 3.40(3)) that for every bilinear Pfister form τ, the form $\tau \otimes [1]$ is strongly multiplicative. $\square$

Definition 3.48. Following established terminology (e.g. [47]) we call a quadratic or bilinear form φ over a field k *round* whenever $N(\varphi) = D(\varphi)$. Analogously we call a quadratic or bilinear module E over k round whenever $N(E) = D(E)$, thus, in finite dimensions, when the (up to isometry unique) associated form φ is round.

By Theorem 3.42 it is clear that a form φ over k is strongly multiplicative if and only if for every field extension L of k the form φ_L is round. For this to hold, it is sufficient to know that $\varphi_{k(t)}$ is round where, as before, $t = (t_1, \ldots, t_n)$ is a tuple of $n = \dim \varphi$ unknowns over k.

Theorem 3.47 tells us, among other things, that a round quasilinear quadratic form φ over a field of characteristic 2 is already strongly multiplicative. In Theorem 3.47 we supposed that φ is anisotropic, to be sure, but this assumption can subsequently be dropped.

We remark that in general there are many more round forms than strongly multiplicative forms among the non quasilinear quadratic forms. For example if $k = \mathbb{R}$, then for every $n \in \mathbb{N}$ the form $n \times [1]$ is round. This form is nonetheless strongly multiplicative only when n is a power of 2.

Now let us assume again that char $k = 2$. Among other things, Theorem 3.47 tells us that every anisotropic quasilinear round quadratic space over k can be considered as a finite field extension K of k with $K \subset k^{1/2}$ in a unique way. We want to give a second, more structural, proof of this fact. This will guide us in the right direction for a classification of the anisotropic round bilinear forms over k.

Thus, let (E, q) be an anisotropic round quasilinear quadratic module over k. Every vector $x \in E$ is uniquely determined by the value $q(x) \in k$. For if $q(x) = q(y)$, then $q(x - y) = 0$ and so $x - y = 0$. {Note: this holds for *every* anisotropic quasilinear space.} Since $D(E)$ is a group, there exists a vector e_0 with $q(e_0) = 1$. Furthermore there exists for every $x \in E$, $x \neq 0$ *exactly one* similarity transformation $\sigma_x : E \to E$ with norm $q(x)$. Hence we have for every vector $y \in E$,

$$q(\sigma_x y) = q(x)q(y).$$

We define a multiplication $(x, y) \mapsto x \cdot y$ on E by $x \cdot y := \sigma_x y$ for $x \neq 0$, $0 \cdot y := 0$. As usual we write xy instead of $x \cdot y$ most of the time. Clearly we then have for arbitrary $x, y \in E$ that

$$q(xy) = q(x)q(y).$$

Since $q(yx) = q(y)q(x) = q(xy)$ it follows that $yx = xy$. If z is a further element in E we have

$$q((xy)z) = q(x)q(y)q(z) = q(x(yz)),$$

so that $(xy)z = x(yz)$. Finally,

$$q(e_0 z) = q(e_0)q(z) = q(z),$$

and so $e_0 z = z$. Our multiplication is thus commutative and associative and has e_0 as identity element. By definition it is k-linear in the second argument and thus also in the first argument. Hence E is a commutative k-algebra. If $x \in E$, $x \neq 0$, then $q(x) \neq 0$ and furthermore $q(x \cdot x) = q(x)^2$, so that

$$q(q(x)^{-1} x \cdot x) = 1.$$

It follows that $q(x)^{-1} x \cdot x = e_0$. Thus x has an inverse $x^{-1} = q(x)^{-1} x$. The ring E is therefore a field. As usual we denote its identity element e_0 from now on by 1 and identify k with the subfield $k \cdot 1$ of E. In this way E becomes a field extension of k and for every $x \in E$ we have

$$x^2 = q(x).$$

If $k^{1/2}$ is the set of all square roots $\sqrt{a}$ of elements $a \in k$ in a given fixed algebraic closure $\tilde{k}$ of k, then we have exactly one field embedding $E \hookrightarrow k^{1/2}$ of E in the field $k^{1/2}$ over k. Thus we have derived the implication (2) $\Rightarrow$ (3) in Theorem 3.47 for the second time. In fact we obtained an analogous result for E of infinite dimension over k, which we actually will not need in what follows.

Now let (E, B) be a round anisotropic *bilinear* module over the field k of characteristic 2. The quadratic form $n(x) := B(x, x)$ on E, associated to B, is round, quasilinear and anisotropic. Thus E carries the structure of a field extension of k with $x^2 = n(x)$ for all $x \in E$.

Definition 3.49. We call the extension field U of k in $k^{1/2}$ which is isomorphic to E over k, the *inseparable field extension of k associated to* (E, B).

The field $U \subset k^{1/2}$ should be considered as a first invariant of the space (E, B). What else do we need in order to describe (E, B) completely up to isometry?

We endow the k-vector space E with the linear form

$$t : E \to k, \quad t(x) := B(1, x).$$

Clearly, $t(1) = n(1) = 1$. Now B is completely determined by t since for every $x \in E$, $x \neq 0$ there is exactly one similarity transformation $\sigma_x : E \xrightarrow{\sim} E$ of norm $n(x)$ and for $y \in E$ we have $\sigma_x(y) = xy$, all of this being justified by our analysis of the round quasilinear space (E, n) above. Hence we have for arbitrary $y, z \in E$

$$B(xy, xz) = n(x)B(y, z). \tag{3.9}$$

This equality remains valid when $x = 0$. Let $y \neq 0$, $x = y^{-1}$, then (3.9) implies that

$$n(y)^{-1}B(y, z) = B(1, y^{-1}z) = t(y^{-1}z),$$

and so

$$B(y, z) = n(y)t(y^{-1}z) = t(n(y) \cdot y^{-1}z) = t(y^2 y^{-1}z) = t(yz).$$

The equality

$$B(y, z) = t(yz)$$

remains valid when $y = 0$.

Conversely, let U be an extension field of k with $k \subset U \subset k^{1/2}$ and let $t : U \to k$ be a linear form on the k-vector space U with $t(1) = 1$. On U we define a symmetric bilinear form

$$\beta_t : U \times U \to k, \quad \beta_t(x, y) := t(xy).$$

We have $\beta_t(x, x) = t(x^2) = x^2 t(1) = x^2$. The quadratic standard form $x \mapsto x^2$ on U is thus associated to β_t. In particular, the bilinear k-module (U, β_t) is anisotropic. For every $x \in U$, $x \neq 0$ we define the k-linear transformation

$$\sigma_x : U \longrightarrow U, \quad y \longmapsto xy.$$

We have

$$\beta_t(\sigma_x y, \sigma_x z) = \beta_t(xy, xz) = t(x^2 yz) = x^2 t(yz) = x^2 \beta_t(y, z).$$

Thus σ_x is a similarity transformation on U of norm x^2. This shows that (U, β_t) is round.

We summarize these considerations in the following theorem.

Theorem 3.50. *The round anisotropic bilinear modules over k are up to isometry the pairs (U, β_t) with U an extension field of k in $k^{1/2}$ and t a k-linear form on U with $t(1) = 1$. For any round anisotropic bilinear module (E, B) over k there is exactly* *one such pair* (U, β_t) *with* $(E, B) \cong (U, \beta_t)$ *and furthermore exactly one isometry from* (E, B) *to* (U, β_t).

Theorem 3.51. *Let φ be an anisotropic round bilinear form over a field k of characteristic 2.*

  (*i*) *φ is a Pfister form.*
  (*ii*) *If τ is a bilinear Pfister form with $\tau < \varphi$, then there exists a further bilinear Pfister form σ with $\varphi \cong \sigma \otimes \tau$.*

Proof. We have $\langle 1 \rangle < \varphi$. Let $\tau < \varphi$ be an arbitrary Pfister form. We choose a Pfister form σ over k of maximal dimension with $\sigma \otimes \tau < \varphi$. {Note: it is possible that $\sigma = \langle 1 \rangle$.} We will show that $\sigma \otimes \tau = \varphi$, which will prove the theorem.

We will use the geometric language. Let $E = (E, B)$ be a bilinear space which arises from φ and let F be a subspace of E arising from $\sigma \otimes \tau$. E carries the structure of a field extension of k with $x^2 = B(x, x)$ for all $x \in E$. Furthermore there is a linear form $t : E \to k$ with $B(x, y) = t(xy)$ for all $x, y \in E$. Finally, F is a subfield of E with $k \subset E$. {If $\sigma \otimes \tau \cong \langle 1 \otimes a_1 \rangle \otimes \cdots \otimes \langle 1, a_m \rangle$, then $F = k(\sqrt{a_1}, \ldots, \sqrt{a_m})$ and $t(\sqrt{a_{i_1} \ldots a_{i_r}}) = 0$ for $r > 0$, $1 \le i_1 < \cdots < i_r \le m$.}

Assume for the sake of contradiction that $F \ne E$. We choose $z \in E$ with $z \ne 0$, $B(z, F) = 0$. For $x, y \in F$ we then have

$$B(x, zy) = t(xyz) = B(xy, z) = 0.$$

Thus zF is a subspace of E, perpendicular to F. Finally we have for $x, y \in F$ with $b := B(z, z)$ the equality
$$B(zx, zy) = bB(x, y).$$

All this shows that $F \perp zF$ is a subspace of E, with associated bilinear form the Pfister form $\langle 1, b \rangle \otimes \sigma \otimes \tau$. This contradicts the maximality of $\dim \sigma$. Thus $F = E$.
$\square$

By Theorems 3.44 and 3.51 it is now clear that over *every* field the anisotropic strongly multiplicative bilinear forms are exactly the anisotropic Pfister forms.

3.7 Divisibility by Pfister Forms

In this section we will deal with the divisibility of a quadratic form by a given quadratic or bilinear Pfister form in the sense of the following general definition.

Definition 3.52. Let φ and τ be quadratic forms and σ a bilinear form over k. We say that φ is *divisible by τ* if there exists a bilinear form ψ over k with $\varphi \cong \psi \otimes \tau$. Similarly we say that φ is *divisible by σ* if there exists a quadratic form χ over k with $\varphi \cong \sigma \otimes \chi$. We then write $\tau \mid \varphi$ resp. $\sigma \mid \varphi$.

First we will study the divisibility by a given (anisotropic) *quadratic* Pfister form.

Theorem 3.53. *Let τ be a quadratic Pfister form over a field k and φ an anisotropic strictly regular quadratic form over k. The following statements are equivalent:*

(i) $\tau \mid \varphi$.

(ii) There exists a nondegenerate bilinear form ψ over k with $\varphi \sim \psi \otimes \tau$.

(iii) $\varphi \otimes k(\tau) \sim 0$.

Proof. The implications $(i) \Rightarrow (ii) \Rightarrow (iii)$ are trivial.

$(iii) \Rightarrow (i)$: Let $\dim \tau = n$ (a 2-power) and let $t = (t_1, \ldots, t_n)$ be a tuple of n unknowns over k. We can think of $\tau(t) \in k[t]$ as a monic polynomial. By the Norm Theorem (cf. §3.5) our assumption $\varphi \otimes k(\tau) \sim 0$ implies that $\tau(t)$ is a similarity norm of $\varphi_{k(t)}$. We choose an element $a \in D(\varphi)$. Then $a\tau(t) \in D(\varphi_{k(t)})$ and thus, by the Subform Theorem (cf. §3.3),

$$\varphi \cong a\tau \perp \varphi_1$$

where φ_1 is a further quadratic form over k. It now follows that $\varphi_1 \otimes k(\tau) \sim 0$. By induction on $\dim \varphi$ we obtain $\varphi \cong \chi \otimes \tau$ for a bilinear form $\chi = \langle a, a_2, \ldots, a_r \rangle$. □

The following theorem is a variation of Theorem 3.53 in the case where φ is also a Pfister form.

Theorem 3.54. *Let τ and ρ be quadratic Pfister forms over a field k. Assume that the form ρ is anisotropic. The following statements are equivalent:*

(i) There exists a bilinear Pfister form σ over k with $\rho \cong \sigma \otimes \tau$.

(ii) $\tau \mid \rho$.

(iii) There exists an element $a \in k^*$ with $a\tau < \rho$.

(iv) $\rho \otimes k(\tau) \sim 0$.

(v) There exists a place $\lambda : k(\rho) \to k(\tau) \cup \infty$ over k.

Proof. The implications $(i) \Rightarrow (ii)$ and $(ii) \Rightarrow (iii)$ are trivial. The implications $(iii) \Rightarrow (v)$ and $(iv) \Leftrightarrow (v)$ both follow from the fact that $k(\rho)$ is a generic zero field of ρ and that every isotropic quadratic Pfister form is already hyperbolic. The implication $(iv) \Rightarrow (ii)$ is clear by Theorem 3.53. To finish the proof we verify the implication $(iii) \Rightarrow (i)$. Thus let $a\tau < \rho$ for some $a \in k^*$. Then $a \in D(\rho) = N(\rho)$. Hence $\tau < a\rho \cong \rho$. Now Theorem 3.44 tells us that there exists a bilinear Pfister form σ over k with $\rho \cong \sigma \otimes \tau$. □

Corollary 3.55. *Let τ_1, τ_2 be quadratic Pfister forms over k such that $k(\tau_1) \sim_k k(\tau_2)$. Then $\tau_1 \cong \tau_2$.*

Note that in the proof of Theorem 3.53 we showed that for a given $a \in D(\varphi)$ we could choose the factor ψ in such a way that $a \in D(\psi)$. Similarly one could ask whether in Theorem 3.54 the Pfister form σ can be chosen so as to be divisible by $\langle 1, a \rangle$, for an appropriately given binary form $\langle 1, a \rangle$. This leads us to the next lemma which is very useful for what follows.

Lemma 3.56. *Let σ be a bilinear and τ a quadratic Pfister form over k. Let $\dim \sigma > 1$ and $a \in D(\sigma' \otimes \tau)$, where—as before (§1.4, §3.2)—σ' denotes the pure part of σ. Then there exists a bilinear Pfister form γ over k with*

$$\sigma \otimes \tau \cong \gamma \otimes \langle 1, a \rangle \otimes \tau.$$

Proof. (a) If $\sigma \otimes \tau$ is isotropic, then $\sigma \otimes \tau$ is hyperbolic and the statement can easily be verified. Namely, if $\dim \sigma > 2$, then we can choose γ to be the hyperbolic Pfister form of the correct dimension. If $\sigma = \langle 1, c \rangle$ however, then $a = cu$ for some $u \in D(\tau)$, so that $\langle 1, c \rangle \otimes \tau \cong \langle 1, a \rangle \otimes \tau$, and $\gamma = \langle 1 \rangle$ will do.

(b) Suppose next that $\sigma \otimes \tau$ is anisotropic. Let $\mu := \langle 1, a \rangle \otimes \tau$ and $K := k(\mu)$. By Theorem 3.54 it is sufficient to show that $(\sigma \otimes \tau)_K \sim 0$.

Since $\mu_K \sim 0$ we have that $\tau_K \cong -a\tau_K$ and

$$(\sigma \otimes \tau)_K = \tau_K \perp (\sigma' \otimes \tau)_K \cong (-a\tau \perp \sigma' \otimes \tau)_K.$$

By assumption the element a is represented by $\sigma' \otimes \tau$. Thus the form $-a\tau \perp \sigma' \otimes \tau$ is isotropic. Hence $(\sigma \otimes \tau)_K$ is isotropic. Since $\sigma \otimes \tau$ is a Pfister form we have $(\sigma \otimes \tau)_K \sim 0$, as required.   □

Now suppose that ρ_1, ρ_2 are two quadratic Pfister forms over k, possibly of different dimension. We are looking for a quadratic Pfister form τ, perhaps of higher dimension, which divides both ρ_1 and ρ_2. In this situation a concept emerges which was coined by Elman and Lam [15, §4] in characteristic $\neq 2$:

Definition 3.57.

(a) Let r be a natural number. We call ρ_1 and ρ_2 *r-linked*, more precisely, *quadratically r-linked*, if over k there exist an r-fold quadratic Pfister form τ and bilinear Pfister forms σ_1, σ_2 such that

$$\rho_1 \cong \sigma_1 \otimes \tau, \quad \rho_2 \cong \sigma_2 \otimes \tau.$$

(b) We define the *quadratic linkage number* $j(\rho_1, \rho_2)$ as follows: if ρ_1 and ρ_2 are 1-linked, then $j(\rho_1, \rho_2)$ is the largest number r such that ρ_1 and ρ_2 are r-linked. Otherwise we set $j(\rho_1, \rho_2) = 0$.

Remark. Elman and Lam only use the term "*r*-linked". In our context the adjective "quadratic" is necessary since in $\operatorname{char} k = 2$ we also can, and should, inquire about *bilinear* Pfister forms which divide both ρ_1 and ρ_2, see below.

Theorem 3.58 (cf. [15, Prop. 4.4] for char $k \neq 2$). [12] *Assume that the quadratic Pfister forms ρ_1, ρ_2 are anisotropic and 1-linked. Then the form $\rho_1 \perp -\rho_2$ has Witt index*

$$\operatorname{ind}(\rho_1 \perp -\rho_2) = 2^r$$

with $r := j(\rho_1, \rho_2) \geq 1$. Furthermore, if τ is a quadratic s-fold Pfister form with $1 \leq s < r$, dividing both ρ_1 and ρ_2, then there exists an $(r - s)$-fold bilinear Pfister form γ and bilinear Pfister forms σ_1, σ_2 such that

$$\rho_1 \cong \sigma_1 \otimes \gamma \otimes \tau, \quad \rho_2 \cong \sigma_2 \otimes \gamma \otimes \tau.$$

---

[12] Elman and Lam do not require the assumption that $j(\rho_1, \rho_2) \geq 1$, cf. our Theorem 3.65 below.

Proof. We assume that we have factorizations $\rho_1 \cong \sigma_1 \otimes \tau$, $\rho_2 \cong \sigma_2 \otimes \tau$ where τ is a quadratic Pfister form and σ_1 and σ_2 are bilinear Pfister forms. We verify the following two statements from which the theorem readily follows:

  (*i*) $\mathrm{ind}(\rho_1 \perp \rho_2) \geq \dim \tau$.
  (*ii*) If $\mathrm{ind}(\rho_1 \perp -\rho_2) > \dim \tau$, then there exists an $a \in k^*$ and bilinear Pfister forms γ_1, γ_2 such that

$$\rho_1 \cong \gamma_1 \otimes \langle 1, a \rangle \otimes \tau, \qquad \rho_2 \cong \gamma_2 \otimes \langle 1, a \rangle \otimes \tau.$$

(*i*): We have $\rho_1 \cong \tau \perp \sigma_1' \otimes \tau$, $\rho_2 \cong \tau \perp \sigma_2' \otimes \tau$, and thus

$$\rho_1 \perp -\rho_2 \cong \tau \otimes \langle 1, -1 \rangle \perp \sigma_1' \otimes \tau \perp (-\sigma_2' \otimes \tau).$$

(*ii*): Now let $\mathrm{ind}(\rho_1 \perp \rho_2) > \dim \tau$. By our calculation this shows that the form $\sigma_1' \otimes \tau \perp (-\sigma_2' \otimes \tau)$ is isotropic. Hence there exists an $a \in k^*$ which is represented by both of the forms $\sigma_1' \otimes \tau$ and $\sigma_2' \otimes \tau$. By Lemma 3.56 we have factorizations

$$\sigma_1 \otimes \tau \cong \gamma_1 \otimes \langle 1, a \rangle \otimes \tau, \qquad \sigma_2 \otimes \tau \cong \gamma_2 \otimes \langle 1, a \rangle \otimes \tau$$

with bilinear Pfister forms γ_1, γ_2, as required.   □

Let us now deal with the problems of divisibility by a given *bilinear* Pfister form. Unfortunately there apparently does not (yet) exist a good counterpart to Theorem 3.53 above. There is one for Theorem 3.54 however. To derive it we need a further lemma.

Lemma 3.59. *Let ρ be a quadratic and σ a bilinear Pfister form over k. Assume that the form ρ is anisotropic. Let $\tau = \begin{bmatrix} 1 & 1 \\ 1 & b \end{bmatrix}$ be a binary Pfister form over k with*

$$\tau \perp \sigma' \otimes [1] < \rho.$$

Then there exists a bilinear Pfister form α over k such that

$$\rho \cong \alpha \otimes \sigma \otimes \tau.$$

Proof. Assume that the dimension of σ is 2^d. Then the Pfister form $\mu := \sigma \otimes \tau$ is of dimension 2^{d+1}. Let $E = (E, q)$ be a quadratic space corresponding to μ and let $K := k(\mu)$. The space $K \otimes E$ is hyperbolic. Hence it contains a 2^d-dimensional vector subspace U with $q(U) = 0$. Let V be a subspace of $K \otimes E$ corresponding to the subform $(\tau \perp \sigma' \otimes [1])_K$ of μ_K. It has dimension $2^d + 1$. From the obvious inequality

$$\dim(V \cap U) + \dim E \geq \dim V + \dim U$$

we see that $\dim(V \cap U) \geq 1$. Hence the quadratic space V is isotropic. Thus $\rho \otimes K$ is isotropic and so $\rho \otimes K \sim 0$. Now Theorem 3.54 tells us that $\rho \cong \alpha \otimes \mu$ for some bilinear Pfister form α.   □

Theorem 3.60. *Let σ be a bilinear Pfister form and ρ an anisotropic quadratic Pfister form over k with $\sigma \otimes [1] < \rho$. Then there exists a quadratic Pfister form τ over k such that $\rho \cong \sigma \otimes \tau$.*

Proof. Let $E = (E, q)$ be a quadratic space over k which corresponds to the form ρ, and let V be a subspace of E which corresponds to the subform $\sigma \otimes [1]$ of ρ. We have an orthogonal decomposition $V = ke_0 \perp V'$ with $q(e_0) = 1$ and V' a quasilinear space corresponding to the form $\sigma' \otimes [1]$. The bilinear form $B = B_q$ vanishes on $V \times V$. There is a linear form s on V with $s(e_0) = 1$, $s(V') = 0$. Furthermore there is a vector $f_0 \in E$ with $s(v) = B(v, f_0)$ for all $v \in V$. The subspace $V + kf_0$ of E corresponds to the quadratic form $\begin{bmatrix} 1 & 1 \\ 1 & b \end{bmatrix} \perp \sigma' \otimes [1]$. This form is thus a subform of ρ. The theorem now follows from Lemma 3.59. $\square$

Definition 3.61. If φ is an anisotropic bilinear form over k and $\dim \varphi > 1$, then we define $k(\varphi) := k(\varphi \otimes [1])$. If $\varphi = \langle a_1, \ldots, a_n \rangle$, then $k(\varphi)$ is the quotient field of the ring $k[t_1, \ldots, t_n]/(a_1 t_1^2 + \cdots + a_n t_n^2)$. Note that the polynomial $\varphi(t) = a_1 t_1^2 + \cdots + a_n t_n^2$ is irreducible since φ is anisotropic.

Theorem 3.62. *Let σ be an anisotropic bilinear Pfister form and ρ an anisotropic quadratic Pfister form over k. Assume that $\sigma \neq \langle 1 \rangle$. The following statements are equivalent:*

(i) There exists a quadratic Pfister form τ over k with $\rho \cong \sigma \otimes \tau$.
(ii) $\sigma \mid \rho$.
(iii) There exists an $a \in k^*$ with $\sigma \otimes [a] < \rho$.
(iv) $\rho \otimes k(\sigma) \sim 0$.
(v) There exists a place $\lambda : k(\rho) \to k(\sigma) \cup \infty$.

Proof. The implications $(i) \Rightarrow (ii) \Rightarrow (iii) \Rightarrow (iv)$ and $(iv) \Leftrightarrow (v)$ are obvious.

$(iv) \Rightarrow (iii)$: Let $t = (t_1, \ldots, t_n)$ be a tuple of $n := \dim \sigma$ unknowns over k. We can treat $\sigma(t) \in k[t]$ as a monic polynomial. By the Norm Theorem, (iv) implies that $\sigma(t)$ is a similarity norm of $\rho_{k(t)}$ and thus that $\sigma(t)$ is represented by $\rho_{k(t)}$. The Subform Theorem then gives us $\sigma \otimes [1] < \rho$.

$(iii) \Rightarrow (i)$: We have $a \in D(\rho)$, hence $\rho \cong a\rho$, and so $\sigma \otimes [1] < \rho$. By Theorem 3.60 we then get $\rho \cong \sigma \otimes \tau$ for some quadratic Pfister form τ. $\square$

Starting from Theorem 3.60 we would also like to establish a counterpart to Theorem 3.58, dealing with bilinear Pfister forms that simultaneously divide two given quadratic Pfister forms. We need a further lemma for this purpose.

Lemma 3.63. *Let σ be a bilinear Pfister form and ρ an anisotropic quadratic Pfister form over k. Furthermore let $a \neq 0$ be an element of k with*

$$\sigma \otimes [1] \perp [a] < \rho.$$

Then there exists a quadratic Pfister form τ over k such that

$$\rho \cong \sigma \otimes \langle 1, a \rangle \otimes \tau.$$

Proof. Let $n := \dim \sigma$ (a 2-power), and let $(t, u) = (t_1, \ldots, t_n, u_1, \ldots, u_n)$ be a tuple of $2n$ unknowns over k. The form $\sigma_{k(t,u)}$ is round (Theorem 3.10). Hence there exists an n-tuple $v \in k(t, u)^n$ with

$$\sigma(t) + a\sigma(u) = \sigma(u)(\sigma(v) + a).$$

By our assumption the elements $\sigma(u)$ and $\sigma(v) + a$ are represented by $\rho_{k(t,u)}$. Since $\rho_{k(t,u)}$ is round as well, $\sigma(t) + a\sigma(u)$ is represented by $\rho_{k(t,u)}$. By the Subform Theorem we then get

$$\sigma \otimes \langle 1, a \rangle \otimes [1] < \rho,$$

and we can conclude the proof by applying Theorem 3.60. $\qquad\qquad\qquad\qquad$ □

Definition 3.64.

(a) Let $r \in \mathbb{N}_0$. We call two quadratic Pfister forms ρ_1, ρ_2 over k *bilinearly r-linked* if there exist a bilinear r-fold Pfister form σ and quadratic Pfister forms τ_1, τ_2 over k such that

$$\rho_1 \cong \sigma \otimes \tau_1, \quad \rho_2 \cong \sigma \otimes \tau_2.$$

(b) We define the *bilinear linkage number* $i(\rho_1, \rho_2)$ of ρ_1 and ρ_2 as follows: If ρ_1 is neither divisible by ρ_2, nor ρ_2 divisible by ρ_1, then $i(\rho_1, \rho_2)$ is the largest number r such that ρ_1 and ρ_2 are bilinearly r-linked. In the cases $\rho_1 \mid \rho_2$ and $\rho_2 \mid \rho_1$ we let $i(\rho_1, \rho_2)$ be the number d such that $2^d = \min(\dim \rho_1, \dim \rho_2)$.

Remark. If char $k \neq 2$, it follows immediately from Definitions 3.57 and 3.64 that $i(\rho_1, \rho_2)$ coincides with the quadratic linkage number $j(\rho_1, \rho_2)$, which we then simply call the *linkage number* of ρ_1 and ρ_2.

Theorem 3.65 ([15, Prop. 4.4] for char $k \neq 2$). *Let ρ_1 and ρ_2 be two anisotropic quadratic Pfister forms over k.*

(i) *The Witt index $\mathrm{ind}(\rho_1 \perp -\rho_2)$ is a 2-power 2^r with $r \geq j(\rho_1, \rho_2)$ and $r \geq i(\rho_1, \rho_2)$. If $j(\rho_1, \rho_2) > 0$, then $r = j(\rho_1, \rho_2)$. Otherwise $r = i(\rho_1, \rho_2)$.*

(ii) *Let σ be a bilinear s-fold Pfister form with $0 \leq s < r$ that divides ρ_1 and ρ_2, i.e.*

$$\rho_1 \cong \sigma \otimes \tau_1, \quad \rho_2 \cong \sigma \otimes \tau_2$$

with quadratic Pfister forms τ_1, τ_2 (cf. Theorem 3.54). Then there exists either an $(r - s)$-fold bilinear Pfister form γ or an $(r - s)$-fold quadratic Pfister form γ such that $\sigma \otimes \gamma$ divides both Pfister forms ρ_1 and ρ_2, i.e. either

$$\rho_1 \cong \sigma \otimes \gamma \otimes \mu_1, \quad \rho_2 \cong \sigma \otimes \gamma \otimes \mu_2$$

with quadratic Pfister forms μ_1, μ_2, or

$$\rho_1 \cong \alpha_1 \otimes \sigma \otimes \gamma, \quad \rho_2 \cong \alpha_2 \otimes \sigma \otimes \gamma$$

with bilinear Pfister forms α_1, α_2.

Proof. We proceed as in the proof of Theorem 3.58 and assume that we have factorizations $\rho_1 \cong \sigma \otimes \tau_1$, $\rho_2 \cong \sigma \otimes \tau_2$ with quadratic Pfister forms τ_1, τ_2 and a bilinear Pfister form σ. {We allow $\sigma = \langle 1 \rangle$.} We now want to verify the following two statements which, together with Theorem 3.60, will establish the proof.

(a) $\mathrm{ind}(\rho_1 \perp -\rho_2) \geq \dim \sigma$.

(b) If $\mathrm{ind}(\rho_1 \perp \rho_2) > \dim \sigma$, then either there exists an $a \in k^*$ and quadratic Pfister forms μ_1, μ_2 such that

$$\rho_1 \cong \sigma \otimes \langle 1, a \rangle \otimes \mu_1, \quad \rho_2 \cong \sigma \otimes \langle 1, a \rangle \otimes \mu_2, \tag{3.10}$$

or there exist over k a quadratic Pfister form μ of dimension 2 and bilinear Pfister forms α_1, α_2 such that

$$\rho_1 \cong \alpha_1 \otimes \sigma \otimes \mu, \quad \rho_2 \cong \alpha_2 \otimes \sigma \otimes \mu. \tag{3.11}$$

Let (E_1, q_1) and (E_2, q_2) be quadratic spaces associated to ρ_1 and ρ_2, and let $(E, q) := (E_1, q_1) \perp (E_2, -q_2)$. The corresponding vector space decomposition $E = E_1 \oplus E_2$ gives rise to projections $p_1 : E \longrightarrow E_1$, $p_2 : E \longrightarrow E_2$.

We have $[1] < \tau_1$ and $[1] < \tau_2$, and thus

$$\sigma \otimes [1, -1] < \sigma \otimes \tau_1 \perp \sigma \otimes (-\tau_2) \cong \rho_1 \perp -\rho_2.$$

Hence E contains a subspace U with $q(U) = 0$, $\dim U = \dim \sigma$, such that the form $\sigma \otimes [1]$ corresponds to $p_1(U)$ as a subspace of (E_1, q_1) as well as to $p_2(U)$ as a subspace of (E_2, q_2). To see this, note that p_i $(i = 1, 2)$ maps the vector space U linearly isomorphic to $p_i(U)$ since E_1 and E_2 are anisotropic as subspaces of (E, q).

Since E contains the totally isotropic subspace U, it is already clear that $\mathrm{ind}(q_1 \perp -q_2) \geq \dim \sigma$.

Assume now that $\mathrm{ind}(q_1 \perp -q_2) > \dim \sigma$. Then (E, q) contains a totally isotropic subspace $V = U + kv$ of dimension $1 + \dim \sigma$. Again we know that p_i maps the vector space V linearly isomorphic to $p_i(V)$ $(i = 1, 2)$, and for arbitrary $x \in V$ we have:

$$q_1(p_1(x)) = q_2(p_2(x)),$$

and thus also for $x, y \in V$:

$$B_1(p_1(x), p_1(y)) = B_2(p_2(x), p_2(y)),$$

where B_i denotes the bilinear form B_{q_i} corresponding to q_i. We distinguish two cases:

Case 1: $B_1(p_1(v), p_1(u)) = 0$. Now $\sigma \perp [a]$ corresponds to $p_1(V)$ with respect to q_1 and also to $p_2(V)$ with respect to q_2 where $a := q_1(p_1(v)) = q_2(p_2(v))$. Thus $\sigma \perp [a]$ is a subform of both ρ_1 and ρ_2. By Lemma 3.63 we obtain (3.10) with quadratic Pfister forms μ_1 and μ_2.

Case 2: $B_1(p_1(v), p_1(u)) \neq 0$. We choose an orthogonal basis $e_1, \ldots, e_n$ of $p_1(U)$ with respect to q_1 with $B_1(e_1, p_1(v)) = 1$, $B_1(e_i, p_1(v)) = 0$ for $2 \leq i \leq n$, which is certainly possible. Then, corresponding to $p_1(V)$ with respect to q_1 and also to

$p_2(V)$ with respect to q_2, we have the form

$$\begin{bmatrix} a_1 & 1 \\ 1 & b_1 \end{bmatrix} \perp [a_2, \ldots, a_n]$$

with $b_1 := q_1(p_1(v))$ and $a_i = q_1(e_i)$, and we have

$$\sigma \otimes [1] \cong [a_1, a_2, \ldots, a_n].$$

Now by Lemma 1.61,

$$\begin{bmatrix} a_1 & 1 \\ 1 & b_1 \end{bmatrix} \cong \langle a_1 \rangle \otimes \begin{bmatrix} 1 & 1 \\ 1 & b \end{bmatrix}$$

with $b := a_1 b_1$. Since the three forms σ, ρ_1, ρ_2 are round, it follows that

$$\begin{bmatrix} 1 & 1 \\ 1 & b \end{bmatrix} \perp [a_2 a_1^{-1}, \ldots, a_n a_1^{-1}] < \rho_1$$

for $i = 1$ and $i = 2$, and also that

$$[1, a_2 a_1^{-1}, \ldots, a_n a_1^{-1}] \cong \sigma \otimes [1].$$

Hence

$$\begin{bmatrix} 1 & 1 \\ 1 & b \end{bmatrix} \perp \sigma' \otimes [1] < \rho_i$$

for $i = 1$ and $i = 2$. Now Lemma 3.59 tells us that the factorizations (3.11) above hold with $\mu = \begin{bmatrix} 1 & 1 \\ 1 & b \end{bmatrix}$ and bilinear Pfister forms α_1 and α_2. $\qquad \square$

Problem 3.66. Let char $k = 2$, $a \in k^*$, and let ρ_1, ρ_2 be 2-fold anisotropic quadratic Pfister forms with $\langle 1, a \rangle \mid \rho_1$ and $\langle 1, a \rangle \mid \rho_2$. Does there always exist a quadratic form $\tau = \begin{bmatrix} 1 & 1 \\ 1 & b \end{bmatrix}$ over k with $\tau \mid \rho_1$ and $\tau \mid \rho_2$?

The author has not yet been able to give either a positive or a negative answer to this question. If the answer is "yes", then by Theorem 3.65 it is immediate that for two anisotropic quadratic Pfister forms ρ_1, ρ_2 over k we *always* have that

$$\operatorname{ind}(\rho_1 \perp -\rho_2) = 2^{j(\rho_1, \rho_2)}.$$

Finally we remark that R. Baeza proved several theorems in this section, or parts of them, more generally over semi-local rings instead of fields; see in particular [6, Chap. IV, §4]. As our proofs rely on the Norm Theorem, the Subform Theorem, and elements of generic splitting, it is until this day not possible to preserve them over semi-local rings. Thus Baeza's proofs differ from the ones we presented here. They are substantially more "constructive" than ours. Therefore they deserve also over fields an interest for their own sake, just as the proofs of Elman and Lam in [15] when the characteristic is different from 2.

3.8 Pfister Neighbours and Excellent Forms

In the following k denotes a field of arbitrary characteristic and a "form" will always mean a *nondegenerate quadratic* form.

Definition 3.67 (cf. [34, Def. 7.4] for char $k \neq 2$). A form φ over k is called a *Pfister neighbour* if there exist a quadratic Pfister form τ over k and an element $a \in k^*$ such that $\varphi < a\tau$ and $\dim \varphi > \frac{1}{2} \dim \tau$. More precisely we then call φ a *neighbour of the Pfister form τ*.

In §3.2 we already introduced the notion of *close Pfister neighbour* for the case char $k = 2$. Close Pfister neighbours are of course Pfister neighbours in the current sense.

If φ is a neighbour of a Pfister form τ, then we must clearly have $\dim \varphi \geq 2$, and in case $\dim \varphi = 2$ we must moreover have $\varphi \cong a\tau$.

Theorem 3.68. *If φ is a neighbour of a Pfister form τ and a is an element of k^* with $\varphi < a\tau$, then the form τ and also the form $a\tau$ are up to isometry uniquely determined by φ. The fields $k(\varphi)$ and $k(\tau)$ are specialization equivalent over k.*

Proof. The form $a\tau_K$ is hyperbolic over the field $K := k(\tau)$. Let $E = (E, q)$ be a quadratic space corresponding to $a\tau_K$ and $F = (F, q|F)$ a quadratic subspace of E corresponding to the subform φ_K of $a\tau_K$. Let $\dim E = 2^d$. There exists a vector subspace U of E of dimension 2^{d-1} with $q(U) = 0$. From the obvious inequality

$$\dim(F \cap U) + \dim E \geq \dim F + \dim U$$

and the assumption that $\dim F > 2^{d-1}$ one sees that $\dim(F \cap U) > 0$, and thus that $F \cap U \neq \{0\}$. Hence F is isotropic and thus the form $\varphi_K = \varphi \otimes k(\tau)$ is isotropic. Therefore there exists a place from $k(\varphi)$ to $k(\tau)$ over k. Since φ is a subform of $a\tau$, the form $\tau \otimes k(\varphi)$ is also isotropic. Thus there also exists a place from $k(\tau)$ to $k(\varphi)$. (A conclusion we have already reached many times!) Hence $k(\tau) \sim_k k(\varphi)$.

By Corollary 3.55, it now follows that τ is uniquely determined by φ up to isometry. If $\varphi < a\tau$ and $c \in D(\varphi)$ is arbitrarily chosen, then $ac \in D(\tau)$, thus $\tau \cong ac\tau$, and so $a\tau \cong c\tau$. This shows that also the form $a\tau$ is uniquely determined by φ up to isometry. $\square$

Definition 3.69. Let φ be a Pfister neighbour, thus $\varphi < a\tau$ where τ is a Pfister form and $\dim \varphi > \frac{1}{2} \dim \tau$. Then we call the by φ up to isometry uniquely determined polar $\mathrm{Pol}_{a\tau}(\varphi)$ of φ in $a\tau$ (cf. §3.2) the *complementary form* of φ. Furthermore we call τ the *neighbouring Pfister form* of φ.

Remark.

(1) Again let $2^d = \dim \tau$. Clearly the complementary form $\eta = \mathrm{Pol}_{a\tau}(\varphi)$ of φ has dimension $2^d - \dim \varphi < 2^{d-1}$. Furthermore φ and η have the same quasilinear part, $QL(\varphi) = QL(\eta)$, as already observed in Lemma 3.16.

(2) In particular η is strictly regular if and only if φ is strictly regular. We then have $\varphi \perp \eta \cong a\tau$. This is always the case when k has characteristic $\neq 2$. In characteristic 2 we also have that η is regular if and only if φ is regular.

(3) By §3.2, φ is a close Pfister neighbour if and only if η is quasilinear. We then have $\eta = QL(\varphi)$. Moreover, if φ is anisotropic, then φ has height 1 by §3.2. If φ is isotropic, then φ splits.

Theorem 3.70. *Let φ be an anisotropic Pfister neighbour over k with complementary form η. Then $\varphi \otimes k(\varphi) \sim (-\eta) \otimes k(\varphi)$.*

Proof. Let τ be the neighbouring Pfister form of φ and $\varphi < a\tau$. If φ is strictly regular, then $a\tau \cong \varphi \perp \eta$ and $\tau \otimes k(\varphi) \sim 0$, and thus certainly $\varphi \otimes k(\varphi) \sim (-\eta) \otimes k(\varphi)$. This holds in particular if k is of characteristic $\neq 2$.

Assume now that $\operatorname{char} k = 2$ and that φ is not strictly regular.[13] Let χ be the quasilinear part of φ, $\chi = QL(\varphi)$. By Scholium 3.17, we have an orthogonal decomposition $\tau = \mu \perp \gamma \perp \rho$ with

$$\chi < \mu, \quad \dim \mu = 2 \dim \chi, \quad \varphi \cong \chi \perp \gamma, \quad \eta = \chi \perp \rho.$$

With $K := k(\varphi)$ we obtain

$$\mu_K \perp \gamma_K \perp \rho_K \sim 0,$$

thus $\gamma_K \sim \mu_K \perp \rho_K$ (note that $\operatorname{char} k = 2$!), and so

$$\varphi_K \sim \chi_K \perp \mu_K \perp \rho_K.$$

Let $\chi \cong [a_1, \ldots, a_r]$. Then

$$\mu_K \cong \begin{bmatrix} a_1 & 1 \\ 1 & b_1 \end{bmatrix} \perp \cdots \perp \begin{bmatrix} a_r & 1 \\ 1 & b_r \end{bmatrix}$$

with elements b_i of k. Hence

$$\chi \perp \mu \cong \chi \perp \begin{bmatrix} 0 & 1 \\ 1 & b_1 \end{bmatrix} \perp \cdots \perp \begin{bmatrix} 0 & 1 \\ 1 & b_r \end{bmatrix} \cong \chi \perp r \times H,$$

and we obtain

$$\varphi_K \sim \chi_K \perp \rho_K = \eta_K. \qquad \square$$

We return to the theory of generic splitting established in §2.4.

For a given form φ, even of small dimension, it is in general very complicated to explicitly determine the splitting pattern $SP(\varphi)$ or even the higher kernel forms. However, we will now specify a larger class of forms for which it is possible to do this, the "excellent forms". We define a class of forms by induction on the dimension (cf. [34, §7] for $\operatorname{char} k \neq 2$).

Definition 3.71. A form φ is called *excellent* if $\dim \varphi \leq 1$, or φ is quasilinear, or φ is a Pfister neighbour with excellent complementary form.

---

[13] The following calculation generalizes the proof of Theorem 3.18, $(i) - (iii)$ in §3.2.

{Note that if the ground field k has characteristic 2 and if $\dim \varphi \leq 1$, then φ is of course quasilinear. In characteristic $\neq 2$ on the other hand, there are no quasilinear forms $\varphi \neq 0$.}

Thus, by this definition, a form φ over k is excellent if and only if there exists a finite sequence of forms

$$\eta_0 = \varphi, \eta_1, \ldots, \eta_t$$

over k such that η_t is quasilinear or at most one-dimensional, and every form η_r with $0 \leq r < t$ is a Pfister neighbour with complementary form η_{r+1}. We call the forms η_r the *higher complementary forms* of φ; more specifically, we call η_r the *rth complementary form* of φ $(0 \leq r \leq t)$.

Theorem 3.72. *Let φ be an anisotropic excellent form over k and let $(\eta_i \mid 0 \leq i \leq t)$ be the sequence of higher complementary forms of φ. For every $r \in \{0, 1, \ldots, t\}$, let K_r denote the free composite of the fields $k(\eta_i)$ with $0 \leq i < r$ over k. {Read $K_0 = k$ for $r = 0$.}*

Claim: $(K_r \mid 0 \leq r \leq t)$ is a generic splitting tower of φ. For every $r \in \{0, 1, \ldots, t\}$, $\varphi \otimes K_r$ has kernel form $(-1)^r \eta_r \otimes K_r$. In particular, φ has height t.

Proof. By induction on r. For $r = 0$ we don't have to show anything, so suppose that $r > 0$. Let τ_0 and τ_1 be the neighbouring Pfister forms of $\varphi = \eta_0$ and η_1 respectively. Thus there are elements a_0, a_1 in k^* with $\varphi < a_0 \tau_0$, $\eta_1 < a_1 \tau_1$, $\eta_1 = \mathrm{Pol}_{a_0 \tau_0}(\varphi)$. Furthermore, $K_1 = k(\varphi)$. By Theorem 3.70 we have $\varphi \otimes K_1 \sim (-\eta_1) \otimes K_1$, and by Theorem 3.68 we have $k(\varphi) \sim_k k(\tau_0)$, $k(\eta_1) \sim_k k(\tau_1)$.

If $\eta_1 \otimes K_1$ were isotropic, then $\tau_1 \otimes k(\tau_0)$ would also be isotropic, and thus hyperbolic. This implies, by Theorem 3.53 (or Theorem 3.54), that the form τ_1 is divisible by τ_0. However, this is absurd since τ_1 has smaller dimension than τ_0. Hence $\eta_1 \otimes K_1$ is anisotropic. Thus $(-\eta_1) \otimes K_1$ is the kernel form of $\varphi \otimes K_1$.

From the definition of excellent forms it is clear that $\eta_1 \otimes K_1$ is excellent and that for every $r \in \{1, \ldots, t\}$ the form $\eta_r \otimes K_1$ is the $(r - 1)$-st complementary form of $\eta_1 \otimes K_1$. If we now apply the induction hypothesis to $\eta_1 \otimes K_1$ we obtain the complete statement of the theorem. With regards to this, one should keep in mind that $K_1(\eta_i \otimes K_1)$ is the free composite of K_1 and $k(\eta_i)$ over k $(1 \leq i \leq t)$ and consequently that K_r is the free composite of the fields $K_1(\eta_i \otimes K_i)$ with $1 \leq i < r$ over K_1 $(1 \leq r \leq t)$. □

In which dimensions do there exist, for a given field k, anisotropic excellent forms over k? We want to discuss this question in the (difficult) case when k has characteristic 2.

Scholium 3.73. *Let char $k = 2$. If φ is an anisotropic excellent form of dimension n over k, then the neighbouring Pfister form τ of φ is again anisotropic, and $\dim \tau$ is clearly the smallest 2-power $2^d \geq n$.*

Conversely, let τ be an anisotropic Pfister form. In this case we construct an excellent form with neighbouring form τ as follows: We choose a sequence $(\tau_r \mid 1 \leq r \leq t)$ of Pfister subforms of τ with $\tau_1 = \tau$, $\tau_r \mid \tau_{r-1}$ for $2 \leq r \leq t$, and a quasilinear subform η of τ_t with $\dim \eta < \frac{1}{2} \dim \tau_t$, which is always possible. {$t \geq 1$. It is allowed that $\eta = 0$.} Then we define successive forms $\eta_t, \eta_{t-1}, \ldots, \eta_0$ in such a way that

$$\eta_t = \eta \,, \ \eta_r < \tau_r \,, \ \eta_{r-1} < \tau_r \,, \ \eta_{r-1} = \mathrm{Pol}_{\tau_r}(\eta_r) \quad for \quad 1 \le r \le t.$$

Now $\varphi := \eta_0$ is an excellent form with neighbour τ and higher complementary forms η_r $(0 \le r \le t)$. The form φ is anisotropic and has quasilinear part η. Up to a scalar factor one can obtain every anisotropic excellent form over k in this manner.

Next we show that over suitable fields of characteristic 2 there exist anisotropic Pfister forms of arbitrarily high dimension.

Theorem 3.74. *Let κ be any field and $k = \kappa(u, u_1, \dots, u_n)$, where $u, u_1, \dots, u_n$ is a tuple of unknowns. The Pfister form $\langle 1, u_1 \rangle \otimes \cdots \otimes \langle 1, u_n \rangle \otimes \begin{bmatrix} 1 & 1 \\ 1 & u \end{bmatrix}$ over k is anisotropic.*

Proof. We assume that κ has characteristic 2 and leave the proof for the case char $\kappa \ne 2$ to the Reader. We work over the iterated power series field

$$K := \kappa((u))((u_1)) \dots ((u_n)),$$

which contains k. Let τ be the Pfister form from above, considered as a form over K. We show by induction on n that τ is anisotropic.

$\underline{n = 0}$: The equation $u = x^2 - x$ is with $x \in \kappa((u))$ unsolvable, so that the form $\begin{bmatrix} 1 & 1 \\ 1 & u \end{bmatrix}$ is anisotropic over $\kappa((u))$.

$\underline{n - 1 \to n}$: Let $F := \kappa((u))((u_1)) \dots ((u_{n-1}))$. We have $\tau = \rho_K \perp u_n \rho_K$ with ρ an anisotropic Pfister form over F by the induction hypothesis. There is a place $\lambda :$ $K \to F \cup \infty$ over F belonging to the obviously discrete valuation of K over F with prime element u_n. The form ρ_K has good reduction with respect to λ and $\lambda_*(\rho_K) = \rho$. We further have that $\lambda(u_n c^2) = 0$ or $= \infty$ for every $c \in K^*$. If τ were isotropic, and thus hyperbolic, then we would have $\rho_K \cong u_n \rho_K$. Applying λ_W would then give $\rho \sim 0$, a contradiction. Therefore τ is anisotropic. $\qquad\square$

Example 3.75. We choose two natural numbers $n > m$, and a sequence of natural numbers

$$s_1 = n > s_2 > \cdots > s_t = m.$$

Let $k = \kappa(u, u_1, \dots, u_n)$, as in Theorem 3.74, and char $\kappa = 2$. We form the—by Theorem 3.74 anisotropic—Pfister forms

$$\tau_r := \langle 1, u_1 \rangle \otimes \cdots \otimes \langle 1, u_{s_r} \rangle \otimes \begin{bmatrix} 1 & 1 \\ 1 & u \end{bmatrix}$$

$(1 \le r \le t)$, and then the quasilinear form

$$\eta := [1, u_1, u_2, \dots, u_s]$$

for some $s < s_t = m$. We have $\eta < \tau_t$ and $\dim \eta < \frac{1}{2} \dim \tau_t$. Using the procedure outlined in the scholium above we obtain an anisotropic excellent form φ of height t over k with $(s + 1)$-dimensional quasilinear part η. $\qquad\square$

We utilize Theorem 3.74 and this example to illustrate principles about the specialization of quadratic forms, obtained earlier (Theorem 3.6, Theorem 2.19). We consider the question of when an excellent form has good reduction with respect to a place, and its specialization is again excellent.

Theorem 3.74 gives us the "generic" $(n+1)$-fold Pfister form of a field κ. Namely, if ρ is an $(n + 1)$-fold Pfister form over κ, and if we have chosen a representation

$$\rho \cong \langle 1, a_1 \rangle \otimes \cdots \otimes \langle 1, a_n \rangle \otimes \begin{bmatrix} 1 & 1 \\ 1 & a \end{bmatrix}$$

$(a \in \kappa, \text{ all } a_i \in \kappa)$, then there exists a place $\lambda : \kappa(u, u_1, \ldots, u_n) \to \kappa \cup \infty$ over κ with $\lambda(u) = a$, $\lambda(u_i) = a_i$ for $i = 1, \ldots, n$. The form

$$\tau := \langle 1, u_1 \rangle \otimes \cdots \otimes \langle 1, u_n \rangle \otimes \begin{bmatrix} 1 & 1 \\ 1 & u \end{bmatrix}$$

has good reduction with respect to every such place λ, and $\lambda_*(\tau) \cong \rho$.

Now suppose that ρ is anisotropic. We choose a binary quadratic subform γ of τ with $1 \in D(\gamma)$, which has good reduction with respect to λ, e.g. $\gamma = \begin{bmatrix} 1 & 1 \\ 1 & u \end{bmatrix}$. If σ is a Pfister form over k with $\gamma < \sigma < \tau$, then σ is clearly obedient with respect to λ. By Theorem 3.6, and Theorem 2.19, it follows that σ has good reduction with respect to λ and that $\lambda_*(\sigma)$ is a subform of $\rho = \lambda_*(\tau)$. Furthermore, $\lambda_*(\gamma)$ is a subform of $\lambda_*(\sigma)$.

We want to obtain that $\lambda_*(\sigma)$ is again a Pfister form. By §2.4, $\lambda_*(\sigma)$ has height ≤ 1. Since $\lambda_*(\sigma)$ is anisotropic, and thus does not split, we must have $h(\lambda_*(\sigma)) = 1$. Furthermore, $[1] < \sigma$ implies, again by Theorem 3.6, etc. that $[1] = \lambda_*([1])$ is a subform of $\lambda_*(\sigma)$, thus $1 \in D(\lambda_*(\sigma))$. Finally Theorem 3.45, tells us that $\lambda_*(\sigma)$ is a Pfister form. It divides $\lambda_*(\tau) = \rho$.

We turn our attention to the construction of excellent forms in the scholium above. There we choose the Pfister form τ_t so that $\gamma < \tau_t$. Now all forms τ_r have good reduction with respect to λ, every $\lambda_*(\tau_r)$ is a Pfister form, and $\lambda_*(\tau_r) \mid \lambda_*(\tau_{r-1})$ for $2 \leq r \leq t$. The quasilinear form η is automatically obedient with respect to λ. Thus, by Theorem 3.6, etc., it has good reduction with respect to λ and $\lambda_*(\eta)$ is a subform of $\lambda_*(\tau_t)$.

If the complementary form η_{t-1} of $\eta = \eta_t$ in τ_t is obedient with respect to λ, then we obtain in the same fashion that η_{t-1} has good reduction with respect to λ, and that $\lambda_*(\eta_{t-1}) < \lambda_*(\tau_t)$. Furthermore, $\lambda_*(\eta_{t-1})$ is now orthogonal to $\lambda_*(\eta_t)$, and by dimension considerations it is now clear that $\lambda_*(\eta_{t-1})$ is the polar of $\lambda_*(\eta_t)$ in $\lambda_*(\tau_t)$.

We can continue this line of reasoning in so far that it is guaranteed that all η_r are obedient with respect to λ, and obtain finally that the excellent form $\varphi = \eta_0$ has good reduction with respect to λ, and that $\lambda_*(\varphi)$ is an excellent subform of $\lambda_*(\tau) = \rho$ with ρ as neighbour.

If one considers $\gamma = \begin{bmatrix} 1 & 1 \\ 1 & u \end{bmatrix}$, and starts with $\eta = [1, u_1, u_2, \ldots, u_s]$, as in the example above, one can reason inductively that all η_r are indeed obedient with respect to λ, and thus actually have good reduction with respect to λ.

Now, if ρ is anisotropic, then it follows from Theorem 3.6, that every subform σ of τ has good reduction with respect to λ, and that $\lambda_*(\sigma)$ is a subform of ρ. If σ is a Pfister form, then $\lambda_*(\sigma)$ has height ≤ 1 by §2.4. Since $\lambda_*(\sigma)$ is anisotropic, and so doesn't split, $\lambda_*(\sigma)$ must have height exactly 1. Moreover, $\lambda_*(\sigma)$ is strictly regular, since this is the case for σ.

Furthermore, $[1] < \sigma$ implies that $[1] = \lambda_*[1]$ is a subform of $\lambda_*(\sigma)$. Now we conclude with Theorem 3.45, that $\lambda_*(\sigma)$ is a Pfister form (and it divides the form ρ since it is a subform of ρ).

Every Pfister subform σ of τ thus has good reduction with respect to λ, and provides a Pfister subform $\lambda_*(\sigma)$ of ρ. By the scholium above it is now clear that the anisotropic excellent form φ in our example above has good reduction with respect to λ, and that $\lambda_*(\varphi)$ is a neighbour of the Pfister form ρ.

Nevertheless we cannot consider φ to be a "generic" excellent form of height 1 with particular splitting pattern for κ since there are too many other possibilities to construct excellent subforms of τ. Our reasoning about the specialization of excellent forms remains valid in the general situation however. We record:

Theorem 3.76. *Let* $\lambda : K \to L \cup \infty$ *be a place and* τ *a Pfister form which has good reduction with respect to* λ. *Assume that the form* $\lambda_*(\tau)$ *is anisotropic. Then* $\lambda_*(\tau)$ *is again a Pfister form. Every excellent subform* φ *of* τ *which is a neighbour of* τ *has good reduction with respect to* λ, *and* $\lambda_*(\varphi)$ *is an excellent form, neighbouring* $\lambda_*(\tau)$, *of the same height as* φ.

In the situation of the last theorem, the form $\lambda_*(\tau)$ is furthermore also a Pfister form when it is isotropic, since it follows already from the generic splitting theory in §2.1 that $\lambda_*(\tau) \cong 2^{d-1} \times H$ with $2^d = \dim \tau$.

3.9 Regular Forms of Height 1

In the following let φ be a *regular* anisotropic quadratic form *of height* 1 over a field k. The purpose of this section is the proof of the following theorem.

Theorem 3.77. φ *is excellent.*

More particularly: if φ is of even dimension, then $\varphi = a\tau$ with $a \in k^*$ and τ a Pfister form. If φ is of odd dimension, then $\varphi = a\,\mathrm{Pol}_\tau([1])$ with $a \in k^*$ and τ a Pfister form of dimension ≥ 4. Is furthermore char $k \neq 2$, then $\mathrm{Pol}_\tau([1])$ is the pure part of τ, cf. §3.2.

For φ of even dimension we already proved the theorem in §3.6, cf. Theorem 3.45. Thus, suppose from now on that the dimension of φ is odd.

We start with the proof for char $k \neq 2$ (cf. [33, p. 81 ff.]). We interpret φ as a symmetric bilinear form. Upon multiplying φ by a scalar factor we may assume without loss of generality that $d(\varphi) = 1$. {The signed determinant $d(\varphi)$ was introduced in §1.2.}

We first show that φ does not represent the element 1. By way of contradiction assume that φ does represent 1. Then we have an orthogonal decomposition $\varphi = \langle 1 \rangle \perp \chi$. Since $h(\varphi) = 1$ we have $\varphi \otimes k(\varphi) \sim \langle 1 \rangle$, and so $\chi \otimes k(\varphi) \sim 0$. As usual the Norm Theorem and the Subform Theorem tell us that $a\varphi < \chi$ for a suitable $a \in k^*$. This is absurd, however, since the dimension of χ is smaller than the dimension of φ. Therefore $1 \notin D(\varphi)$. The form $\tau := \langle 1 \rangle \perp (-\varphi)$ is thus anisotropic.

Let L/k be an arbitrary field extension of k. If $\varphi \otimes L$ is anisotropic, then $h(\varphi \otimes L) = 1$ since $h(\varphi \otimes L) \leq h(\varphi)$ and $h(\varphi \otimes L) \neq 0$. If we apply what we just proved to $\varphi \otimes L$ instead of φ, we see that $\tau \otimes L$ is anisotropic. On the other hand, if $\varphi \otimes L$ is isotropic, then $\varphi \otimes L \sim \langle 1 \rangle$, and so $\tau \otimes L \sim 0$. This shows that $\tau \otimes L$ is either anisotropic or hyperbolic. Thus τ has height 1. Furthermore $1 \in D(\tau)$. As already established before (Theorem 3.45), τ is a Pfister form. It follows that $\varphi = -\tau'$.

Let us now continue by proving Theorem 3.77 for char $k = 2$ and dim φ odd, dim $\varphi = 2m + 1$ with $m > 0$. Without loss of generality we may assume that φ has quasilinear part $QL(\varphi) = [1]$. We have to show that φ is a close neighbour (cf. §3.2) of a Pfister form.

We deal with the case $m = 1$ separately. In this case we have

$$\varphi \cong [1] \perp c \begin{bmatrix} 1 & 1 \\ 1 & b \end{bmatrix}$$

with $c \in k^*$, $b \in k$, and so φ is a close neighbour of the Pfister form $\langle 1, c \rangle \otimes \begin{bmatrix} 1 & 1 \\ 1 & b \end{bmatrix}$.

From now on let $m \geq 2$, and thus dim $\varphi \geq 5$. We require an easy lemma of a general nature that we will prove at a later stage.

Lemma 3.78. *Let ψ and χ be quadratic forms over k, and assume that $\psi \perp \chi$ is nondegenerate. Then*

$$\operatorname{ind} \psi + \dim \chi \geq \operatorname{ind}(\psi \perp \chi). \qquad \square$$

We have an orthogonal decomposition

$$\varphi \cong [1] \perp \rho$$

with ρ strictly regular, dim $\rho = 2m$. Since the height of φ is 1 we have $\operatorname{ind}(\varphi \otimes k(\varphi)) = m$. The lemma implies that

$$\operatorname{ind}(\rho \otimes k(\varphi)) \geq m - 1 > 0.$$

So, $\rho \otimes k(\varphi)$ is in particular isotropic. Since $\varphi \otimes k(\rho)$ is trivially also isotropic it follows that $k(\varphi) \sim_k k(\rho)$ and then

$$\operatorname{ind}(\rho \otimes k(\rho)) \geq m - 1.$$

Let us assume for the sake of contradiction that $\operatorname{ind}(\rho \otimes k(\rho)) > m - 1$, thus $\rho \otimes k(\rho) \sim 0$. By earlier work we have that $\rho = a\sigma$ where σ is a Pfister form. Hence $\varphi = [1] \perp a\sigma$ is a neighbour of the Pfister form $\tau := \langle 1, a \rangle \otimes \sigma$. It follows from Theorem 3.68 that $k(\varphi) \sim_k k(\tau)$. This implies $k(\rho) \sim_k k(\tau)$, since $k(\sigma)$ is also specialization equivalent to $k(\varphi)$ over k. By Corollary 3.55, this means $\sigma \cong \tau$, which is not possible since $\dim \tau = 2 \dim \sigma$. Therefore

$$\mathrm{ind}(\rho \otimes k(\rho)) = m - 1.$$

Thus $\rho \otimes k(\rho)$ is the orthogonal sum of $m - 1$ hyperbolic planes and an anisotropic binary form. Hence $\rho \otimes k(\rho)$ has non-vanishing Arf-invariant. Thus we certainly have $\mathrm{Arf}(\rho) \neq 0$. Let $\mathrm{Arf}(\rho) = a + \wp k$ and

$$\sigma := \begin{bmatrix} 1 & 1 \\ 1 & a \end{bmatrix}.$$

Consider the form $\tau := \sigma \perp \rho$. Then $\varphi < \tau$ and $\mathrm{Arf}(\tau) = 0$. {Note that we are now in a situation which is similar to the proof involving the form τ in char $k \neq 2$.} Assume that $\varphi \otimes k(\sigma)$ is anisotropic. This form then has height 1 and

$$\varphi \otimes k(\sigma) = [1] \perp \rho \otimes k(\sigma).$$

Thus, we should get $\mathrm{Arf}(\rho \otimes k(\sigma)) \neq 0$, which is false. Therefore $\varphi \otimes k(\sigma)$ is isotropic and it follows that

$$\varphi \otimes k(\sigma) \cong [1] \perp m \times H.$$

Hence, $m \times H < \tau \otimes k(\sigma)$. Since $\dim \tau = 2m + 2$ and $\mathrm{Arf}(\tau \otimes k(\sigma)) = 0$ we obtain $\tau \otimes k(\sigma) \sim 0$ and so $\rho \otimes k(\sigma) \sim 0$. By Theorem 3.53, there exist $a_1, \ldots, a_m \in k^*$ with

$$\rho \cong \langle a_1, \ldots, a_m \rangle \otimes \sigma.$$

Consequently,

$$\tau \cong \langle 1, a_1, \ldots, a_m \rangle \otimes \sigma.$$

We will now show that τ is anisotropic. Assume for the sake of contradiction that this is not so. Then there exist vectors $x_0, x_1, \ldots, x_m \in k^2$ with

$$\sigma(x_0) + \sum_{i=1}^{m} a_i \sigma(x_i) = 0$$

and not all $x_i = 0$. Since ρ is anisotropic, we must have $x_0 \neq 0$ and thus also $\sigma(x_0) \neq 0$. Thus there are vectors $y_1, \ldots, y_m \in k^2$ with $\sigma(y_i) = \sigma(x_0)^{-1} \sigma(x_i)$ for $1 \leq i \leq m$. Hence,

$$1 + \sum_{i=1}^{m} a_i \sigma(y_i) = 0.$$

This contradicts the anisotropy of φ. Thus τ is indeed anisotropic.

Now it follows easily that $\varphi \otimes k(\tau)$ is isotropic. Indeed, if $\varphi \otimes k(\tau)$ were anisotropic, then we could apply what we just proved to the form $\varphi \otimes k(\tau)$ of height 1 and we would conclude that $\tau \otimes k(\tau)$ would be anisotropic, which is of course false.

Hence there exists a place from $k(\varphi)$ to $k(\tau)$ over k, and it follows that

$$\varphi \otimes k(\tau) \cong [1] \perp m \times H.$$

Thus $\tau \otimes k(\tau) > m \times H$ and, since $\mathrm{Arf}(\tau) = 0$ and $\dim \tau = 2m + 2$, even $\tau \otimes k(\tau) \sim 0$. Furthermore, $1 \in D(\tau)$ and τ is anisotropic. By earlier work τ is a Pfister form. The form φ is a close neighbour of this Pfister form.

Let us finally supply the proof of the lemma above. Then our theorem will be completely proved.

The field k may have arbitrary characteristic. Let ψ and χ be quadratic forms over k. If $\mathrm{ind}(\psi \perp \chi) \leq \dim \chi$, there is nothing to prove. Thus, suppose $\mathrm{ind}(\psi \perp \chi) > \dim \chi$. Let E be the quadratic space of ψ and G the quadratic space of χ. The space $E \perp G$, nondegenerate by assumption, contains a totally isotropic subspace V with $\dim V = \mathrm{ind}(\psi \perp \chi)$. Let $\pi : V \to G$ be the restriction of the natural projection $E \perp G \to G$ to V. The kernel of π is then $V \cap E$. This space is totally isotropic and

$$\dim(V \cap E) + \dim G \geq \dim V.$$

Therefore,

$$\dim(V \cap E) \geq \mathrm{ind}(\psi \perp \chi) - \dim \chi,$$

which establishes the assertion of the lemma.

3.10 Some Open Problems in Characteristic 2

We subdivide this section into parts (A) – (D), posing four problems which seem to be open.

As before, the word "form" stands for nondegenerate quadratic form and the ground field k is arbitrary, unless indicated otherwise.

(A) Let char $k = 2$. *Which nonregular forms of height 1 are there?*

If φ is anisotropic and $\dim \varphi = 2 + \dim QL(\varphi)$, then φ surely has height 1. Not every such form is excellent! As a simple example we take an anisotropic four-dimensional form

$$\varphi = [a_1, a_2] \perp \begin{bmatrix} a_3 & 1 \\ 1 & b_3 \end{bmatrix}.$$

If φ were excellent, it would have to be a Pfister form up to a scalar factor, which is not the case.

We give another example. Let

$$\varphi = [a_1, a_2, a_3, a_4] \perp \begin{bmatrix} a_5 & 1 \\ 1 & b_5 \end{bmatrix}$$

be anisotropic. {E.g., let $a_1, \ldots, a_5, b_5$ be unknowns over a field κ and let $k = \kappa(a_1, \ldots, a_5, b_5)$.} We have $h(\varphi) = 1$ and $\dim \varphi = 6$. Let $\varphi < \tau$ and τ a strictly regular hull of φ (cf. §3.2). Let $\chi := QL(\varphi)$. By Scholium 3.17, we have

$$\varphi = \chi \perp \gamma, \quad \tau = \mu \perp \gamma$$

with $\chi < \mu$, $\dim \mu = 2 \dim \chi = 8$. Thus, $\dim \tau = 8 + 2 = 10$. The form φ is certainly not a Pfister neighbour.

Let us ask a more precise question: *Do there exist any anisotropic forms of height 1, other than close Pfister neighbours and anisotropic forms φ with $\dim(\varphi) = 2 + \dim QL(\varphi)$?*

(B) Let φ be a neighbour of a Pfister form τ and let η be the complementary form. Then $\varphi \otimes k(\varphi) \sim (-\eta) \otimes k(\varphi)$ (Theorem 3.42). *Is $\eta \otimes k(\varphi)$ anisotropic?*

If the characteristic of the ground field is different from two, this is true, as shown originally by R. Fitzgerald [18]. Hence $(-\eta) \otimes k(\varphi)$ is the first higher kernel form of φ. Later on, D. Hoffmann proved the following far-reaching theorem:

Theorem 3.79 ([19]). *Let* char $k \neq 2$ *and let φ and η be anisotropic forms over k. Assume that there exists a 2-power 2^m such that $\dim \eta \leq 2^m < \dim \varphi$. Then $\eta \otimes k(\varphi)$ is anisotropic.*

Does this statement remain true in characteristic $2?$[14]

One could simply try to extend the proofs of Fitzgerald and Hoffmann to characteristic 2. This approach, however, leads to problems very soon since these proofs make extensive use of the following argument: let ψ_1 and ψ_2 be anisotropic forms over k whose orthogonal sum $\psi_1 \perp \psi_2$ is isotropic. Then there exists an $a \in k^*$ such that $\psi_1 = [a] \perp \psi_1'$ and $\psi_2 = [-a] \perp \psi_2'$. But this is only correct when k has characteristic $\neq 2$.[15] In characteristic 2 there does not seem to exist anything that achieves results similar to this simple but effective argument.

(C)

Definition 3.80. Let L/k be a field extension and ψ a form over L. We say that ψ is *definable over k* if there exists a form η over k such that $\psi \cong \eta \otimes L$.

Let φ be a form over k. Let $(K_r \mid 0 \leq r \leq h)$ be a generic splitting tower of φ and let $\varphi_r := \ker(\varphi \otimes K_r)$, the rth higher kernel form of φ ($0 \leq r \leq h$). We are interested in the situation where all φ_r are definable over k. One should observe that this property is independent of the choice of the generic splitting tower $(K_r \mid 0 \leq r \leq h)$.

Problem. *For which anisotropic forms φ over k are all higher kernel forms of φ definable over k?*

Theorem 3.72 tells us that this is the case for excellent forms. For char $k \neq 2$ it is known that there are no other forms with this property [34, Th. 7.14]. {For this reason many texts, e.g. [39, §13], *define* excellence via this property when char $k \neq 2$. We will *not* do that here.}

---

[14] See Addendum at the end of this section!

[15] In characteristic 2 the argument remains valid only for *symmetric bilinear forms*.

Obviously every anisotropic form of height 1 satisfies this property. Among such forms there are non-excellent ones, as seen under (A). In the case that the first problem under (B) has an affirmative answer,[16] it is not so hard to obtain further examples of anisotropic forms whose higher kernel forms are definable over the ground field. Simply embed η into an anisotropic Pfister form τ, of dimension $> 2 \dim \eta$ (we may assume without loss of generality that $1 \in D(\eta)$). Then all higher kernel forms of the polar φ of η in τ are also definable over k.

(D) Let $(K_r \mid 0 \le r \le h)$ be a generic splitting tower of a form φ over k and let $(\varphi_r \mid 0 \le r \le h)$ be the associated higher kernel forms of φ. We fix a number r with $0 < r < h$. We assume that φ_r is definable over k. *Is there, up to isometry, only one form η over k such that $\varphi_r \cong \eta \otimes K_r$?* {Note that this is trivially true for $r = 0$ and $r = h$.}

It is known that the answer is "yes" for char $k \ne 2$. A proof can be found in [34, §7]. Fortunately this proof remains valid in characteristic 2 when the form φ is *strictly regular*. We will reproduce it below. In slightly more generality we can show:

Theorem 3.81. *Let φ be strictly regular. Let η_1 and η_2 be anisotropic forms over k with $\eta_1 \otimes K_r \sim \eta_2 \otimes K_r \sim \varphi \otimes K_r$ and $\dim \eta_1 < \dim \varphi_{r-1}$, $\dim \eta_2 < \dim \varphi_{r-1}$. Then $\eta_1 \cong \eta_2$.*

Proof ([34, p.1 ff.]). We assume without loss of generality that the form η_1 has minimal dimension among all anisotropic forms δ over k with $\delta \otimes K_r \sim \varphi \otimes K_r$.

We first settle the case $r = 1$. Assume that η_1 and η_2 are not isometric. Then $\eta_1 \perp (-\eta_2)$ has a kernel form $\zeta \ne 0$ with $\zeta \otimes k(\varphi) \sim 0$. By the Norm Theorem and the Subform Theorem there exist a form ψ over k and an element $a \in k^*$ such that

$$\varphi \perp \psi \cong a\zeta.$$

It follows that $\varphi \otimes k(\varphi) \sim (-\psi) \otimes k(\varphi)$. Now

$$\dim \eta_1 + \dim \eta_2 \ge \dim \zeta = \dim \varphi + \dim \psi,$$

and so

$$\dim \eta_1 - \dim \psi \ge \dim \varphi - \dim \eta_2 > 0.$$

This contradicts the minimality of $\dim \eta_1$. Therefore, $\eta_1 \cong \eta_2$.

Assume next that $r \ge 2$. We proceed by induction on r. Assume that η_1 and η_2 are not isometric, and thus $\eta_1 \not\cong \eta_2$. Let s be the largest integer between 0 and $r - 1$ such that $\eta_1 \otimes K_s \not\cong \eta_2 \otimes K_s$. If $s > 0$ we apply the induction hypothesis for $r - s$ to the kernel forms of $\varphi \otimes K_s$, $\eta_1 \otimes K_s$, $\eta_2 \otimes K_s$, and obtain a contradiction. Hence $s = 0$. For the kernel form ζ of $\eta_1 \perp (-\eta_2)$ we have again that $\zeta \otimes k(\varphi) \sim 0$ and, just as in the case $r = 1$ above, we obtain a contradiction with the minimality of $\dim \eta_1$. □

---

[16] See Addendum at the end of this section!

Let φ be a neighbour of a Pfister form τ and let η be the complementary form of φ. By §3.6 we have $\varphi \otimes k(\varphi) \sim (-\eta) \otimes k(\varphi)$. We further assume that $\dim \eta \geq \dim QL(\eta) + 2$. {In the case char $k \neq 2$ this implies $\dim \eta \geq 2$.} Now we can form the function field $k(\eta)$. We have $k(\varphi) \sim_k k(\tau)$ and $\tau \otimes k(\eta) \sim 0$. Hence there exists a place from $k(\varphi)$ to $k(\eta)$ over k, and so we also have $\varphi \otimes k(\eta) \sim (-\eta) \otimes k(\eta)$. Conversely we now ask:

Let φ and η be anisotropic forms over k with $\dim \varphi > \dim \eta \geq 2 + \dim QL(\eta)$. Assume that $\varphi \otimes k(\varphi) \sim \eta \otimes k(\varphi)$ and $\varphi \otimes k(\eta) \sim \eta \otimes k(\eta)$. Is then φ a Pfister neighbour with complementary form $-\eta$?

If char $k \neq 2$, this is true, cf. [34, Th. 8.9]. The proof of this fact again utilizes the argument described in (B), which breaks down in characteristic 2.

Addendum (2009)

Problem (B) is now solved. Hoffmann and Laghribi have established Theorem 3.79 above also in characteristic 2, without any restrictions [23], [14, §26].

3.11 Leading Form and Degree Function

Let k be a field of arbitrary characteristic and let φ be a *form* over k, which is just as before understood to be a *nondegenerate quadratic* form.

In case the characteristic of k is different from 2, we alternatively interpret φ as a bilinear form. In fact we identify φ with the bilinear form β on k^n ($n = \dim \varphi$) which satisfies $\varphi(x) = \beta(x, x)$. {Thus $\beta = \frac{1}{2} B_\varphi$, where B_φ is the bilinear form associated to φ.} In particular we make the identification

$$\langle a_1, \ldots, a_n \rangle = [a_1, \ldots, a_n] \qquad (a_i \in k^*).$$

One should bear in mind that with respect to this identification quadratic Pfister forms over k correspond to bilinear Pfister forms over k.

In the rest of this section we will always assume that φ is strictly regular.

Theorem 3.82. *Assume that φ is not hyperbolic. Then the minimum of the dimensions of the kernel forms* $\ker(\varphi \otimes L)$, *where L runs through all field extensions of k with $\varphi \otimes L \nsim 0$, is a 2-power 2^d.*

Proof. If $\dim \varphi$ is odd (and thus char $K \neq 2$), this minimum is clearly 1, and the statement holds with $d = 0$.

Thus assume that $\dim \varphi$ is even. Let $(K_r \mid 0 \leq r \leq h)$ be a generic splitting tower of φ. Since φ is not hyperbolic we have $h \geq 1$.

The kernel form of $\varphi \otimes K_{h-1}$ has height 1. By Theorem 3.45 it is thus of the form $a\tau$ with $a \in K_{h-1}^*$ and τ a d-fold Pfister form over K_{h-1} for some $d \geq 1$. By our generic splitting theory (cf. the end of §2.1 or Theorem 2.12) it is clear that 2^d is the required minimum. $\square$

Definition 3.83. We call the number d, appearing in Theorem 3.82, the *degree* of the form φ and denote it by $\deg \varphi$. If φ is hyperbolic, we set $\deg \varphi = \infty$.

According to the proof of Theorem 3.82, we have $\deg \varphi = 0$ in case $\dim \varphi$ is odd (and so $\operatorname{char} k \neq 2$), and $\deg \varphi \geq 1$ otherwise. An anisotropic n-fold Pfister form clearly has degree n.

The notion of degree of a form is closely connected to the notion of a "leading form", which we introduce next.

Definition 3.84. Let φ be nondegenerate and let $(K_r \mid 0 \leq r \leq h)$ be a generic splitting tower of φ.
(a) We call any field extension of k, which is specialization equivalent to K_{h-1} over k, a *leading field for φ*.
(b) Let F be a leading field for φ. By §3.9 the kernel form ψ of $\varphi \otimes F$ is of the form $a\tau$ for $\dim \varphi$ even, and of the form $a\tau'$ with $a \in F^*$ and τ a Pfister form for $\dim \varphi$ odd. {Recall: τ' denotes the pure part of τ.} We call this Pfister form τ over F the *leading form of φ over F*.

From the proof of Theorem 3.82 it is clear that in even dimension the degree of φ corresponds to the degree of any leading form τ of φ. In odd dimension there is a connection between the degree of τ and the degree of the even-dimensional form $\varphi \perp -d(\varphi)$, where $d(\varphi)$ denotes the signed determinant of φ, interpreted as a one-dimensional quadratic form, cf. §1.2.

Theorem 3.85. *Assume that φ is not split and that $\dim \varphi$ is odd (and so $\operatorname{char} k \neq 2$ since φ is strictly regular). Let τ be a leading form of φ. Then the forms τ and $\varphi \perp -d(\varphi)$ have the same degree.*

Proof. The form $\psi := \varphi \perp -d(\varphi)$ is not split. It has signed determinant[17] $d(\psi) = 1$. Let F be a leading field for φ and E a leading field for ψ. We have

$$\varphi \otimes F \sim a\tau', \quad \psi \otimes E \sim b\sigma,$$

where a and b are scalars, τ is the leading form of φ and σ is the leading form of ψ. Let $n := \deg \tau$, $m := \deg \sigma = \deg \psi$. Since $d(\psi) = 1$ we have $d(\sigma) = 1$, and so $m \geq 2$. Now $d(\tau') = \langle -1 \rangle$, so that $d(\varphi) \otimes F = d(\varphi \otimes F) = \langle -a \rangle$. Hence

$$\psi \otimes F \sim a\tau' \perp \langle a \rangle = a\tau.$$

Therefore the kernel form of $\psi \otimes F$ has dimension 2^n and it follows that $m \leq n$. On the other hand we have

$$\varphi \otimes E \sim b\sigma \perp (d(\varphi) \otimes E).$$

The form on the right-hand side is of dimension $2^m + 1$. It cannot be split since σ is anisotropic and has dimension ≥ 4. From the generic splitting theory (see for instance Scholium 1.43) it follows that $2^m + 1 \geq \dim(a\tau') = 2^n - 1$, and so $2^m + 2 \geq 2^n$. Since $m \leq n$ and $m \geq 2$, we must have $m = n$. $\qquad\square$

---

[17] We write $d(\psi) = 1$ instead of $d(\psi) = \langle 1 \rangle$.

We now describe a situation where the leading form of φ can be determined.

Theorem 3.86 (cf. [33, Th. 6.3] for char $k \neq 2$). *Let $n \in \mathbb{N}$. Let φ be an orthogonal sum $a\tau \perp \psi$ with $a \in k^*$, τ a Pfister form over k of degree $n \geq 1$, and ψ a form over k with $\deg \psi \geq n + 1$. Let F be a leading field for φ.*

(i) The leading form of φ over F is $\tau \otimes F$.

(ii) If $\deg \psi \geq n + 2$, then $\varphi \otimes F$ has kernel form $(a\tau) \otimes F$.

Proof (as in [33, p. 88 ff.]). (*i*) We may assume that ψ is not split. We will show that φ has degree n.

Let $(L_i \,|\, 0 \leq i \leq e)$ be a generic splitting tower of ψ. Assume that $\tau \otimes L_e \sim 0$. Let s be maximal in[18] $[0, e-1]$ with $\tau \otimes L_s \not\sim 0$, so that $\tau \otimes L_s$ is anisotropic. Let ψ_s be the kernel form of $\psi \otimes L_s$. Then $\tau \otimes L_s(\psi_s) \sim 0$. It then follows from the Norm Theorem and the Subform Theorem in the usual manner that $b\psi_s < \tau \otimes L_s$ for some $b \in L_s^*$. Thus, $\deg \psi = \deg \psi_s \leq n$, which contradicts our assumption about ψ. Therefore $\tau \otimes L_e$ is anisotropic. Hence, $\varphi \otimes L_e$ has kernel form $(a\tau) \otimes L_e$, and it follows that $\deg \varphi \leq n$.

Assume that φ has degree $m < n$. Let $(K_j \,|\, 0 \leq j \leq h)$ be a generic splitting tower of φ. Now $\varphi \otimes K_{h-1}$ is of the form $b\rho$ with $b \in K_{h-1}^*$ and ρ a Pfister form of degree m over K_{h-1}. We have

$$\psi \otimes K_{h-1} \sim b\rho \perp (-a)(\tau \otimes K_{h-1}).$$

The form on the right has dimension $2^m + 2^n < 2^{n+1}$. From $\deg \psi \geq n + 1$ it follows that $\psi \otimes K_{h-1} \sim 0$.

We obtain $b\rho \sim a(\tau \otimes K_{h-1})$. Since $\dim \rho < \dim \tau$, it follows that $\tau \otimes K_{h-1} \sim 0$, and so $\rho \sim 0$, a contradiction. Thus φ has degree n.

(*ii*) We have $\psi \otimes K_h \sim (-a\tau) \otimes K_h$ and $\deg \psi > n$. Thus $\psi \otimes K_h \sim 0$ and so $\tau \otimes K_h \sim 0$. Let s be maximal in $[0, h-1]$ with $\tau \otimes K_s \not\sim 0$, so that $\tau \otimes K_s$ is anisotropic. From the Norm Theorem and the Subform Theorem it follows again that $b\varphi_s < \tau \otimes K_s$ for some $b \in K_s^*$. Since φ has degree n we must have $s = h - 1$ and

$$\varphi_{h-1} \cong b(\tau \otimes K_{h-1}).$$

Thus $\tau \otimes K_{h-1}$ is the leading form of φ over K_{h-1}. Furthermore,

$$\psi \otimes K_{h-1} \sim [\varphi \perp (-a\tau)] \otimes K_{h-1} \sim \langle b, -a \rangle \otimes \tau \otimes K_{h-1}.$$

If $\deg \psi \geq n + 2$, then $\psi \otimes K_{h-1} \sim 0$ and so

$$\varphi_{h-1} \cong (a\tau) \otimes K_{h-1}. \qquad \square$$

It is already clear from Definition 3.83 that every form which is Witt equivalent to φ is of the same degree as φ. Therefore we have a degree function

$$\deg : Wq(k) \longrightarrow \mathbb{N}_0 \cup \{\infty\}$$

---

[18] For $n \in \mathbb{N}$ we denote the set of numbers $0, 1, 2, \ldots, n$ by $[0, n]$.

on the Witt group $Wq(k)$ of strictly regular forms over k, given by $\deg(\{\varphi\}) := \deg \varphi$. In char $k = 2$ this function has values in $\mathbb{N} \cup \{\infty\}$.

The degree function motivates a further definition.

Definition 3.87. For every $n \in \mathbb{N}$ we denote by $J_n(k)$ the set of all Witt classes ξ of strictly regular forms with $\deg(\xi) \geq n$.

What does this notation mean for $n = 1$? Clearly $J_1(k)$ is the set of Witt classes of strictly regular forms of even dimension. If char $k = 2$, we thus have $J_1(k) = Wq(k)$. If char $k \neq 2$, then $Wq(k) = W(k)$ is a ring and $J_1(k)$ is an ideal of $W(k)$, namely the kernel of the dimension index

$$\nu : W(k) \longrightarrow \mathbb{Z}/2, \quad \{\varphi\} \longmapsto \dim \varphi + 2\mathbb{Z},$$

which already made an appearance in §1.2. This homomorphism ν is also available in characteristic 2. Its kernel is denoted by $I(k)$ and is called the *fundamental ideal* of $W(k)$. Thus $W(k)/I(k) = \mathbb{Z}/2$. In char $k \neq 2$ we have $I(k) = J_1(k)$.

In more generality we have:

Theorem 3.88. *For every* $n \in \mathbb{N}$, $J_n(k)$ *is a* $W(k)$-*submodule of* $Wq(k)$.

Proof (as in [33, p. 89 ff.]).

(a) We want to show that $J_n(k)$ is closed under addition in $Wq(k)$. We will prove the following equivalent statement: for any two strictly regular forms φ_1, φ_2 over k we have

$$\deg(\varphi_1 \perp \varphi_2) \geq \min(\deg \varphi_1, \deg \varphi_2). \tag{3.12}$$

This statement is trivial when φ_1 or φ_2 has odd dimension, and also when $\varphi_1 \perp \varphi_2 \sim 0$. In the following we exclude these cases.

Let $n := \deg(\varphi_1 \perp \varphi_2)$. There exists a field extension L of k such that

$$\ker(\varphi_1 \otimes L \perp \varphi_2 \otimes L) = a\rho,$$

for some $a \in L^*$ and a Pfister form ρ of degree n. By definition of the degree function we have

$$\deg(\varphi_i \otimes L) \geq \deg \varphi_i \quad (i = 1, 2).$$

Therefore it suffices to prove the statement for the forms $\widetilde{\varphi}_i := \varphi_i \otimes L$ instead of φ_i. If $\deg \widetilde{\varphi}_2 > n$, then the Witt equivalence

$$\widetilde{\varphi}_1 \sim a\rho \perp (-\widetilde{\varphi}_2)$$

and Theorem 3.86 give us that $\widetilde{\varphi}_1$ has degree n. Thus we always have

$$\min(\deg \widetilde{\varphi}_1, \deg \widetilde{\varphi}_2) \leq n.$$

(b) $W(k)$ is additively generated by the one-dimensional bilinear forms $\langle a \rangle$. Since for every strictly regular form φ, the form $a\varphi$ is of the same degree as φ, it is clear that every $J_n(k)$ is stable under multiplication with elements from $W(k)$. $\square$

An immediate consequence of Theorem 3.88 is

Corollary 3.89. *Let φ and ψ be strictly regular forms over k with $\deg \varphi \neq \deg \psi$. Then*

$$\deg(\varphi \perp \psi) = \min(\deg \varphi, \deg \psi).$$

It is commonplace to denote the nth power of the ideal $I(k)$ by $I^n(k)$ (instead of $I(k)^n$). It is also customary to leave out the curly brackets $\{,\}$, which indicate that not the form itself, but rather its Witt class, is considered. For example, one writes $\gamma \in I^n(k)$ to indicate that the Witt class of a nondegenerate bilinear form γ is in $I^n(k)$. We will follow this custom.

The ideal $I(k)$ is additively generated by the bilinear forms $a\langle 1, b \rangle$ where $a, b \in k^*$. Thus $I^n(k)$ is additively generated by the forms $a\tau$ where $a \in k^*$ and τ is an n-fold bilinear Pfister form which we may in addition assume to be anisotropic. This is also true for $n = 0$, if we formally write $I^0(k) = W(k)$.

The $W(k)$-module $J_1(k)$ is additively generated by the binary forms $a \begin{bmatrix} 1 & 1 \\ 1 & b \end{bmatrix}$. Thus the $W(k)$-module $I^n(k)J_1(k)$ is additively generated by the forms $a\tau$, where $a \in k^*$ and τ is a quadratic anisotropic Pfister form of degree $n + 1$. Thus, by Theorem 3.88, every element of $I^n(k)J_1(k)$ is of degree $\geq n + 1$. Hence we obtain

Corollary 3.90. *For every $n \in \mathbb{N}_0$ we have*

$$I^n(k)J_1(k) \subset J_{n+1}(k).$$

Arason and Pfister already showed in 1970, without a generic splitting theory (which did not exist yet at the time), that in characteristic $\neq 2$ every anisotropic form $\varphi \neq 0$ which lies in $I^n(k)$ is of dimension at least 2^n (the so-called Arason–Pfister Hauptsatz [1]). The proof of Arason–Pfister can be modified in such a way that it also works in characteristic 2, but we will not elaborate that here. {One should of course replace $I^n(k)$ by $I^{n-1}(k)J_1(k)$.} The statement of our Corollary 3.90 is equivalent to the Arason–Pfistert Hauptsatz, in every characteristic.

Next we wish to characterize the strictly regular even-dimensional forms φ of degree 1 by employing the discriminant algebra $\Delta(\varphi)$ (cf. the beginning of §2.5). We begin with some comments on this invariant of φ.

If L/k is a quadratic separable field extension, there exists a generator ω of L/k with minimal equation $\omega^2 + \omega = a$, where $1 + 4a \in k^*$. The norm form $N_{L/k}(x)$ of L/k has the value matrix $\begin{bmatrix} 1 & 1 \\ 1 & a \end{bmatrix}$ with respect to $1, \omega$. If $\operatorname{char} k \neq 2$ then $1, 1 + 2\omega$ is an orthogonal basis with respect to the norm form, which is thus of the form $\langle 1, -1 - 4a \rangle$. Also, $(1 + 2\omega)^2 = 1 + 4a$.

All of this remains basically true for $a = 0$ when L is equal to the split quadratic separable algebra $k \times k$. In this case the norm form is hyperbolic instead of anisotropic as before.

This can all be verified easily and so we obtain straight away the following elementary lemma.

Lemma 3.91. *The isomorphism classes of quadratic separable k-algebras L are in one-one correspondence with the isometry classes of binary $(= 2$-dimensional$)$ Pfister forms τ via $L = \Delta(\tau)$, $\tau = N_{L/k}$.*

We recall further that in characteristic 2 the discriminant algebra $\Delta(\varphi)$ is just the Arf-invariant in disguise (see the beginning of §2.5), while in characteristic $\neq 2$ it corresponds to the signed determinant $d(\varphi)$ since then $\langle 1, -d(\varphi) \rangle$ is the binary Pfister form associated to $\Delta(\varphi)$. Thus one verifies immediately:

Lemma 3.92. *Let φ_1 and φ_2 be two regular forms of even dimension.*
(a) $\Delta(\varphi_1 \perp \varphi_2) = 1$ if and only if $\Delta(\varphi_1) = \Delta(\varphi_2)$.
(b) $\Delta(\varphi_1 \perp \varphi_2) = \Delta(\varphi_1)$ if and only if $\Delta(\varphi_2) = 1$.

Remark 3.93. These statements can be placed on a more conceptual footing by endowing the set of isomorphism classes of quadratic separable k-algebras with a multiplication, which turns it into an abelian group $QS(k)$ of exponent 2, and observing that for regular forms φ_1, φ_2 of even dimension, $\Delta(\varphi_1 \perp \varphi_2) = \Delta(\varphi_1) \cdot \Delta(\varphi_2)$, cf. [8, Chap. IV], [9], [42, §10].

Theorem 3.94. *Let φ be a strictly regular form of even dimension. Let τ be the binary Pfister form over k with $\Delta(\varphi) = \Delta(\tau)$ (cf. Lemma 3.91).*
(a) If $\Delta(\varphi) = 1$, then $\deg \varphi \geq 2$.
(b) If $\Delta(\varphi) \neq 1$ (thus τ is anisotropic), then $\deg(\varphi) = 1$ and $\varphi \sim \tau \perp \psi$ for some strictly regular form ψ of degree ≥ 2. Furthermore, if F is a leading field for φ, then $\tau \otimes F$ is the associated leading form of φ.

Proof. Let F be a leading field for φ, and ρ the associated leading form of φ. Thus $\ker(\varphi \otimes F) = a\rho$ for some scalar $a \in F^*$. If $\Delta(\varphi) = 1$, then $\Delta(a\rho) = 1$. It follows that ρ, and thus also φ, is of degree ≥ 2.

Assume now that $\Delta(\varphi) \neq 1$. By Lemma 3.92 the form $\psi := \varphi \perp \tau$ has discriminant $\Delta(\psi) = 1$. Thus $\deg \psi \geq 2$, by the first part of the proof. From $\varphi \sim -\tau \perp \psi$ and Theorem 3.86 it follows that φ has degree 1 and leading form $\tau \otimes F$. $\qquad \square$

We note that only part (a) of the lemma was used. Part (b) will be needed for the first time only in §3.13.

It follows immediately from this theorem that $J_2(k)$ is the set of all Witt classes $\{\varphi\}$ of regular forms φ with $\dim \varphi$ even and $\Delta(\varphi) = 1$. By Remark 3.93 it follows further that the group $QS(k)$ is isomorphic to the additive group $J_1(k)/J_2(k)$. We will not need this in the sequel, but it is nevertheless good to know.

We return to an examination of the modules $J_n(k)$ for arbitrary n.

Theorem 3.95. *Let φ be a strictly regular quadratic form and α a nondegenerate bilinear form. Assume that α has odd dimension. Then*

$$\deg(\alpha \otimes \varphi) = \deg \varphi.$$

Proof. This is clear when the dimension of φ is odd, and also when $\varphi \sim 0$. Therefore we will exclude those cases. Thus let $h(\varphi) > 0$ and $n := \deg \varphi \geq 1$. By Theorem 3.88 we have $\deg(\alpha \otimes \varphi) \geq n$.

We choose a leading field F for φ and obtain $\varphi \otimes F \sim a\tau$ for some $a \in F^*$ and some n-fold anisotropic Pfister form τ. We also consider some decomposition

$$\alpha \cong \langle b \rangle \perp b_1\rho_1 \perp \cdots \perp b_r\rho_r$$

where $b, b_1, \ldots, b_r \in k^*$ and $\rho_1, \ldots, \rho_r$ are binary bilinear Pfister forms. Hence,

$$(\alpha \otimes \varphi) \otimes F \sim b\tau \perp b_1(\rho_1 \otimes \tau) \perp \cdots \perp b_r(\rho_r \otimes \tau),$$

where we shortened $(\rho_i \otimes F) \otimes \tau$ to $\rho_i \otimes \tau$. The $\rho_i \otimes \tau$ are $(n + 1)$-fold quadratic Pfister forms. By Theorem 3.88, $b_1(\rho_1 \otimes \tau) \perp \cdots \perp b_r(\rho_r \otimes \tau)$ has degree $\geq n + 1$. By Theorem 3.86 (or Corollary 3.89), $(\alpha \otimes \varphi) \otimes F$ has degree n. It follows that $\deg(\alpha \otimes \varphi) \leq n$ and so $\deg(\alpha \otimes \varphi) = n$. $\square$

We record a much weaker, but still interesting, version of this theorem:

Corollary 3.96. *Let α be a bilinear form of odd dimension and φ a quadratic strictly regular form with $\varphi \nsim 0$. Then $\alpha \otimes \varphi \nsim 0$.*

Theorem 3.97. *For arbitrary $m \geq 1$, $n \geq 1$ we have*

$$I^m(k)J_n(k) \subset J_{m+n}(k).$$

Proof (cf. [33, p. 91]). Of course it suffices to show that $I(k)J_n(k) \subset J_{n+1}(k)$. Since $I(k)$ is generated by the forms $\langle 1, a \rangle$ as an ideal, the theorem boils down to the following statement: if φ is a strictly regular quadratic form of even dimension over k which is not hyperbolic, then, for given $a \in k^*$, the degree of $\langle 1, a \rangle \otimes \varphi$ is larger than the degree of φ.

We prove this statement by induction on $h(\varphi)$. The case $h(\varphi) = 1$ is clear. Let $h(\varphi) > 1$. If the form $\alpha := \langle 1, a \rangle \otimes \varphi$ splits, the statement is true. Assume now that $\alpha \nsim 0$. We choose a leading field F for α. Let ρ be the associated leading form of α over F. Since $\deg(\varphi \otimes F) \geq \deg \varphi$ it suffices to show that $\deg \rho > \deg(\varphi \otimes F)$. We have $h(\varphi \otimes F) \leq h(\varphi)$. If we replace k by F and φ by $\varphi \otimes F$, we are reduced to the case

$$\langle 1, -a \rangle \otimes \varphi \sim \rho \tag{3.13}$$

for some anisotropic Pfister form ρ. Now we must show that $\deg \rho > \deg \varphi$.

Assume next that $\rho \otimes k(\varphi)$ is anisotropic. We have

$$\langle 1, -a \rangle \otimes (\varphi \otimes k(\varphi)) \sim \rho \otimes k(\varphi)$$

and $\deg(\varphi \otimes k(\varphi)) = \deg \varphi$, $\deg(\rho \otimes k(\varphi)) = \deg(\rho)$, $h(\varphi \otimes k(\varphi)) = h(\varphi) - 1$. Thus $\deg \rho > \deg \varphi$ by the induction hypothesis.

The remaining case is when $\rho \otimes k(\varphi)$ splits. By the Norm Theorem and the Subform Theorem we have

$$\rho \cong b\varphi \perp \zeta, \tag{3.14}$$

for some $b \in k^*$ and some form ζ over k. Assume that $\zeta = 0$. Then

$$\langle 1, -a, -b \rangle \otimes \varphi \sim 0$$

by (3.13) and (3.14), contradicting the corollary to the previous theorem. Hence $\zeta \neq 0$, and so $\dim \varphi < \dim \rho$. Since ρ is a Pfister form, it follows that $\deg \varphi < \deg \rho$. $\square$

Nowadays it is known that for every natural number n,

$$J_n(k) = I^n(k)$$

when char $k \neq 2$ [51], and

$$J_n(k) = I^{n-1}(k)Wq(k)$$

when char $k = 2$ [2]. Thus, in Theorem 3.97 we have, *de facto*, $I^m(k)J_n(k) = J_{n+m}(k)$.

The proofs of these statements are very dissimilar, and go far beyond the scope of this book. The case characteristic $\neq 2$ seems to be the more difficult one.

3.12 The Companion Form of an Odd-dimensional Regular Form

In this section k is an arbitrary field and φ *is always a regular form of odd dimension over k*.

If char $k \neq 2$, then φ is strictly regular. In characteristic 2, however, φ has a one-dimensional quasilinear part, $QL(\varphi) = [a]$. In characteristic $\neq 2$ we also interpret φ as a bilinear form in the manner indicated at the beginning of §3.11.

Theorem 3.98. *Up to isometry there is a unique form ψ over k with the following four properties:*
(1) *ψ is strictly regular.*
(2) *$\Delta(\psi) = 1$.*
(3) *$\varphi < \psi$.*
(4) *$\dim \psi = \dim \varphi + 1$.*

Proof. If char $k \neq 2$, we need $\psi = \varphi \perp \langle b \rangle$ for some $b \in k^*$ such that the condition $d(\psi) = 1$ is satisfied. So we take b such that $\langle b \rangle = -d(\varphi)$.

Now let char $k = 2$. We choose an orthogonal decomposition $\varphi := \chi \perp [a]$. Thus χ is strictly regular and $QL(\varphi) = [a]$. We need $\psi = \chi \perp \alpha$ for some binary strictly regular form α, to be determined. The condition $\chi \perp [a] < \psi$ is equivalent with $[a] < \alpha$ (cf. Theorem 3.33). Thus, by Lemma 3.92(a), α should be chosen in such a way that $a \in D(\alpha)$ and $\Delta(\alpha) = \Delta(\chi)$. By Lemma 3.91, there is exactly one binary Pfister form τ with $\Delta(\tau) = \Delta(\chi)$. We must take $\alpha = a\tau$. $\square$

Definition 3.99. We call the form ψ, described in Theorem 3.98, the *companion* of φ. We denote this form, which is uniquely determined by φ, by $Com(\varphi)$.

Remark. The mapping $\varphi \mapsto Com(\varphi)$ satisfies the following rules:
(1) If α is a strictly regular form with $\Delta(\alpha) = 1$, then $Com(\varphi \perp \alpha) = Com(\varphi) \perp \alpha$. In particular, $Com(\gamma) \sim Com(\varphi)$ for every regular form $\gamma \sim \varphi$.

(2) $\text{Com}(a\varphi) = a\text{Com}(\varphi)$ for every $a \in k^*$.

(3) $\text{Com}(\varphi \otimes L) = \text{Com}(\varphi) \otimes L$ for every field extension $L \supset k$.

(4) If char $k \neq 2$, then $\text{Com}(\varphi) = \varphi \perp -d(\varphi)$.

Proof. The rules (1) – (3) follow immediately from the definition of a companion. Rule (4) is clear from the proof of Theorem 3.98 above. $\square$

Conversely, every strictly regular form ψ of even dimension and discriminant 1 generally acts as the companion of many regular forms φ. More precisely the following theorem holds, which can easily be verified upon consulting the explanations about polars in § 3.2.

Theorem 3.100.

(i) *Let ψ be a strictly regular form of even dimension over k with $\Delta(\psi) = 1$. For every $a \in D(\psi)$ the polar $\varphi := \text{Pol}_\psi([a])$ (cf. §3.2) is a strictly regular form of odd dimension with $\text{Com}(\varphi) = \psi$. If k is of characteristic 2, then $[a] = QL(\varphi)$. If k is of characteristic $\neq 2$, then $\langle a \rangle = -d(\varphi)$.*

(ii) *The isometry classes (φ) of strictly regular forms φ of odd dimension over k are in one–one correspondence with the pairs $\big((\psi), [a]\big)$, where (ψ) is the isometry class of a strictly regular form of even dimension and discriminant 1 (thus $\{\psi\} \in J_2(k)$) and $a \in D(\psi)$, via $\varphi = \text{Pol}_\psi[a]$ and $\psi = \text{Com}(\varphi)$, $[a] = QL(\varphi)$ for char $k = 2$, $\langle a \rangle = -d(\varphi)$ for char $k \neq 2$.*

The following theorem gives another method for retrieving φ from ψ and $d(\varphi)$ from $QL(\varphi)$, respectively, by means of the Cancellation Theorem (Theorem 1.67— for fields).

Theorem 3.101. *Let $\psi = \text{Com}(\varphi)$.*

(a) *If char $k \neq 2$, then $\varphi \perp H \cong \psi \perp d(\varphi)$.*

(b) *If char $k = 2$, then $\varphi \perp H \cong \psi \perp QL(\varphi)$.*

Proof. (a) This is clear since $\psi = \varphi \perp -d(\varphi)$ in this case.

(b) Since $QL(\varphi) = [a]$ we have

$$\varphi \perp H \cong \text{Pol}_\psi([a]) \perp H \cong \text{Pol}_{\psi \perp H}([a]).$$

Furthermore, $\psi \perp H \cong \psi \perp \left[\begin{smallmatrix} a & 1 \\ 1 & 0 \end{smallmatrix}\right]$. Thus,

$$\text{Pol}_{\psi \perp H}([a]) \cong \psi \perp [a]. \qquad \square$$

Next we present necessary and sufficient conditions for the companion $\text{Com}(\varphi)$ to be isotropic or split.

Theorem 3.102. *The following statements are equivalent:*

(i) $\text{Com}(\varphi)$ *is isotropic.*

(ii) *There exists an element $b \in k^*$ and a strictly regular form χ of even dimension with $\Delta(\chi) = 1$ and $\varphi \cong \chi \perp [b]$.*

(iii) *There exists a strictly regular form* χ *of even discriminant with* $\Delta(\chi) = 1, \chi < \varphi$ *and* $1 + \dim \chi = \dim \varphi$.

If (i) – (iii) *hold, then* $\mathrm{Com}(\varphi) = \chi \perp H$ *and, furthermore,* $[b] = QL(\varphi)$ *when* char $k = 2$ *and* $\langle b \rangle = d(\varphi)$ *when* char $k \neq 2$.

Proof. If χ is a strictly regular form with $\chi < \varphi$, $\dim \chi = \dim \varphi - 1$, then we have a decomposition $\varphi \cong \chi \perp [b]$ for some $b \in k^*$. This establishes the equivalence (ii) $\Leftrightarrow$ (iii). We also see that then $QL(\varphi) = [b]$ in characteristic 2 and $d(\varphi) = \langle b \rangle$ in characteristic $\neq 2$.

Let $\psi := \mathrm{Com}(\varphi)$. By Theorem 3.101 we have an isometry $\varphi \perp H \cong \psi \perp [b]$. If ψ is isotropic, we have in addition a decomposition $\psi \cong \chi \perp H$ and we have $\Delta(\chi) = 1$. It follows that $\varphi \cong \chi \perp [b]$. Conversely, if $\varphi \cong \chi \perp [b]$ with $\Delta(\chi) = 1$, then the form $\chi \perp H \cong \chi \perp \begin{bmatrix} b & 1 \\ 1 & 0 \end{bmatrix}$ satisfies the conditions of Theorem 3.98, and so $\chi \perp H \cong \psi$.   □

Theorem 3.103. $\mathrm{Com}(\varphi)$ *splits if and only if* φ *splits.*

Proof. If $\varphi \cong r \times H \perp [a]$, then the form $\psi := r \times H \perp \begin{bmatrix} a & 1 \\ 1 & 0 \end{bmatrix} \cong (r+1) \times H$ satisfies the four conditions of Theorem 3.98. Hence, $\psi = \mathrm{Com}(\varphi)$. Conversely, if $\mathrm{Com}(\varphi)$ is hyperbolic, then $\varphi \perp H$ splits by Theorem 3.101. This implies that φ splits.   □

As a digression we remark that the unicity statement contained in Theorem 3.102 leads to a new kind of cancellation theorem:

Theorem 3.104. *Let* $a \in k$ *and let* χ_1, χ_2 *be two strictly regular forms of even dimension with* $\Delta(\chi_1) = \Delta(\chi_2)$ *and* $\chi_1 \perp [a] \cong \chi_2 \perp [a]$. *Then* $\chi_1 \cong \chi_2$.

Proof. This is trivial when $a = 0$. Thus let $a \neq 0$. If $\Delta(\chi_1) = \Delta(\chi_2) = 1$, the statement follows immediately from Theorem 3.102. If only $\Delta(\chi_1) = \Delta(\chi_2)$, then $\Delta(\chi_1 \perp \chi_2) = \Delta(\chi_2 \perp \chi_2) = 1$ by Lemma 3.92. We have $\chi_1 \perp \chi_2 \perp [a] \cong \chi_2 \perp \chi_2 \perp [a]$, and deduce that $\chi_1 \perp \chi_2 \cong \chi_2 \perp \chi_2$, and finally that $\chi_1 \cong \chi_2$ with the old Cancellation Theorem, Theorem 1.67.   □

We make a general statement about the Witt indices of φ and $\mathrm{Com}(\varphi)$:

Theorem 3.105. *We have*

$$\mathrm{ind}(\varphi) \leq \mathrm{ind}(\mathrm{Com}(\varphi)) \leq 1 + \mathrm{ind}(\varphi).$$

Proof. Let $\psi := \mathrm{Com}(\varphi)$. Since $\varphi < \psi$ we have $\mathrm{ind}(\varphi) \leq \mathrm{ind}(\psi)$. On the other hand we have an isometry $\varphi \perp H \cong \psi \perp [a]$ by Theorem 3.101. Hence

$$\mathrm{ind}(\psi) \leq \mathrm{ind}(\psi \perp [a]) = 1 + \mathrm{ind}(\varphi).$$ □

Theorem 3.106. $h(\varphi) \leq 2h(\mathrm{Com}(\varphi))$.

Proof. Let $\psi := \mathrm{Com}(\varphi)$, $h := h(\varphi)$, $e := h(\psi)$. Then $h + 1$ is the number of possible Witt indices of $\varphi \otimes K$, where K traverses all field extensions of k, thus $h + 1 = \mathrm{card}\,\mathrm{SP}(\varphi)$. Similarly, $e + 1 = \mathrm{card}\,\mathrm{SP}(\psi)$. Furthermore we have $\mathrm{Com}(\varphi \otimes K) = \psi \otimes K$. If $\mathrm{ind}(\psi \otimes K) = j$, then $\mathrm{ind}(\varphi \otimes K) = j$ or $j + 1$ by Theorem 3.105. However, if $\mathrm{ind}(\psi \otimes K)$ has the highest possible value $\frac{1}{2} \dim \psi$, then $\mathrm{ind}(\varphi \otimes K)$ also has the highest possible value $\frac{1}{2} \dim \psi - 1$ by Theorem 3.103. It follows that $h + 1 \leq 2e + 1$, and so $h \leq 2e$. $\square$

In §3.11 we associated to every strictly regular form, which is not split, a *leading form*. We will now also introduce a leading form for regular forms of odd dimension, in characteristic 2. First some preliminaries.

Definition 3.107. Let τ be a (quadratic) Pfister form over k. Since $1 \in D(\tau)$ we can form the polar $\mathrm{Pol}_\tau[1]$, which we call the *unit polar of* τ and denote by τ'.

Remark 3.108.
(1) If $\dim \tau = 2$ and $\mathrm{char}\,k = 2$, then $\tau' = [1]$, and forming the unit polar is not interesting. On the other hand, if $\dim \tau \geq 4$, then clearly τ is the strictly regular companion of τ' and τ' is the regular form of odd dimension associated to the pair $(\tau, [1])$ in the sense of Theorem 3.100.
(2) For an arbitrary Pfister form σ over k we introduced the *pure part* of σ, denoted by σ', earlier (§3.2). In $\mathrm{char}\,k \neq 2$ this pure part is identical to the unit polar.

It stands to reason to look for a relationship between the generic splitting behaviour of the form φ and its companion $\mathrm{Com}(\varphi) = \psi$, and in particular to compare the leading forms of φ and ψ (cf. Definition 3.84). A result in this vein, presented below, can already be found in [33, §5] for the characteristic $\neq 2$ case. A partial result in characteristic $\neq 2$ was anticipated in Theorem 3.85.

Theorem 3.109. *Let φ be a regular form of odd dimension over k and $\psi := \mathrm{Com}(\varphi)$ its companion. Let $(K_r \mid 0 \leq r \leq h)$ be a generic splitting tower of φ and $(L_s \mid 0 \leq s \leq e)$ a generic splitting tower of ψ. We assume that φ, and so (by Theorem 3.103) also ψ, does not split. Let τ be the leading form of φ over K_{h-1} and σ the leading form of ψ over L_{e-1}. Then we claim:*
   (i) $K_h \sim_k L_e$.
   (ii) There is a place $\lambda : L_{e-1} \to K_{h-1} \cup \infty$ over k.
   (iii) σ has good reduction with respect to every such place λ, and $\lambda_*(\sigma) \cong \tau$. In particular, $\deg \tau = \deg \sigma = \deg \psi$.

Proof. For every field extension L of k we have $\mathrm{Com}(\varphi \otimes L) = \psi \otimes L$. Thus $\psi \otimes L$ splits if and only if $\varphi \otimes L$ splits by Theorem 3.103. Hence K_h and L_e are specialization equivalent over k.

The kernel form φ_{h-1} of $\varphi \otimes K_{h-1}$ is of the form $a\tau'$ for some $a \in K_{h-1}^*$. {Incidentally, a can be chosen in k, but we do not need that now.} We have

$$\psi \otimes K_{h-1} = \mathrm{Com}(\varphi \otimes K_{h-1}) \sim \mathrm{Com}(\varphi_{h-1}) = a\tau. \tag{3.15}$$

It follows that $\deg \tau \geq \deg \psi$. On the other hand, the kernel form ψ_{e-1} of $\psi \otimes L_{e-1}$ is of the form $b\sigma$ for some $b \in L_{e-1}^*$. Hence, $\varphi \otimes L_{e-1} \sim b\sigma \perp [a]$, where $[a] = QL(\varphi)$ in characteristic 2 and $[a] = d(\varphi)$ in characteristic $\neq 2$. Since the form σ is anisotropic and of dimension at least 4, $b\sigma \perp [a]$ cannot split.

It follows from the general generic splitting theory that $\dim \sigma + 1 \geq \dim \tau'$, hence $\dim \sigma + 2 \geq \dim \tau$. Earlier we established that $\dim \sigma \leq \dim \tau$. Since $\dim \sigma \geq 4$ and the integers $\dim \sigma$ and $\dim \tau$ are both 2-powers, it follows that $\dim \sigma = \dim \tau$. Now (3.15) tells us, by the generic splitting theory, that there exists a place λ from L_{e-1} to K_{h-1} over k, and furthermore that $b\sigma$ has good reduction with respect to every such place λ and that $\lambda_*(b\sigma) = a\tau$. Since $b\sigma$ has good reduction with respect to λ, $b\sigma$ will certainly represent a unit c in the valuation ring $\mathfrak{o}_\lambda$. Then we have $b\sigma \cong c\sigma$, and it follows that σ has good reduction with respect to λ, and that $\lambda_*(\sigma) = \lambda(c)a\tau$. This forces $\lambda_*(\sigma) = \tau$.   □

As an illustration of the results obtained so far, we classify the anisotropic regular forms of odd dimension whose strictly regular companion is of height 1.

Theorem 3.110. *Let φ be anisotropic. The companion $\psi := \mathrm{Com}(\varphi)$ is of height 1 if and only if one of the following two cases happens:*

(A) $\varphi \cong a\tau'$ *for some anisotropic Pfister form τ of degree ≥ 2 and some $a \in k^*$.*
(B) $\varphi \cong a\tau \perp [b]$ *for some anisotropic Pfister form τ of degree ≥ 2, some $a \in k^*$ and $b \in k^*$ with $-ab \notin D(\tau)$.*

In case (A) we have $\psi \cong a\tau$, and so ψ is anisotropic. In case (B) we have $\psi \cong a\tau \perp H$, and so ψ is isotropic. In case (A) φ is of height 1, in case (B) of height 2. In both cases φ is excellent.

Proof. (1) Let τ be a Pfister form of degree ≥ 2 and let a and b be elements of k^*. If $\varphi \cong a\tau'$, then $\psi = a\tau$, and so $h(\psi) = 1$. Furthermore we have $h(\varphi) = 1$ in this case. If $\varphi \cong a\tau \perp [b]$ and $-ab \notin D(\tau)$, then φ is anisotropic, and $\psi \cong a\tau \perp \begin{bmatrix} b & 1 \\ 1 & 0 \end{bmatrix} \cong a\tau \perp H$. Therefore ψ is also of height 1 in this case. Furthermore,

$$\varphi \perp b\tau' \cong a\langle 1, ab \rangle \otimes \tau.$$

Thus φ is a neighbour of the anisotropic Pfister form $\langle 1, ab \rangle \otimes \tau$ with complementary form $b\tau'$. Hence φ is excellent of height 2.

(2) Assume now that $h(\psi) = 1$. By Theorem 3.105 we have $\mathrm{ind}(\psi) \leq 1$.

Case (A): ψ is anisotropic. Then $\psi \cong c\tau$ for some $c \in k^*$ and some Pfister form τ of degree ≥ 2. If $\mathrm{char}\, k = 2$, and $QL(\varphi) = [a]$, then we may choose $c = a$ and have $\varphi = \mathrm{Pol}_\psi([a]) \cong a\tau'$. If $\mathrm{char}\, k \neq 2$, then φ is a strictly regular subform of ψ, and so we obtain again that $\varphi \cong a\tau'$ for some $a \in k^*$.

Case (B): ψ is isotropic. Then $\psi \cong a\tau \perp H$ for some $a \in k^*$ and some Pfister form τ of degree 2. By Theorem 3.101 we have $\varphi \perp H \cong \psi \perp [b]$ for some $b \in k^*$. Hence, $\varphi \cong a\tau \perp [b]$. Since φ is anisotropic it follows that $-ab \notin D(\tau)$.   □

Next we will more closely consider and compare the generic splitting towers of φ and $\psi := \mathrm{Com}(\varphi)$. Assume as before that $\dim \varphi$ is odd. Let further ak^{*2} be the square class with $d(\varphi) = \langle -a \rangle$ in case char $k \neq 2$, and $QL(\varphi) = [a]$ in case char $k = 2$. Thus $a \in D(\psi)$ and $\varphi = \mathrm{Pol}_\psi([a])$. We need two lemmata.

Lemma 3.111. *Let* $\varphi_0 := \ker(\varphi)$ *and* $\psi_0 := \ker(\psi)$. *The following statements are equivalent:*

   (i) $\mathrm{Com}(\varphi_0) = \psi_0$.
   (ii) $\mathrm{Com}(\varphi_0)$ *is anisotropic.*
   (iii) $\mathrm{ind}(\psi) = \mathrm{ind}(\varphi)$.
   (iv) $a \in D(\psi_0)$.

Proof. Let $r := \mathrm{ind}(\varphi)$, hence $\varphi \cong \varphi_0 \perp r \times H$ and $\chi := \mathrm{Com}(\varphi_0)$. From Remark 3.108(1) above it follows that

$$\psi \cong \chi \perp r \times H.$$

This is the Witt decomposition of ψ if and only if $\mathrm{ind}(\psi) = r$ or, equivalently, if and only if χ is anisotropic. This already proves the equivalence of $(i) - (iii)$.

   Furthermore, $d(\varphi_0) = d(\varphi) = \langle -a \rangle$ if char $k \neq 2$, and $QL(\varphi_0) = QL(\varphi) = [a]$ if char $k = 2$. Thus $a \in D(\chi)$ and $\varphi_0 = \mathrm{Pol}_\chi([a])$. If $\chi = \psi_0$, then certainly $a \in D(\psi_0)$. This establishes $(i) \Rightarrow (iv)$.

   Finally, assume that $a \in D(\psi_0)$. Of course $\psi_0 \sim \chi$. Hence, $\mathrm{Pol}_{\psi_0}([a]) \sim \mathrm{Pol}_\chi[a] = \varphi_0$, and so $\varphi_0 \cong \mathrm{Pol}_{\psi_0}([a])$, since the forms φ_0 and $\mathrm{Pol}_{\psi_0}([a])$ are both anisotropic. Therefore, $\psi_0 = \mathrm{Com}(\varphi_0)$. This shows $(iv) \Rightarrow (i)$.   □

   We assume from now on that φ, and thus ψ, is not split.

Lemma 3.112. *Let* $\psi = \mathrm{Com}(\varphi)$ *be anisotropic. Let* ψ_1 *denote the kernel form of* $\psi \otimes k(\psi)$.
   (i) *There exists a place from* $k(\psi)$ *to* $k(\varphi)$ *over* k.
   (ii) $k(\varphi) \sim_k k(\psi)$ *if and only if* $a \in D(\psi_1)$ *or* $\mathrm{ind}(\psi \otimes k(\psi)) \geq 2$.
   (iii) *If* $k(\varphi)$ *is not specialization equivalent to* $k(\psi)$ *over* k, *then* $\varphi \otimes k(\psi)$ *is anisotropic.*

Proof. (i) is clear since φ is a subform of ψ. Let $K := k(\psi)$. We have $\mathrm{Com}(\varphi \otimes K) = \mathrm{Com}(\psi) \otimes K$. By the general theory of generic splitting it is clear that $k(\varphi)$ is specialization equivalent to K over k if and only if $\varphi \otimes K$ is isotropic.

   A priori we have $\mathrm{ind}(\varphi \otimes K) = \mathrm{ind}(\psi \otimes K)$ or $\mathrm{ind}(\varphi \otimes K) = \mathrm{ind}(\psi \otimes K) - 1$ (by Theorem 3.105). If $\mathrm{ind}(\psi \otimes K) \geq 2$, then $\varphi \otimes K$ is definitely isotropic. Assume now that $\mathrm{ind}(\psi \otimes K) = 1$, thus $\psi_K \cong \psi_1 \perp H$. The form $\varphi \otimes K$ is of index 0 or 1. By Lemma 3.111, $\varphi \otimes K$ is of index 1 if and only if $a \in D(\psi_1)$. This happens if and only if $\varphi \otimes K$ is isotropic.   □

   Let $h := h(\varphi)$ and $e := h(\psi)$. Further, let $(L_s \mid 0 \leq s \leq e)$ be a generic splitting tower of $\psi = \mathrm{Com}(\varphi)$, and let $\psi_s := \ker(\psi \otimes L_s)$, $\eta_s := \ker(\varphi \otimes L_s)$. We know that $\varphi \otimes L_e$ splits, so that η_e is one-dimensional and thus $\eta_e = [-a]$.

   We are looking for an integer $m \in [0, e]$, as large as possible, such that $(L_s \mid 0 \leq s \leq m)$ is a *truncated generic splitting tower* of φ, which signifies that there is a generic splitting tower $(K_r \mid 0 \leq r \leq h)$ of φ with $h \geq m$ and $K_r = L_r$ for $0 \leq r \leq m$.

Theorem 3.113. *Let* $a \in D(\psi_m)$ *for some* $m \in [0, e-1]$.

(a) $(L_s \mid 0 \le s \le m)$ *is a truncated generic splitting tower of* φ. *Furthermore,* $\mathrm{ind}(\varphi \otimes L_s) = \mathrm{ind}(\psi \otimes L_s)$ *and* $\mathrm{Com}(\eta_s) = \psi_s$ *for* $0 \le s \le m$.

(b) *If, in addition,* $a \notin D(\psi_{m+1})$, *then* $\mathrm{ind}(\varphi \otimes L_{m+1}) = \mathrm{ind}(\psi \otimes L_{m+1}) - 1$ *and* $\mathrm{Com}(\eta_{m+1}) \cong \psi_{m+1} \perp H$. *The following alternative also holds: if* $\mathrm{ind}(\psi_m \otimes L_{m+1}) \ge 2$, *then* $(L_s \mid 0 \le s \le m+1)$ *is a truncated generic splitting tower of* φ. *On the other hand, if* $\mathrm{ind}(\psi_m \otimes L_{m+1}) = 1$, *then* $\eta_m \otimes L_{m+1}$ *is anisotropic, and so* $\eta_m \otimes L_{m+1} = \eta_{m+1}$ *and*

$$\mathrm{ind}(\varphi \otimes L_{m+1}) = \mathrm{ind}(\varphi \otimes L_m) = \mathrm{ind}(\psi \otimes L_m).$$

Proof. By induction on m. Lemma 3.111 and Lemma 3.112 give us the case $m = 0$. Thus let $m > 0$. From the induction hypothesis we know that $(L_s \mid 0 \le s \le m - 1)$ is a truncated generic splitting tower of φ and that $\mathrm{ind}(\varphi \otimes L_s) = \mathrm{ind}(\psi \otimes L_s)$ and $\mathrm{Com}(\eta_s) = \psi_s$ for $0 \le s \le m - 1$. Let $K := L_{m-1}$. By assumption we have $L_m \sim_K K(\psi_{m-1})$ and $a \in D(\psi_m)$. It follows that the element a is also represented by $\ker(\psi_{m-1} \otimes K(\psi_{m-1}))$. Furthermore, $\psi_{m-1} = \mathrm{Com}(\eta_{m-1})$. From Lemma 3.112 it follows that $K(\psi_{m-1}) \sim_K K(\eta_{m-1})$, and so $L_m \sim_K K(\eta_{m-1})$. Hence, $(L_s \mid 0 \le s \le m)$ is a truncated generic splitting tower of φ. Applying Lemma 3.111 to $\eta_{m-1} \otimes L_m$ further shows that $\mathrm{Com}(\eta_m) = \psi_m$, and finally that $\eta_{m-1} \otimes L_m$ and $\psi_{m-1} \otimes L_m$ have the same index. By the induction hypothesis $\varphi \otimes L_{m-1}$ and $\psi \otimes L_{m-1}$ also have the same index. Therefore $\varphi \otimes L_m$ and $\psi \otimes L_m$ have the same index. This proves part (a) of the theorem. Applying Lemma 3.112 again, but this time to η_m and $\psi_m = \mathrm{Com}(\eta_m)$, gives part (b). $\square$

Theorem 3.114. *We have* $h \ge e$. {*Recall that* $h := h(\varphi)$, $e := h(\psi)$. *We assume that* $h > 0$, $e > 0$.} *If* $a \notin D(\psi_0)$, *then we even have* $h \ge e + 1$. *The following statements are equivalent:*

(i) $a \in D(\psi_{e-1})$.

(ii) $(L_s \mid 0 \le s \le e)$ *is a generic splitting tower of* φ. *(In particular,* $h = e$.)

(iii) $\mathrm{Com}(\eta_s) = \psi_s$ *for* $0 \le s \le e - 1$.

(iv) $\mathrm{ind}(\varphi \otimes L_s) = \mathrm{ind}(\psi \otimes L_s)$ *for* $0 \le s \le e - 1$.

(v) L_{e-1} *is a leading field for* φ.

Proof. We distinguish three cases.

Case 1: $a \notin D(\psi_0)$. For every $s \in [0, e]$ we establish the following. We have $a \notin D(\psi_s)$, and so $\psi_s \perp [-a]$ is anisotropic. By Theorem 3.101 we have $\psi \perp [-a] \cong \varphi \perp H$. Thus $\psi_s \perp [-a] \sim \eta_s$ by the anisotropy of both forms, hence $\eta_s \cong \psi_s \perp [-a]$. For $s = e - 1$ we obtain $\eta_{e-1} \cong b\sigma \perp [-a]$ for some $b \in L_{e-1}^*$ and σ the leading form of ψ over L_{e-1}.

This form η_{e-1} is a neighbour of the Pfister form $\langle 1, -ab \rangle \otimes \sigma$ (which is thus anisotropic) and has $\zeta := -a\sigma'$ as complementary form, and thus $(-a\sigma') \otimes L_{e-1}(\eta_{e-1})$ as first higher kernel form. The kernel form of $\varphi \otimes L$ has a different dimension for each of the fields $L = L_0, L_1, \ldots, L_e, L_{e-1}(\eta_{e-1})$. Hence $h > e$.

Case 2: $a \in D(\psi_{e-1})$. Since $\psi_e = 0$, and thus $a \notin D(\psi_e)$, we may apply Theorem 3.113. By part (a) of said theorem, $(L_s \mid 0 \le s \le e - 1)$ is a truncated generic splitting tower of φ and $\text{Com}(\eta_s) = \psi_s$ for every $s \in [0, e-1]$. In particular, η_{e-1} is not split. But $\eta_{e-1} \otimes L_e$ splits, since $\varphi \otimes L_e$ splits. Thus, by part (b) of Theorem 3.113, $(L_s \mid 0 \leq s \leq e)$ is a truncated generic splitting tower of φ. Since $\varphi \otimes L_e$ splits, this tower is actually a full generic splitting tower of φ.

Case 3: The rest. Now there exists a largest integer $m \in [0, e-2]$ such that $a \in D(\psi_m)$. By Theorem 3.113(a), $(L_s \mid 0 \leq s \leq m)$ is a truncated generic splitting tower of φ and $\text{Com}(\eta_s) = \psi_s$ for $0 \leq s \leq m$. Furthermore, by Theorem 3.113(b), $\text{Com}(\eta_{m+1}) = \psi_{m+1} \perp H$, and we have the following alternatives:

Case 3(a): $\text{ind}(\psi_m \otimes L_{m+1}) \geq 2$. Now $(L_s \mid 0 \leq s \leq m+1)$ is a truncated generic splitting tower of φ.

Case 3(b): $\text{ind}(\psi_m \otimes L_m) = 1$. Now $\eta_m \otimes L_{m+1} = \eta_{m+1}$.

$(L_s \mid m+1 \leq s \leq e)$ is a generic splitting tower of $\text{Com}(\eta_{m+1}) \sim \psi_{m+1} \perp H$, and we have $a \notin D(\psi_{m+1})$. For this form ψ_{m+1} we are in Case 1, discussed above, and conclude that $h(\eta_{m+1}) > h(\psi_{m+1}) = e - (m+1)$, and so $h(\eta_{m+1}) \geq e - m$. In Case 3(a) we have $h(\eta_{m+1}) = h - (m+1)$, and we obtain $h \geq e + 1$. In case 3(b) we have $\eta_{m+1} \cong \eta_m \otimes L$ and we may conclude that $h(\eta_{m+1}) \leq h(\eta_m) = h - m$. Therefore, $e - m \leq h - m$ and so $e \leq h$.

Thus $e \leq h$ in all cases, and in Case 2 we even have $e < h$. Our analysis shows furthermore that all the conditions $(i) - (iv)$ of the theorem hold in Case 2, and are thus equivalent. The implication $(ii) \Rightarrow (v)$ is trivial. Assume (v), then $\eta_{e-1} \cong a\tau'$ where τ is the leading form of φ over L_{e-1}. Thus $\psi_{e-1} \sim \text{Com}(a\tau') = a\tau$ and then $\psi_{e-1} \cong a\tau$, since the forms ψ_{e-1} and τ are both anisotropic. Hence $a \in D(\psi_{e-1})$. This proves the implication $(v) \Rightarrow (i)$. $\square$

Note that the analysis carried out above also establishes the following:

Corollary 3.115. *If there exists an $m \in [0, e-1]$ with $a \in D(\psi_m)$, $a \notin D(\psi_{m+1})$, and* $\text{ind}(\psi_m \otimes L_m) \geq 2$, *then $h > e$.*

3.13 Definability of the Leading Form over the Base Field

In the following φ denotes a regular (quadratic) form over an arbitrary field k, F denotes a leading field for φ and σ denotes the associated leading form of φ.

Definition 3.116. We say that the leading form of φ is *defined over k* if there exists a form τ over k such that $\sigma \cong \tau \otimes F$.

Obviously this property is independent of the choice of the leading field F. Moreover, τ is to a large extent uniquely determined by σ. We namely have the following

Theorem 3.117 ([34, Prop. 9.2] for char $\neq 2$). *Assume that the leading form of φ is defined over k.*

(i) *If the dimension of φ is even, then there exists up to isometry precisely one form τ over k such that $\tau \otimes F \cong \sigma$.*

    (ii) If the dimension of φ is odd, then there exists up to isometry precisely one form τ over k such that $1 \in D(\tau)$ and $\tau \otimes F \cong \sigma$.
    (iii) In both cases τ is a Pfister form.

Proof. We choose a generic splitting tower $(K_r \mid 0 \leq r \leq h)$ of φ and assume without loss of generality that $F = K_{h-1}$. Let τ_1 and τ_2 be forms over k such that

$$\tau_1 \otimes K_{h-1} \cong \tau_2 \otimes K_{h-1} \cong \sigma.$$

Assume that $\tau_1 \not\cong \tau_2$. We must have $h \geq 2$. Let s be the largest integer in $[0, h-2]$ such that $\tau_1 \otimes K_s \not\cong \tau_2 \otimes K_s$, and let ζ be the kernel form of $[\tau_1 \perp (-\tau_2)] \otimes K_s$. Then $\zeta \neq 0$ and the dimension of ζ is even. For $0 \leq r \leq h$ let φ_r denote the kernel form of $\varphi \otimes K_r$. The form ζ is split by $K_s(\varphi_s)$. By the usual argument (the Norm and Subform Theorems) we thus have a decomposition

$$\varphi_s \perp \psi \cong c\zeta$$

for some $c \in K_s^*$ and some form ψ over K_s. It follows that

$$\varphi_s \otimes K_{s+1} \sim (-\psi) \otimes K_{s+1},$$

and thus $\dim \psi \geq \dim \varphi_{s+1}$. We obtain

$$2 \dim \tau_1 \geq \dim \zeta \geq \dim \varphi_s + \dim \varphi_{s+1} \geq 2 \dim \varphi_{s+1} + 2. \qquad (3.16)$$

If $\dim \varphi$ is even, this gives a contradiction and so $\tau_1 \cong \tau_2$.

Assume now that $\dim \varphi$ is odd and moreover that $[1] < \tau_1$ and $[1] < \tau_2$. Then $\tau_1 \perp (-\tau_2)$ is isotropic and we obtain from (3.16) that

$$2 \dim \tau_1 \geq \dim \zeta + 2 \geq 2 \dim \varphi_{s+1} + 4,$$

which is again a contradiction, so that $\tau_1 \cong \tau_2$ as before.

In addition there always exists a form τ over k such that $1 \in D(\tau)$ and $\tau \otimes K_{h-1} \cong \sigma$ (also when $\dim \varphi$ is odd). Namely, if η is a form over k such that $\eta \otimes K_{h-1} \cong \sigma$, and if an element $c \in D(\eta)$ is chosen, then $c\eta$ is such a form τ.

Now let τ be the uniquely determined form over k with $1 \in D(\tau)$ and $\tau \otimes K_{h-1} \cong \sigma$. Further, let $t = (t_1, \ldots, t_n)$ be an n-tuple of unknowns over k with $n := \dim \tau$. We consider the forms $\tilde{\varphi} := \varphi \otimes k(t)$ and $\tilde{\tau} := \tau \otimes k(t)$. By Theorem 2.58, $(K_r(t) \mid 0 \leq r \leq h)$ is a generic splitting tower of $\tilde{\varphi}$ and $\tilde{\tau} \otimes K_{h-1}(t)$ is the associated leading form of $\tilde{\varphi} \otimes k(t)$ (also in the event that $\dim \varphi$ is odd). Since this leading form is a Pfister form we have

$$\tilde{\tau} \otimes K_{h-1}(t) \cong [\tau(t)\tilde{\tau}] \otimes K_{h-1}(t).$$

The form $\tau(t)\tilde{\tau}$ also represents 1. It follows from what we already proved that $\tau(t)\tilde{\tau} \cong \tilde{\tau}$. This indicates that τ is strongly multiplicative, and thus is a Pfister form (cf. §3.6).

                                                            □

Remark. If dim φ is odd, we cannot dispense with the assumption that $1 \in D(\tau)$ in assertion *(ii)*, as the following example shows. Let ρ be a Pfister form, $a \in k^*$ and $\langle 1, a \rangle \otimes \rho$ anisotropic. Then the form $\varphi := \rho \perp [a]$ is a neighbour of $\langle 1, a \rangle \otimes \rho$ with complementary form

$$\eta := \mathrm{Pol}_{\rho \perp a\rho}(\rho \perp [a]) = \mathrm{Pol}_{a\rho}([a]) = a\mathrm{Pol}_\rho([1]) = a\rho'.$$

Thus φ has height 2 and leading form $\rho \otimes F$, for instance with $F = k(\varphi)$. However, we do have $\rho \otimes F \cong (-a\rho) \otimes F$, which does *not* contradict our theorem since $1 \notin D(-a\rho)$.

Definition 3.118. Assume that $h(\varphi) > 0$ and that the leading form σ of φ is defined over k. Let τ be the Pfister form over k with $\sigma \cong \tau \otimes F$. Then we say that the leading form of φ is *defined over k by τ*.

Example.
(1) If φ is excellent, then the kernel form of $\varphi \otimes F$ is definable over k by Theorem 3.72 (for this terminology, see §3.10(C)). This implies, also when dim φ is odd, that the leading form of φ is defined over k.
(2) Let $\varphi \sim a\tau \perp \psi$ with $a \in k^*$, τ a Pfister form of degree $n \geq 1$, and ψ strictly regular with $\deg \psi \geq n + 1$. By Theorem 3.86 the leading form of φ is defined over k by τ.
(3) Let dim φ be odd and $\Delta(\varphi) \neq 1$. Let τ be the anisotropic binary Pfister form with $\Delta(\tau) = \Delta(\varphi)$ (cf. Lemma 3.91). The form $\psi := \varphi \perp (-\tau)$ has discriminant $\Delta(\psi) = 1$, and thus degree ≥ 2 (cf. Theorem 3.94). We have $\varphi \sim \tau \perp \psi$ and $\deg \tau = 1$. Thus, by the previous example, the leading form of φ is defined over k by τ.

In the search for forms whose leading form is defined over the ground field, we may restrict ourselves to forms of even dimension by the following theorem.

Theorem 3.119 ([34, Prop. 9.4] for char $k \neq 2$). *Assume that φ is not split and that* dim φ *is odd. Let τ be an anisotropic Pfister form over k (of degree ≥ 2). The leading form of φ is defined over k by τ if and only if this is the case for the companion* Com(φ).

Proof. Let $\psi := \mathrm{Com}(\varphi)$. If the leading form of ψ is defined over k by τ, then Theorem 3.109 shows that the same is true for φ {although the leading fields of φ and ψ might not be equivalent, see Theorem 3.122 below}.

Assume now that the leading form of φ is defined over k by τ. We must show that the same is true for ψ. Let F be a leading field for φ, and let $QL(\varphi) = [a]$ for char $k = 2$, $d(\varphi) = \langle -a \rangle$ for char $k \neq 2$, where $a \in k^*$. In both cases we have $\varphi \perp H \cong \psi \perp [-a]$ by Theorem 3.101.

Furthermore, $\ker(\varphi \otimes F) = c(\tau' \otimes F)$ for some $c \in F^*$. We see that $QL(\varphi \otimes F) = [c]$ for char $k = 2$, $d(\varphi \otimes F) = \langle -c \rangle$ for char $k \neq 2$. Therefore c and a represent the same square class of F, and we may write

$$\ker(\varphi \otimes F) = (a\tau') \otimes F.$$

Let E be a leading field for ψ and σ the leading form of ψ over E. By Theorem 3.109, σ and τ have the same dimension ≥ 4. The kernel form of $\psi \otimes E$ is $b\sigma$ for some $b \in E^*$.

Suppose for the sake of contradiction that σ is not isomorphic to $\tilde{\tau} := \tau \otimes E$. Then $\sigma \otimes E(\tilde{\tau})$ is anisotropic. We introduce the form

$$\alpha := b(\sigma \otimes E(\tilde{\tau})) \perp [-a]$$

over $E(\tilde{\tau})$ and know from above that $\varphi \otimes E(\tilde{\tau}) \sim \alpha$ and $\dim \alpha = \dim \tau + 1$.

And now for a little thought experiment. Let T be a field extension of $E(\tilde{\tau})$ for which $\alpha \otimes T$ is isotropic. Then the kernel form of $\varphi \otimes T$, which coincides with the kernel form of $\alpha \otimes T$, has dimension $\leq \dim \tau'$. Thus there exists a place from F to T over k. Since $\varphi \otimes F \sim (a\tau') \otimes F$ and $\tau \otimes T$ splits, it follows that

$$\alpha \otimes T \sim \varphi \otimes T \sim a(\tau' \otimes T) \sim [-a].$$

By the definition of α and the Cancellation Theorem, Theorem 3.104, we obtain $\sigma \otimes T \sim 0$.

This thought experiment shows for starters that α itself must be anisotropic since $\sigma \otimes E(\tilde{\tau})$ is anisotropic. (Take $T = E(\tilde{\tau})$.) Secondly, it shows that α splits over every field extension of $E(\tilde{\tau})$, over which α is isotropic. Thus α has height 1. By §3.9 α is up to a scalar factor the unit polar of a Pfister form. Hence $\dim \alpha + 1$ is a 2-power. This is the contradiction we were looking for because $\dim \alpha + 1 = \dim \tau + 2$ cannot be a 2-power since $\dim \tau$ is a 2-power ≥ 4. This proves $\sigma \cong \tau \otimes E$. □

We come to the main theorem of this section.

Theorem 3.120 ([34, Th. 9.6] for char $k \neq 2$). *Let $n \in \mathbb{N}$, let τ be a Pfister form of degree n over k and let φ be a non-split (strictly) regular form of even dimension over k. The following statements are equivalent:*

(i) The leading form of φ is defined over k by τ.
(ii) $\varphi \equiv \tau \mod J_{n+1}(k)$.

{*Recall: $J_r(k)$ is the module of Witt classes of strictly regular forms over k of degree $\geq r$.*}

Proof. The implication $(ii) \Rightarrow (i)$ was already established in Theorem 3.86. Assume now that (i) holds. We want to show that the form $\psi := \varphi \perp (-\tau)$ has degree $> n$. In order to achieve this we will for the first time make serious use of the theory developed in §2.5.

We assume without loss of generality that ψ is not hyperbolic. Let $(K_r \mid 0 \leq r \leq h)$ be a regular generic splitting tower of φ (cf. Definition 2.56) and let $\varphi_r := \ker(\varphi \otimes K_r)$. For every $r \in [0, h]$ we consider the field compositum $K_r \cdot k(\tau)$ over k in accordance with Definition 2.57. Let J be the set of all $r \in [0, h]$ such that $\varphi_r \otimes K_r \cdot k(\tau)$ is anisotropic and let $r_0 < r_1 < \cdots < r_e$ be the elements of J. We know from Theorem 2.58, that $(K_{r_i} \cdot k(\tau) \mid 0 \leq i \leq e)$ is a generic splitting tower of $\varphi \otimes k(\tau)$ and, since $\varphi \otimes k(\tau) \sim \psi \otimes k(\tau)$, also of $\psi \otimes k(\tau)$. Since $\psi \not\sim 0$ we have $e \geq 1$.

Furthermore $\varphi \otimes (K_{h-1} \cdot k(\tau)) \sim 0$, since $\varphi \otimes K_{h-1}$ has kernel form $b(\tau \otimes K_{h-1})$ for some $b \in K_{h-1}^*$ and $\tau \otimes k(\tau) \sim 0$. Thus $h - 1 \notin J$. Therefore

$$\ker(\psi \otimes K_{r_{e-1}} \cdot k(\tau)) = \varphi_{r_{e-1}} \otimes K_{r_{e-1}} \cdot k(\tau)$$

has dimension $> 2^n$. This means that $\psi \otimes k(\tau)$ has degree $> n$.

Let E be a leading field for ψ and ρ the associated leading form of ψ. By Theorem 2.58, ρ splits over $E \cdot k(\tau) = E(\tau \otimes E)$. This implies by Theorem 3.53, and since $\dim \rho = \dim \tau$, that $\rho \cong \tau \otimes E$, and so $\psi \otimes E \sim a(\tau \otimes E)$ for some $a \in E^*$. We therefore obtain

$$\varphi \otimes E \sim \tau \otimes E \perp a(\tau \otimes E) = \langle 1, a \rangle \otimes (\tau \otimes E).$$

In particular we have $\deg(\varphi \otimes E) = n + 1$ or $= \infty$. We now make use of the lemma below. We apply it to φ and the field extension E of k and obtain $\tau \otimes E \sim 0$. But this is nonsense because $\tau \otimes E \cong \rho$. Thus we must have $\deg \psi > n$.   □

Lemma 3.121. *Continuing with the assumptions of Theorem 3.120, let L be a field extension of k with $\deg(\varphi \otimes L) > \deg \varphi$. Then $\tau \otimes L$ splits.*

Proof. Let $(K_r \mid 0 \le r \le h)$ be the affine standard tower of φ (Definition 2.51). Thus, $K_0 = k$ and $K_{r+1} = K_r(\varphi_r)$ where $\varphi_r = \ker(\varphi \otimes K_r)$. By Theorem 2.58, $\tau \otimes K_{h-1} \cdot L$ splits.

Assume for the sake of contradiction that $\tau \otimes L$ is *not* split. We certainly have $h \ge 2$. Let $s \in [0, h - 2]$ be the largest integer with $\tau \otimes K_s \cdot L$ anisotropic. Let $\bar{\tau} := \tau \otimes K_s \cdot L$ and $\bar{\varphi}_s := \varphi \otimes K_s \cdot L$. Then $K_{s+1} \cdot L = (K_s \cdot L)(\bar{\varphi}_s)$ and $\bar{\tau} \otimes (K_s \cdot L)(\bar{\varphi}_s) = 0$. By the usual argument it follows that $\bar{\varphi}_s < c\bar{\tau}$ for some $c \in (K_s \cdot L)^*$. In particular, $\dim \varphi_s \le \dim \tau = \dim \varphi_{h-1}$, which is nonsense. Therefore $\tau \otimes L$ must be split.   □

Remark. We proved the converse of statement *(i)* in Theorem 3.86. For characteristic $\ne 2$ the converse of statement *(ii)* of this theorem was also proved in [34, p. 23 ff.]: If $\varphi \equiv a\tau \mod J_{n+2}(k)$, then the $(h - 1)$-st kernel form φ_{h-1} is defined over k by $a\tau$, $\varphi_{h-1} \cong (a\tau) \otimes K_{h-1}$. So far it has not been possible to extend the proof in [34] to characteristic 2. One runs into the problem formulated at the end of §3.10, Part (C), which must be solved at least in a special case.

By means of Theorem 3.120 we can improve upon Theorem 3.119:

Theorem 3.122. *Let φ again be a nondegenerate regular form of odd dimension over k, and let ψ be the companion of φ. Let ak^{*2} be the square class with $[a] = QL(\varphi)$, in case $\operatorname{char} k = 2$, and $\langle a \rangle = -d(\varphi)$, in case $\operatorname{char} k \ne 2$. Let $(K_r \mid 0 \le r \le h)$ be a generic splitting tower of φ with associated kernel forms $\varphi_r := \ker(\varphi \otimes K_r)$ and let $(L_s \mid 0 \le s \le e)$ be a generic splitting tower of ψ with associated kernel forms $\psi_s := \ker(\psi \otimes L_s)$. Assume furthermore that the leading form of φ—therefore also the leading form of ψ—is defined over k by τ. Finally, let $h > 1$. {Note that otherwise $\varphi \cong a\tau'$, $\psi \cong a\tau$, which is not of interest here.} Then the following statements are equivalent:*

(i) $\dim \varphi_{h-2} > \dim \tau + 1$.

(ii) $L_{e-1} \sim_k K_{h-1}$.

(iii) $a \in D(\psi_{e-1})$, and so $\psi_{e-1} \cong (a\tau) \otimes L_{e-1}$.

If these statements do not hold, then $L_{e-1} \sim_k K_{h-2}$.

Proof. In Theorem 3.109, we already determined that there exists a place from L_{e-1} to K_{h-1} over k. Furthermore we have $\varphi \perp H \cong \psi \perp [-a]$, and so $\varphi \otimes L_{e-1} \sim \psi_{e-1} \perp [-a]$. Hence there is certainly a place from K_{h-2} to L_{e-1} over k. This place can be extended to a place from K_{h-1} to L_{e-1} if and only if the form $\psi_{e-1} \perp [-a]$ is isotropic, so that $a \in D(\psi_{e-1})$. This establishes the equivalence (ii) $\Leftrightarrow$ (iii).

If the form $\psi_{e-1} \perp [-a]$ is anisotropic, then it is the kernel form of $\varphi \otimes L_{e-1}$. If $\dim \varphi_{h-2} > \dim \tau + 1$, this cannot happen for dimension reasons. Therefore $\psi_{e-1} \perp [-a]$ is isotropic in this case. This proves (i) $\Rightarrow$ (iii).

Now we assume that (i) does *not* hold, in other words that $\dim \varphi_{h-2} = \dim \tau + 1$. We want to show that then $L_{e-1} \sim_k K_{h-2}$. Since K_{h-1} is not equivalent to K_{h-2} over k, this will imply that (iii) does *not* hold. This will prove the implication (iii) $\Rightarrow$ (i), in addition to the last statement of the theorem.

We have already established above that there is a place from K_{h-2} to L_{e-1} over k. We have to show that there is also a place over k in the reverse direction. By the generic splitting theory this means that the kernel form of $\psi \otimes K_{h-2}$ must have dimension $\leq 2^n$.

Let $m := \operatorname{ind}(\varphi \otimes K_{h-2})$. Then

$$2^n + 1 = \dim \varphi_{h-2} = \dim \varphi - 2m.$$

Furthermore we have $\operatorname{ind}(\psi \otimes K_{h-2}) \geq m$. Hence,

$$\dim \ker(\psi \otimes K_{h-2}) \leq \dim \psi - 2m = 2^n + 2.$$

If the inequality is strict, then $\dim \ker(\psi \otimes K_{h-2}) \leq 2^n$, and we are done.

We assume then that the form $\alpha := \ker(\psi \otimes K_{h-2})$ has dimension $2^n + 2$, and want to obtain a contradiction. We have $\deg \alpha \geq \deg \psi = n$, but $\dim \alpha < 2^{n+1}$, since $n \geq 2$. This forces $\deg \alpha = n$. By Theorem 2.58 it is now clear that α has a leading form over K_{h-2}, defined by $\tau \otimes K_{h-2}$. (The form $\tau \otimes K_{h-2}$ is in particular anisotropic.) By Theorem 3.120 this means that

$$\alpha \sim (\tau \otimes K_{h-2}) \perp \chi$$

for some form χ over K_{h-2} of degree $> n$. Let $\tilde{\tau} := \tau \otimes K_{h-2}$. We have $\chi \otimes K_{h-2}(\tilde{\tau}) \sim \alpha \otimes K_{h-2}(\tilde{\tau})$. Therefore $\alpha \otimes K_{h-2}(\tilde{\tau})$ has degree $> n$. However, $\dim \alpha < 2^{n+1}$. This forces $\alpha \otimes K_{h-2}(\tilde{\tau}) \sim 0$. The theory of divisibility by Pfister forms (Theorem 3.53) then tells us that $\alpha \cong \gamma \otimes \tilde{\tau}$ for a suitable form γ over K_{h-2}. Now $\dim \alpha$ is in particular divisible by $\dim \tau = 2^n$. This is the contradiction we were looking for, since $\dim \alpha = 2^n + 2 < 2^{n+1}$. This establishes that L_{e-1} is specialization equivalent to K_{h-2} over k. $\square$

In the proof of Theorem 3.122 we did not use Theorem 3.114. (Would the proof have been simpler if we had?) If (*i*)–(*iii*) in Theorem 3.122 are valid, this theorem does give us the additional information that the tower $(L_s \mid 0 \le s \le e)$ is a generic splitting tower of φ, and in particular that $h = e$.

Chapter 4
Specialization with Respect to Quadratic Places

4.1 Quadratic Places; Specialization of Bilinear Forms

In this section we mean by a "form" always a nondegenerate symmetric bilinear form.

To begin with we recall a result from §1.3. Let $\lambda : K \to L \cup \infty$ be a place and $\mathfrak{o} = \mathfrak{o}_\lambda$ the associated valuation ring. We defined an additive map

$$\lambda_W : W(K) \longrightarrow W(L),$$

which has the following values for one-dimensional forms (= square classes):

$$\lambda_W \langle \varepsilon \rangle = \langle \lambda(\varepsilon) \rangle \quad \text{for } \varepsilon \in \mathfrak{o}^*, \tag{4.1}$$

$$\lambda_W(\alpha) = 0 \quad \text{for } \alpha \in Q(K) \setminus Q(\mathfrak{o}). \tag{4.2}$$

Here we identified the square class group $Q(\mathfrak{o}) = \mathfrak{o}^*/\mathfrak{o}^{*2}$ with its image in $Q(K)$ under the natural map $\varepsilon \mathfrak{o}^{*2} \mapsto \varepsilon K^{*2}$.

More generally we have that the natural map $\{M\} \mapsto \{K \otimes_\mathfrak{o} M\}$ from $W(\mathfrak{o})$ to $W(K)$ is injective. (Here $\{M\}$ denotes the Witt class of a nondegenerate $\mathfrak{o}$-module M). Thus we can also regard $W(\mathfrak{o})$ as a subring of $W(K)$. The restriction of λ_W to $W(\mathfrak{o})$ can then be interpreted as the ring homomorphism $W(\mathfrak{o}) \to W(L)$, induced by the ring homomorphism $\lambda|\mathfrak{o} : \mathfrak{o} \to L$. The restriction $\lambda_W|W(\mathfrak{o})$ is thus also multiplicative. Furthermore, $\lambda_W : W(K) \to W(L)$ is semilinear with respect to $W(\mathfrak{o}) \to W(L)$. In other words, we have for $\xi \in W(\mathfrak{o})$ and $\eta \in W(K)$, that

$$\lambda_W(\xi \eta) = \lambda_W(\xi) \cdot \lambda_W(\eta).$$

Starting from the map λ_W we developed a specialization theory for bilinear forms in §1.3. In particular we associated to a form φ with good reduction with respect to λ (i.e. a form over K that comes from a nondegenerate bilinear module over $\mathfrak{o}$) a form $\lambda_*(\varphi)$ over L such that $\{\lambda_*(\varphi)\} = \lambda_W(\{\varphi\})$ and $\dim \lambda_*(\varphi) = \dim \varphi$. The form $\lambda_*(\varphi)$ is only determined up to stable isometry by the form φ, however. In truth we thus

M. Knebusch, *Specialization of Quadratic and Symmetric Bilinear Forms*,
Algebra and Applications 11, DOI 10.1007/978-1-84882-242-9_4,
© Springer-Verlag London Limited 2010

associated to φ only a stable isometry class of forms over L. Nonetheless we will disregard this fact and just speak of $\lambda_*(\varphi)$ as if it were a form over L.

Note that if char $L \neq 2$, then stable isometry is the same as isometry by Witt's Cancellation Theorem and so $\lambda_*(\varphi)$ is uniquely determined—up to isomorphism— by λ and φ.

Now we would like to associate to a "quadratic place" $\Lambda : K \to L \cup \infty$ (see the following definition) a map $\Lambda_W : W(K) \to W(L)$ in a similar manner as was done above and, using this foundation, develop a specialization theory for quadratic places that generalizes the theory developed in §1.3.

Definition 4.1. Let K and L be fields. A *quadratic place* (= Q-place) from K to L is a triple (λ, H, χ) consisting of a place $\lambda : K \to L \cup \infty$ (in the usual sense), a subgroup H of $Q(K)$ that contains $Q(\mathfrak{o})$ (with $\mathfrak{o} := \mathfrak{o}_\lambda$) and a group homomorphism $\chi : H \to Q(L)$ with $\chi(\langle\varepsilon\rangle) = \langle\lambda(\varepsilon)\rangle$ for every $\varepsilon \in \mathfrak{o}^*$, and thus $\chi(\alpha) = \lambda_*(\alpha) = \lambda_W(\alpha)$ for every $\alpha \in Q(\mathfrak{o})$.

We will often denote the triple (λ, H, χ) by a Greek capital letter, for instance Λ, and use the symbolic notation $\Lambda : K \to L \cup \infty$. We call a group homomorphism from a subgroup H of $Q(K)$ to $Q(L)$ a ($Q(L)$-valued) *character* of H.

Example. To every place $\lambda : K \to L \cup \infty$ we can associate a quadratic place $\widehat{\lambda} :=$ $(\lambda, H_\lambda, \chi_\lambda)$ by letting $H_\lambda = Q(\mathfrak{o})$ and taking for χ_λ the character $\langle\varepsilon\rangle \mapsto \langle\lambda(\varepsilon)\rangle$ given by λ. This Q-place $\widehat{\lambda}$ can then be "extended" to a Q-place $\Lambda := (\lambda, H, \chi)$ if we choose a subgroup H of $Q(K)$ which contains $Q(\mathfrak{o})$ and somehow continue the character χ_λ to a character $\chi : H \to Q(L)$. Such a continuation is always possible since H and $Q(L)$ are elementary abelian of exponent 2, and can thus be interpreted as $\mathbb{F}_2$-vector spaces.

In this setup we imagine that $\widehat{\lambda}$ is the same object as λ, but just dressed up, so that the notion of quadratic place generalizes the notion of place.

Theorem 4.2. *Let $\Lambda = (\lambda, H, \chi)$ be a Q-place from K to L.*
(a) There is precisely one additive map

$$\Lambda_W : W(K) \longrightarrow W(L)$$

with the following properties:

$$\Lambda_W(\alpha) = \chi(\alpha) \quad \text{if } \alpha \in H, \tag{4.3}$$
$$\Lambda_W(\alpha) = 0 \quad \text{if } \alpha \in Q(K) \setminus H. \tag{4.4}$$

(b) The restriction of Λ_W to the ring $W(K, H)$, generated by H in $W(K)$, is a ring homomorphism from $W(K, H)$ to $W(L)$, and Λ_W is semilinear with respect to this ring homomorphism.

Proof. The uniqueness of Λ_W is clear since $W(K)$ is additively generated by $Q(K)$. In order to prove the existence, we choose a subgroup H_0 of H with $H = H_0 \times Q(\mathfrak{o})$ (i.e., a "complement" H_0 of $Q(\mathfrak{o})$ in H). Then we define

$$\Lambda_W(\xi) := \sum_{\alpha \in H_0} \chi(\alpha)\lambda_W(\alpha\xi) \qquad (\xi \in W(K)). \tag{4.5}$$

This map $\Lambda_W : Q(K) \to Q(L)$ is clearly additive. It satisfies conditions (4.3) and (4.4), and also possesses the properties stated in part (b), as can be verified instantly.
□

Assume now that $\Lambda = (\lambda, H, \chi) : K \to L \cup \infty$ is a Q-place and that $\mathfrak{o} := \mathfrak{o}_\lambda$. We choose a subgroup H_0 of H with $H := H_0 \times Q(\mathfrak{o})$. Let φ be a form over K.

Definition 4.3. We say that φ has *good reduction* (abbreviated GR) *with respect to* Λ, when φ has a decomposition (modulo stable isometry)

$$\varphi \approx \perp_{\alpha \in H_0} \alpha\varphi_\alpha, \tag{4.6}$$

in which the forms φ_α are λ-unimodular. We then also say that φ is Λ-*unimodular*. Here we used the notation $\alpha\varphi_\alpha := \alpha \otimes \varphi_\alpha$. (If $\alpha = \langle a \rangle$, then $\alpha\varphi_\alpha = a\varphi_\alpha$ up to isometry.)

One should note that Λ-unimodularity is independent of the choice of the complement H_0 of $Q(\mathfrak{o})$ in H.

Example. If $H = Q(K)$, then *every* form φ over K has good reduction with respect to Λ. Namely, if $\varphi \cong r \times \begin{pmatrix} 0 & 1 \\ 1 & 0 \end{pmatrix}$ is hyperbolic, then φ is actually λ-unimodular. Otherwise we choose a diagonalization $\varphi \cong \langle a_1, \ldots, a_n \rangle$ and take those elements $\langle a_j \rangle \in Q(K)$ together that are in the same coset of $Q(\mathfrak{o})$.

Definition 4.4. Assume that φ has GR with respect to Λ. We choose a decomposition (4.6) of Λ and let

$$\Lambda_*(\varphi) := \perp_{\alpha \in H_0} \chi(\alpha)\lambda_*(\varphi_\alpha). \tag{4.7}$$

We call $\Lambda_*(\varphi)$ a *specialization of φ with respect to* Λ.

Theorem 4.5. *If φ has* GR *with respect to Λ then, up to stable isometry, the form $\Lambda_*(\varphi)$ does not depend on the choice of the decomposition (4.6), and also not on the choice of the complement H_0 of $Q(\mathfrak{o})$ in H.*

Proof. We clearly have

$$\Lambda_W(\{\varphi\}) = \{\Lambda_*(\varphi)\}. \tag{4.8}$$

Moreover, $\dim \Lambda_*(\varphi) = \dim \varphi$. Noting that a form over L is uniquely determined up to stable isometry by its Witt class and dimension finishes the proof.
□

In the future we will call $\Lambda_*(\varphi)$ "the" specialization of φ with respect to Λ. Although sloppy, we hope this notation is harmless and helpful.

Remark. This notion of specialization generalizes the one from §1.3: if λ is a place from K to L, then a form φ over K has GR with respect to $\widehat{\lambda}$ if and only if it has GR with respect to λ, and then we have $\widehat{\lambda}_*(\varphi) \approx \lambda_*(\varphi)$.

Our goal is now to use these notions of good reduction and specialization to build a theory, analogous to the theory developed in §1.3 and then—in the case char $L \neq 2$—to obtain results about the generic splitting of $\Lambda_*(\varphi)$ from the generic splitting behaviour of φ, analogous to §1.4; see §4.3 below.

It is obvious that for forms φ, ψ over K which have GR with respect to Λ, also $\varphi \perp \psi$ and $\varphi \otimes \psi$ have GR with respect to Λ and that

$$\Lambda_*(\varphi \perp \psi) \approx \Lambda_*(\varphi) \perp \Lambda_*(\psi), \quad \Lambda_*(\varphi \otimes \psi) \approx \Lambda_*(\varphi) \otimes \Lambda_*(\psi). \tag{4.9}$$

It is not clear however that the analogue of Theorem 1.26 remains valid. When φ and $\varphi \perp \psi$ have GR with respect to Λ, why should ψ then also have GR with respect to Λ?

Theorem 1.26 was crucial for the development of the generic splitting theory of §1.4. In order to save this theorem we will now broaden the notion of "good reduction" to the new notion of "almost good reduction".

Given a Q-place $\Lambda = (\lambda, H, \chi)$, we choose a subgroup S of $Q(K)$ with $Q(K) = S \times H$. For every form φ over K we then have a decomposition

$$\varphi \approx \underset{s \in S}{\perp} s\varphi_s \tag{4.10}$$

with Λ-unimodular forms S (even with $\cong$ instead of $\approx$). This is obvious for $\varphi \cong \langle a \rangle$ one-dimensional and for $\varphi \cong \left(\begin{smallmatrix} 0 & 1 \\ 1 & 0 \end{smallmatrix} \right)$, and so clear for every form φ over K. We call the decomposition (4.10) a Λ-*modular decomposition* of φ.

Lemma 4.6. *For every* $s \in S$, *the Witt class of* $\Lambda_*(\varphi_s)$ *in a* Λ-*modular decomposition* (4.10) *of* φ *is uniquely determined by* φ, Λ *and* s.

Proof. For every $s \in S$ we choose a decomposition (see (4.6) above)

$$\varphi_s \approx \underset{\alpha \in H_0}{\perp} \alpha\varphi_{s,\alpha}$$

with λ-unimodular forms $\varphi_{s,\alpha}$. Then

$$\varphi \approx \underset{(s,\alpha) \in S \times H_0}{\perp} s\alpha\varphi_{s,\alpha}.$$

Clearly, $\lambda_W(\{s\alpha\varphi\}) = \{\lambda_*(\varphi_{s,\alpha})\}$. Hence the Witt class of $\lambda_*(\varphi_{s,\alpha})$ is uniquely determined by φ and Λ for fixed s and α. It follows that the Witt class of

$$\Lambda_*(\varphi_s) \approx \underset{\alpha \in H_0}{\perp} \chi(\alpha)\lambda_*(\varphi_{s,\alpha})$$

is uniquely determined by φ and Λ for fixed s and H_0. By Theorem 4.5, $\Lambda_*(\varphi_s)$ does not change, up to stable isometry, if we pass from H_0 to a different complement of $Q(\mathfrak{o})$ in H. $\qquad \square$

Definition 4.7. Consider a Λ-modular decomposition (4.10) of φ. We say that φ has *almost good reduction* (abbreviated AGR) *with respect to* Λ if all forms $\Lambda_*(\varphi_s)$ with

$s \in S$, $s \neq 1$, are metabolic. We then define

$$\Lambda_*(\varphi) := \Lambda_*(\varphi_1) \perp \frac{\dim \varphi - \dim \varphi_1}{2} \times \widetilde{H},$$

with $\widetilde{H} := \begin{pmatrix} 0 & 1 \\ 1 & 0 \end{pmatrix}$, and call $\Lambda_*(\varphi)$ "the" *specialization of φ with respect to Λ.*.

It is clear from Lemma 4.6 that the property AGR is independent of the choice of Λ-modular decomposition (4.10), and that also the Witt class of $\Lambda_*(\varphi)$ does not depend on this choice. Since $\dim \Lambda_*(\varphi) = \dim \varphi$ we have that $\Lambda_*(\varphi)$ is uniquely determined by φ up to stable isometry. It is now also clear that the property AGR and the stable isometry class of $\Lambda_*(\varphi)$ are independent of the choice of the complement S of H in $Q(K)$.

One should note that for the property AGR all three components λ, H, χ of $\Lambda = (\lambda, H, \chi)$ play a role, whereas GR depends only on λ and H.

One easily verifies the following:

Theorem 4.8. *Let φ and ψ be forms over K. If φ and ψ have almost good reduction with respect to Λ, this is also the case for $\varphi \perp \psi$ and $\varphi \otimes \psi$, and we have*

$$\Lambda_*(\varphi \perp \psi) \approx \Lambda_*(\varphi) \perp \Lambda_*(\psi), \tag{4.11}$$

$$\Lambda_*(\varphi \otimes \psi) \approx \Lambda_*(\varphi) \otimes \Lambda_*(\psi). \tag{4.12}$$

In addition, the analogue of Theorem 1.26 also holds for almost good reduction. The proof of this analogue—in the current state of affairs—will turn out to be trivial.

Theorem 4.9. *Let φ and ψ be forms over K. Assume that the forms φ and $\varphi \perp \psi$ have almost good reduction with respect to Λ. Then ψ also has almost good reduction with respect to Λ.*

Proof. We choose Λ-modular decompositions

$$\varphi \approx \underset{s \in S}{\perp} s\varphi_s, \quad \psi \approx \underset{s \in S}{\perp} s\psi_s.$$

Then

$$\varphi \perp \psi \approx \underset{s \in S}{\perp} s(\varphi_s \perp \psi_s)$$

is a Λ-modular decomposition of $\varphi \perp \psi$. For every $s \in S$ we have

$$\Lambda_*(\varphi_s \perp \psi_s) \approx \Lambda_*(\varphi_s) \perp \Lambda_*(\psi_s).$$

If $s \neq 1$, then $\Lambda_*(\varphi_s) \sim 0$ and $\Lambda_*(\varphi_s \perp \psi_s) \sim 0$, and thus also $\Lambda_*(\psi_s) \sim 0$. $\qquad \square$

4.2 Almost Good Reduction with Respect to Extensions of Quadratic Places

In this section a "form" will again always mean a nondegenerate (symmetric) bilinear form.

Definition 4.10. Let $\Lambda : K \to L \cup \infty$ and $\Lambda' : K \to L \cup \infty$ be two quadratic places from K to L, $\Lambda = (\lambda, H, \chi)$, $\Lambda' = (\lambda', H', \chi')$. We call Λ' an *expansion* of Λ, and write $\Lambda \subset \Lambda'$, when $\lambda = \lambda'$, $H \subset H'$, and $\chi'|H = \chi$.

Note that every Q-place $\Lambda = (\lambda, H, \chi)$ is an expansion of the Q-place $\widehat{\lambda} = (\lambda, Q(\mathfrak{o}), \lambda_W | Q(\mathfrak{o}))$ with $\mathfrak{o} = \mathfrak{o}_\lambda$.

Remark. If $\Lambda \subset \Lambda'$ and if some form φ over K has good reduction (respectively, almost good reduction) with respect to Λ, then it obviously has GR (respectively, AGR) with respect to Λ' and we have $\Lambda'_*(\varphi) = \Lambda_*(\varphi)$.

Consider a Q-place $\Lambda = (\lambda, H, \chi)$ from K to L. If k is a subfield of K, we obtain from Λ a Q-place $\Gamma = (\gamma, D, \psi)$ from k to L as follows: let γ be the restriction $\lambda|k$ of λ. The group $D \subset Q(k)$ is the preimage of H under the homomorphism $j : Q(k) \to Q(K)$, which maps a square class $\langle a \rangle = ak^{*2}$ of k to the square class $\langle a \rangle_K = aK^{*2}$. (Note that $\langle a \rangle_K = \langle a \rangle \otimes K$ when we interpret square classes as one-dimensional forms.) Finally ψ is the character $\chi \circ j : D \to Q(L)$.

Definition 4.11. We call Γ the *restriction* of the Q-place Λ to k and write $\Gamma = \Lambda|k$.

Caution! If $\Lambda = \widehat{\lambda}$ for a place $\lambda : K \to L \cup \infty$ and $\gamma = \lambda|k$, then $\Lambda|k \supset \widehat{\gamma}$, but $\Lambda|k$ is in general different from $\widehat{\gamma}$.

The following important theorem for quadratic places has no counterpart for places.

Theorem 4.12. *Let k be a subfield of K and φ a form over k. Let $\Lambda : K \to L \cup \infty$ be a Q-place and $\Gamma = \Lambda|k$. The form $\varphi \otimes K$ has AGR with respect to Λ if and only if φ has AGR with respect to Γ, and then $\Lambda_*(\varphi \otimes K) = \Gamma_*(\varphi)$.*

Proof. As above we write $\Lambda = (\lambda, H, \chi)$, $\Gamma = (\gamma, D, \psi)$ with $\gamma = \lambda|k$ etc. We choose a subgroup S of $Q(k)$ with $Q(k) = S \times D$. The homomorphism $j : Q(k) \to Q(K)$ maps S isomorphically to a subgroup of $Q(K)$, which we denote by S_K. We further choose a subgroup $T \supset S_K$ of $Q(K)$ such that $Q(K) = T \times H$. It is always possible to do this. We now start with a Γ-modular decomposition of φ,

$$\varphi \approx \underset{s \in S}{\perp} s \, \varphi_s,$$

for Γ-unimodular forms φ_s over k. We can interpret

$$\varphi \otimes K \approx \underset{s \in S}{\perp} s_K(\varphi_s \otimes K)$$

as a Λ-modular decomposition of $\varphi \otimes K$ for the decomposition $Q(K) = T \times H$, in which the components for $t \in T \setminus S_K$ are all zero. By definition φ has AGR with respect to Γ if and only if $\Gamma_*(\varphi_s) \sim 0$ for all $s \neq 1$ in S, and $\varphi \otimes K$ has AGR with respect to Λ if and only if $\Lambda_*(\varphi_s \otimes K) \sim 0$ for all $s \neq 1$ in S. But for every $s \in S$ we have $\Lambda_*(\varphi_s \otimes K) \approx \Gamma_*(\varphi_s)$. Thus $\varphi \otimes K$ has AGR with respect to Λ if and only if φ has AGR with respect to Γ, and in this case we have, with $r := \frac{1}{2}(\dim \varphi - \dim \varphi_1)$, that

$$\Lambda_*(\varphi \otimes K) = \Lambda_*(\varphi_1 \otimes K) \perp r \times \widetilde{H} \approx \Gamma_*(\varphi_1) \perp r \times \widetilde{H} = \Gamma_*(\varphi). \qquad \square$$

Definition 4.13. As before, let K be a field and k a subfield of K. Consider Q-places $\Lambda : K \to L \cup \infty$ and $\Gamma : k \to L \cup \infty$. We call Λ an *extension of* Γ when $\Lambda|k \supset \Gamma$, and a *strict extension of* Γ when actually $\Lambda|k = \Gamma$.

Theorem 4.14. *Let* $\Lambda : K \to L \cup \infty$ *be an extension of* $\Gamma : k \to L \cup \infty$. *Let* φ *be a form over* k *which has AGR with respect to* Γ. *Then* $\varphi \otimes K$ *has AGR with respect to* Λ *and* $\Lambda_*(\varphi \otimes K) = \Gamma_*(\varphi)$.

Proof. By Theorem 4.12 the statement holds when Λ is a strict extension of Γ. Thus it suffices to consider the case where $K = k$ and $\Lambda = (\lambda, H, \chi)$ is an expansion of $\Gamma = (\gamma, D, \psi)$. Thus, let $K = k, D \subset H, \chi|D = \psi$.

We choose subgroups S_1 of H and T of $Q(k)$ with $H = S_1 \times D$, $Q(k) = T \times H$. Then $S := T \times S_1$ is a subgroup of $Q(k)$ with $Q(k) = S \times D$. Let

$$\varphi \approx \underset{(\alpha, \beta) \in T \times S_1}{\perp} \alpha\beta \, \varphi_{\alpha, \beta}$$

be a Γ-modular decomposition of φ. We let

$$\varphi_\alpha := \underset{\beta \in S_1}{\perp} \beta \, \varphi_{\alpha, \beta} \qquad (\alpha \in T).$$

For every $\alpha \in T$ the form φ_α is Λ-unimodular and

$$\Lambda_*(\varphi_\alpha) \approx \underset{\beta \in S_1}{\perp} \chi(\beta) \, \Gamma_*(\varphi_{\alpha, \beta}). \tag{4.13}$$

It follows that

$$\varphi \approx \underset{\alpha \in T}{\perp} \alpha \, \varphi_\alpha$$

is a Λ-modular decomposition of φ. If $\alpha \neq 1$, then $\alpha\beta \neq 1$ for every $\beta \in S_1$, and so $\Gamma_*(\varphi_{\alpha, \beta}) \sim 0$ for every $\beta \in S_1$. From (4.13) we then see that $\Lambda_*(\varphi_\alpha) \sim 0$. Therefore φ has AGR with respect to Λ.

Furthermore,

$$\Lambda_*(\varphi) \approx \Lambda_*(\varphi_1) \perp \frac{\dim \varphi - \dim \varphi_1}{2} \times \widetilde{H}. \tag{4.14}$$

If $\beta \in S_1, \beta \neq 1$, then $\Gamma_*(\varphi_{1,\beta}) \sim 0$, and we see from the expression (4.13) with $\alpha = 1$ that

$$\Lambda_*(\varphi_1) \approx \Gamma_*(\varphi_{1,1}) \perp \frac{\dim \varphi_1 - \dim \varphi_{1,1}}{2} \times \widetilde{H}. \tag{4.15}$$

From (4.14) and (4.15) we obtain

$$\Lambda_*(\varphi) \approx \Gamma_*(\varphi_{1,1}) \perp \frac{\dim \varphi - \dim \varphi_{1,1}}{2} \times \widetilde{H} \approx \Gamma_*(\varphi). \qquad \square$$

Theorem 4.15. *Let* $\Gamma : k \to L \cup \infty$ *be a Q-place and* φ *a form over k which has AGR with respect to* Γ. *Assume that there exists a field extension* $K \supset k$ *such that* $\varphi \otimes K$ *is isotropic and* Γ *has an extension to a Q-place* $\Lambda : K \to L \cup \infty$. *Then* $\Gamma_*(\varphi)$ *is isotropic. More precisely,*

$$\mathrm{ind}\,\Gamma_*(\varphi) \geq \mathrm{ind}(\varphi \otimes K).$$

Proof. We have $\varphi \otimes K \approx \psi_0 \perp r \times \widetilde{H}$ with ψ_0 anisotropic, $r = \mathrm{ind}(\varphi \otimes K) > 0$. By Theorem 4.14, $\varphi \otimes K$ has AGR with respect to Λ and $\Gamma_*(\varphi) \approx \Lambda_*(\varphi \otimes K)$. Theorem 4.9 tells us that ψ_0 has AGR with respect to Λ. Therefore

$$\Gamma_*(\varphi) \approx \Lambda_*(\psi_0) \perp r \times \widetilde{H}. \qquad \square$$

Theorem 4.15 inspires the hope that for almost good reduction there are theorems about the generic splitting of $\Gamma_*(\varphi)$, analogous to those in §1.4. As in §1.4 one will have to assume that the characteristic of L is different from two though.

4.3 Realization of Quadratic Places; Generic Splitting of Specialized Forms in Characteristic $\neq$ 2

We consider fields of arbitrary characteristic and want to investigate "realizations" of a quadratic place in the sense of the following definition.

Definition 4.16. Let $\Lambda : K \to L \cup \infty$ be a quadratic place. A *realization of* Λ (as an ordinary place) is a pair (i, μ) consisting of a field extension $i : K \hookrightarrow E$ and a place $\mu : E \to L \cup \infty$ with the property $\Lambda \subset \widehat{\mu}|K$. {Here $\widehat{\mu}$ denotes the quadratic place determined by μ, cf. §4.1.} The realization is called *strict* when actually $\Lambda = \widehat{\mu}|K$.

Strict realizations are not easy to find, but in general the following theorem holds which suffices for our generic splitting theory.

Theorem 4.17. *For every quadratic place* $\Lambda : K \to L \cup \infty$ *there exists a realization* $(i : K \hookrightarrow E, \mu)$ *with E purely transcendental over K.*

For the proof we need a lemma, which follows easily from general principles pertaining to the extension of valuations, cf. [11, §2, n°4, Prop. 3 and §8, n°3, Thm. 1].

Lemma 4.18. *Let K be a quadratic field extension of a field E,* $K = E(\alpha)$ *with* $\alpha^2 = a \in E, \alpha \notin E$. *Let* $\rho : E \to L \cup \infty$ *be a place with* $\rho(ac^2) = 0$ *or* ∞ *for all* $c \in E$. *Then* ρ *extends uniquely to a place* $\lambda : K \to L \cup \infty$.

Proof of Theorem 4.17. Let $\Lambda = (\lambda, H, \chi)$, $\mathfrak{o} := \mathfrak{o}_\lambda$ and $\mathfrak{m}$ the maximal ideal of the valuation ring $\mathfrak{o}$. We choose a subgroup P of $Q(K)$ with $Q(K) = P \times Q(\mathfrak{o})$ and a family $(\pi_i \mid i \in I)$ in $\mathfrak{m}$ such that the square classes $\langle \pi_i \rangle$, $i \in I$, form a basis of the $\mathbb{F}_2$-vector space P. Finally we choose families of indeterminates $(t_i \mid i \in I)$ and $(u_i \mid i \in I)$ over the fields K and L respectively.

Let $E' := K(t_i \mid i \in I)$ and $F := L(u_i \mid i \in I)$. The place $\lambda : K \to L \cup \infty$ extends uniquely to a place $\tilde{\lambda} : E' \to F \cup \infty$ with $\tilde{\lambda}(u_i) = t_i$ for every $i \in I$, cf. [11, §10, Prop. 2]. {Note: with $\tilde{\mathfrak{o}} := \mathfrak{o}_{\tilde{\lambda}}$ and $\tilde{\mathfrak{m}}$ the maximal ideal of $\tilde{\mathfrak{o}}$ we have $\tilde{\mathfrak{o}}/\tilde{\mathfrak{m}} \cong (\mathfrak{o}/\mathfrak{m})(\bar{t}_i \mid i \in I)$, where $\bar{t}_i$ denotes the image of t_i in $\tilde{\mathfrak{o}}/\tilde{\mathfrak{m}}$. The family $(\bar{t}_i \mid i \in I)$ is algebraically free over $\mathfrak{o}/\mathfrak{m}$. The natural map $K^*/\mathfrak{o}^* \to E'^*/\tilde{\mathfrak{o}}^*$ is an isomorphism.}

For every $i \in I$ and $f \in E'$ we have $\tilde{\lambda}(\pi_i f^2) = 0$ or ∞. Let $E := E'(\sqrt{\pi_i t_i^{-1}} \mid i \in I)$. This field is again purely transcendental over K. It follows from Lemma 4.18 (possibly with Zorn's lemma) that $\tilde{\lambda}$ extends uniquely to a place $\rho : E \to F \cup \infty$. (We have $\rho(\sqrt{\pi_i t_i^{-1}}) = 0$ for every $i \in I$.) Now $\langle \pi_i \rangle_E = \langle t_i \rangle_E$. Thus $\langle \pi_i \rangle_E$ has GR with respect to ρ and $\rho_*(\langle \pi_i \rangle_E) = \langle u_i \rangle$ for every $i \in I$.

Finally we choose for every $i \in I$ an element a_i of L with $\chi(\langle \pi_i \rangle) = \langle a_i \rangle$. There exists a place $\sigma : F \to L \cup \infty$ over L, in general not uniquely determined, with $\sigma(u_i) = a_i$ for every i. For every such place σ we have $\widehat{\sigma \circ \rho} | K \supset \Lambda$, which can be verified easily. $\square$

In what follows we require that *all fields have characteristic $\neq 2$*. As before, a "form" is a nondegenerate (symmetric) bilinear form, which is now the same as a nondegenerate quadratic form.

As indicated at the end of §4.2, we want to study the generic splitting of quadratically specialized forms $\Lambda_*(\varphi)$ of forms with almost good reduction in a manner, similar to what was done in §1.4 for specialized forms $\lambda_*(\varphi)$ of forms with good reduction. For this purpose we need a supplement to Theorem 1.41 in §1.4 (which itself is a special case of Theorem 2.5 in §2.1) about generic zero fields.

Theorem 4.19 ([33, Thm. 3.3.ii]). *Let $\gamma : k \to L \cup \infty$ be a place and φ a form over k with bad (= not good) reduction with respect to γ. Assume there exists an element $c \in D(\varphi)$ with $\gamma(c) \neq 0, \infty$. Then γ extends to a place $\lambda : k(\varphi) \to L \cup \infty$.*

Proof. We may assume that $\varphi = \langle a_1, \ldots, a_n \rangle$ with $n \geq 2$, $\gamma(a_i) \neq \infty$ for all i, $\gamma(a_1) \neq 0$, $\gamma(a_n c^2) = 0$ or ∞ for every $c \in k$. We can write $k(\varphi) = k(x_1, \ldots, x_n)$ with the only relation $\sum_1^n a_i x_i^2 = 0$. The elements $x_1, \ldots, x_{n-1}$ are algebraically independent over k and $k(\varphi)$ is a quadratic extension $E(x_n)$ of the purely transcendental extension $E = k(x_1, \ldots, x_{n-1})$ of k.

Let $u_1, \ldots, u_{n-1}$ be indeterminates over L and $F := L(u_1, \ldots, u_{n-1})$. Then γ has a unique extension $\mu : E \to F$ with $\mu(x_i) = u_i$ ($i = 1, \ldots, n-1$). We want to show that $\mu(x_n^2 z^2) = 0$ or ∞ for every $z \in E$. By Lemma 4.18 above we then know that μ extends to a place $\rho : k(\varphi) \to F \cup \infty$. By composing ρ with a place $\sigma : F \to L \cup \infty$ over L, we obtain a place $\lambda = \sigma \circ \rho : k(\varphi) \to L \cup \infty$, which extends γ.

Let $z \in E^*$. We write $z = dfg^{-1}$ with $d \in k$ and polynomials $f, g \in k[x_1, \ldots, x_{n-1}]$ whose coefficients are all in the valuation ring $\mathfrak{o}$ of γ, but not all in the maximal ideal of $\mathfrak{o}$. Then $\mu(f)$ and $\mu(g)$ are finite and nonzero. We obtain

$$\mu(x_n^2 z^2) = -[\gamma(a_1)u_1^2 + \cdots + \gamma(a_{n-1})u_{n-1}^2]\mu(f)^2\mu(g)^{-2}\gamma(a_n^{-1}d^2).$$

On the right-hand side all factors, except the last one, are finite and nonzero. In contrast, $\gamma(a_n^{-1}d^2) = 0$ or ∞. Therefore the same is true for $\mu(x_n^2 z^2)$. $\square$

Theorem 4.20. *Let* $\gamma : k \to L \cup \infty$ *be a place and* φ *a form over* k *which has AGR with respect to* $\widehat{\gamma}$. *Assume that the form* $\widehat{\gamma}_*(\varphi)$ *is isotropic. Then there exists a place* $\lambda : k(\varphi) \to L \cup \infty$, *which extends* γ.

Proof. Let $\Gamma := \widehat{\gamma}$. If φ has GR with respect to γ then our statement is covered by Theorem 1.41, since $\widehat{\gamma}_*(\varphi) = \gamma_*(\varphi)$ in this case. Assume now that φ has bad reduction with respect to γ. We choose a subgroup P of $Q(K)$ with $Q(k) = Q(\mathfrak{o}_\gamma) \times P$. Let

$$\varphi \cong \underset{\alpha \in P}{\perp} \alpha\,\varphi_\alpha$$

be a Γ-modular decomposition of φ. This implies that all forms φ_α are γ-unimodular. If $\varphi_1 \neq 0$ then Theorem 4.19 tells us that γ extends to a place from $k(\varphi)$ to L.

Finally assume that $\varphi_1 = 0$. We have dim $\varphi = \dim \Gamma_*(\varphi) \geq 2$. Thus there certainly exists an $\alpha_0 \in P$ with $\varphi_{\alpha_0} \neq 0$. Since φ has AGR, we have $\gamma_*(\varphi_{\alpha_0}) \sim 0$. The form $\widetilde{\varphi} := \alpha_0 \varphi$ has AGR with respect to Γ, since $\varphi_1 = 0$ and thus $\gamma_*(\varphi_1) = 0$. $(\gamma_*(\varphi_1) \sim 0$ would suffice.) We have $k(\varphi) = k(\widetilde{\varphi})$. $\Gamma_*(\widetilde{\varphi})$ has an orthogonal summand $\gamma_*(\varphi_{\alpha_0})$ and is thus isotropic. By the above γ extends to a place from $k(\widetilde{\varphi})$ to L. $\square$

Now we can prove a theorem, crucial for a generic splitting theory of specialized forms. It is the converse of Theorem 4.15 in characteristic $\neq 2$.

Theorem 4.21. *Let* $\Gamma : k \to L \cup \infty$ *be a Q-place and* φ *a form over* k *which has AGR with respect to* Γ. *Assume that the form* $\Gamma_*(\varphi)$ *is isotropic. Let* $K \supset k$ *be a generic zero field. Then there exists a Q-place* $\Lambda : K \to L \cup \infty$, *which extends* Γ {*i.e., with* $\Lambda|k \supset \Gamma$, *cf. Definition 4.13*}.

Proof. Following Theorem 4.17 we choose a purely transcendental extension $E \supset k$ together with a place $\mu : E \to L \cup \infty$ such that $\widehat{\mu}|k \supset \Gamma$. By Theorem 4.14, $\varphi \otimes E$ has AGR with respect to $\widehat{\mu}$ and $\widehat{\mu}_*(\varphi \otimes E) = \Gamma_*(\varphi)$. Therefore $\widehat{\mu}_*(\varphi \otimes E)$ is isotropic. The free compositum $K \cdot E$ of K and E over k is specialization equivalent to $k(\varphi) \cdot E = E(\varphi \otimes E)$ over E. By Theorem 4.20 we can extend μ to a place $\rho : K \cdot E \to L \cup \infty$. Let $\Lambda : K \to L \cup \infty$ be the restriction of the Q-place $\widehat{\rho}$ to K. Clearly,

$$\Lambda|k = (\widehat{\rho}|K)|k = \widehat{\rho}|k = (\widehat{\rho}|E)|k,$$

and also $\widehat{\rho}|E \supset \widehat{\mu}$. Therefore

$$\Lambda|k \supset \widehat{\mu}|k \supset \Gamma.$$ $\square$

With this theorem and Theorem 4.15, we have all the ingredients at our disposal to replicate the arguments of §1.4 about generic splitting of a specialized form. If we traverse the proof of Theorem 1.42 and replace the words "place" and "good reduction" everywhere by "Q-place" and "almost good reduction" respectively, we obtain the following theorem about the Witt decomposition of a specialized form.

Theorem 4.22. *Let φ be a form over a field k. Let $(K_r \mid 0 \leq r \leq h)$ be a generic splitting tower of φ with associated higher kernel forms φ_r and indices i_r. Let Γ : $k \to L \cup \infty$ be a Q-place, with respect to which φ has almost good reduction. Finally, consider a Q-place $\Lambda : K_m \to L \cup \infty$ for some m, $0 \leq m \leq h$, which extends Γ and which in the case $m < h$ cannot be extended to K_{m+1}. Then φ_m has good reduction with respect to Λ. The form $\Gamma_*(\varphi)$ has kernel form $\Lambda_*(\varphi)$ and Witt index $i_0 + \cdots + i_m$.*

A subtle point to make here is that φ_m has GR with respect to Λ, not just AGR. Namely, if φ_m only had AGR, then $\Lambda_*(\varphi_m)$ would certainly be isotropic.

We now also obtain a detailed proposition about the generic splitting of $\Gamma_*(\varphi)$, which parallels word for word Scholium 1.46 of §1.4, and whose formulation we may thus leave to the Reader.

4.4 Stably Conservative Reduction of Quadratic Forms

The splitting theory of a quadratically specialized form, developed above, required the assumption that the characteristic is different from two. We would like to establish such a theory also for characteristic 2, generalizing our theory in Chapter 1, from §1.5 onwards, and in Chapter 2.

Thence we will consider quadratic forms instead of bilinear forms. In the process we will endeavour to stay as close as possible to the specialization concept of "almost good reduction", developed in §4.1–§4.3. Nevertheless important new problems will occur, which all arise from the lack of obedience of many forms and the existence of quasilinear parts. An obvious, immediate generalization of the concept of "almost good reduction" will not suffice.

We first recall some definitions from §1.7 and §2.3.

Let $\lambda : K \to L \cup \infty$ be a place between fields of arbitrary characteristic, $\mathfrak{o} = \mathfrak{o}_\lambda$, $\mathfrak{m} = \mathfrak{m}_\lambda$, $k = \mathfrak{o}/\mathfrak{m}$. The place λ yields a field embedding $\bar{\lambda} : k \to L$.

We choose a subgroup T of $Q(K)$ with $Q(K) = T \times Q(\mathfrak{o})$. A quadratic space $(E, q) = E$ over K is called *weakly obedient with respect to* λ if it has a decomposition

$$E = \underset{\alpha \in T}{\perp} F_\alpha, \quad F_\alpha \cong \langle \alpha \rangle \otimes E_\alpha \tag{4.16}$$

in which every E_α is a quadratic space over K of the form

$$E_\alpha = K \otimes_\mathfrak{o} M_\alpha = K M_\alpha, \tag{4.17}$$

for some quadratic module M_α over $\mathfrak{o}$, which is free of finite rank, and for which the quadratic module $M_\alpha/\mathfrak{m}M_\alpha$ over k is nondegenerate. (M_α is then called "reduced nondegenerate".) We also called (4.16) a *weakly λ-modular decomposition* of E. We described spaces of the form (4.17) as having *fair reduction* (FR), and denoted the space $M_\alpha/\mathfrak{m}M_\alpha \otimes_{\bar{\lambda}} L$ by $\lambda_*(E_\alpha)$. We saw in §2.3 that for every α the Witt class $\{\lambda_*(E_\alpha)\}$ is determined solely by E, λ and α.

We should keep in mind that the quasilinear part $QL(M_\alpha/\mathfrak{m}M_\alpha)$ of $M_\alpha/\mathfrak{m}M_\alpha$ has to be anisotropic, but that $QL(\lambda_*(E_\alpha))$ could be isotropic. {We defined Witt equivalence over fields also for forms with isotropic quasilinear part, cf. §1.6.}

If E actually has the decomposition (4.16), (4.17), in which every M_α is a nondegenerate quadratic $\mathfrak{o}$-module, then we said that E is *obedient* with respect to λ, and further that E_α has *good reduction* (GR) with respect to Λ. Finally we then called (4.16) a λ-*modular decomposition* of E.

Assume now that λ is enlarged to a quadratic place $\Lambda = (\lambda, H, \chi) : K \to L \cup \infty$. We choose direct decompositions $H = H_0 \times Q(\mathfrak{o})$, $Q(K) = S \times H$. The group T from above is now $S \times H_0 = S H_0$.

We switch to the language of quadratic forms instead of quadratic modules, and thus use φ, φ_α etc. instead of E, E_α etc. A "form" will now always be a quadratic form.

Let φ be a nondegenerate form over K.

Definition 4.23. We say that φ has *fair reduction* (= FR) with respect to Λ, or that φ is *weakly Λ-unimodular*, if φ has a decomposition

$$\varphi \cong \bigperp_{\alpha \in H_0} \alpha \otimes \varphi_\alpha = \bigperp_{\alpha \in H_\alpha} \alpha \varphi_\alpha \tag{4.18}$$

in which every φ_α has FR with respect to λ, and then set

$$\Lambda_*(\varphi) := \bigperp_{\alpha \in H_0} \chi(\alpha) \lambda_*(\varphi_\alpha). \tag{4.19}$$

If there actually exists a decomposition (4.18) in which all φ_α have GR with respect to λ, then we say that φ has GR with respect to Λ, or that φ is Λ-*unimodular*.

As in §4.1 we see that $\Lambda_*(\varphi)$ is determined by φ up to isometry, and also that it is independent of the choice of the complement H_0 of $Q(\mathfrak{o})$ in H, if we utilize the weak specialization λ_W from §2.3 instead of the map λ_W from §1.3.[1] {In the case of good reduction, the operator λ_W from §1.7 suffices.}

It goes without saying that φ must be weakly obedient (resp. obedient) with respect to λ in order for φ to have FR (resp. GR) with respect to Λ. It can happen that $\Lambda_*(\varphi)$ is degenerate; for fair reduction even in the case where λ is the canonical place $K \to \mathfrak{o}/\mathfrak{m} \cup \infty$ associated to $\mathfrak{o}$.

Remark.

(1) If φ, ψ have FR (resp. GR) with respect to Λ, and if φ is strictly regular, then $\varphi \perp \psi$ has FR (resp. GR) with respect to Λ and

$$\Lambda_*(\varphi \perp \psi) \cong \Lambda_*(\varphi) \perp \Lambda_*(\psi), \tag{4.20}$$

which can be seen easily.

---

[1] In order to be able to repeat the conclusion of §4.1, it is important that for our definition of Witt equivalence over fields (Definition 1.76), the statement "If $E \sim F$, $\dim E = \dim F$, then $E \cong F$" also holds for degenerate spaces.

(2) If σ is a bilinear form which has GR with respect to Λ, φ is a quadratic form which has FR (resp. GR) with respect to Λ, and *if $\Lambda_*(\varphi)$ is strictly regular*, then $\sigma \otimes \varphi$ has FR (resp. GR) with respect to Λ and

$$\Lambda_*(\sigma \otimes \varphi) \cong \Lambda_*(\sigma) \otimes \Lambda_*(\varphi), \tag{4.21}$$

which can also be verified easily. Note that $\Lambda_*(\sigma)$ is determined by Λ, σ only up to stable isometry. Nevertheless this formula (with true isometry) is correct.

(3) It is important that $\Lambda_*(\varphi)$ is strictly regular in formula (4.21), at any rate when $\Lambda_*(\sigma)$ is isotropic. This follows already from the following observation. Let char $L = 2$ and let $a, b, c \in L^*$. Then

$$\langle a, a \rangle \otimes [c] \cong [ac, ac] \cong [ac, 0],$$

and likewise $\langle b, b \rangle \otimes [c] \cong [bc, 0]$. Furthermore

$$\langle a, a \rangle \cong \begin{pmatrix} a & 1 \\ 1 & 0 \end{pmatrix} \approx \begin{pmatrix} b & 1 \\ 1 & 0 \end{pmatrix} \cong \langle b, b \rangle,$$

but in general $[ac, 0] \not\cong [bc, 0]$.   □

If φ is weakly obedient with respect to λ and if

$$\varphi = \underset{(s,\beta) \in S \times H_0}{\perp} s\beta \varphi_{s,\beta} \tag{4.22}$$

is a weakly λ-unimodular decomposition of φ, in which thus every form $\varphi_{s,\beta}$ has FR with respect to λ, then we can combine the forms $\varphi_{s,\beta}$, for fixed s, in a form

$$\varphi_s := \underset{\beta \in H_0}{\perp} \beta \varphi_{s,\beta} \tag{4.23}$$

which is weakly Λ-unimodular. This results in a coarser decomposition

$$\varphi = \underset{s \in S}{\perp} s \varphi_s. \tag{4.24}$$

We call such a decomposition (4.24) a *weakly Λ-modular decomposition* of φ. If φ is actually obedient with respect to λ, then we have such decompositions (4.23), (4.24), in which the $\varphi_{s,\beta}$ have GR with respect to λ and the φ_s have GR with respect to Λ, and then call (4.24) a *Λ-modular decomposition* of φ.

Definition 4.24.

(a) We say that φ has *almost fair reduction* (= AFR) with respect to Λ if φ is weakly obedient with respect to λ, and if in a Λ-modular decomposition (4.24) of φ the specializations $\Lambda_*(\varphi_s)$, with $s \in S \setminus \{1\}$, are all hyperbolic.

(b) We say that φ has *almost good reduction* (= AGR) with respect to Λ if there is a Λ-modular decomposition (4.24) in which $\Lambda_*(\varphi_s)$ is hyperbolic for each $s \in S \setminus \{1\}$.

Note that these properties are independent of the choice of decomposition (4.24) since every $\Lambda_*(\varphi_s)$ is up to Witt equivalence uniquely determined by φ, Λ and s.

As in §4.1 we see that by almost fair reduction the form

$$\Lambda_*(\varphi) := \Lambda_*(\varphi_1) \perp \frac{\dim \varphi - \dim \varphi_1}{2} \times H$$

is determined up to isometry (instead of only Witt equivalence) by Λ and φ alone. We call $\Lambda_*(\varphi)$ the *specialization of φ with respect to Λ*.

Next there follow two theorems which can be proved in exactly the same way as the analogous statements in §4.1 and §4.2.

Theorem 4.25. *Let φ and ψ be quadratic forms over K. Assume that φ is strictly regular and has AGR with respect to Λ. Assume that ψ has AFR (resp. AGR) with respect to Λ. Then $\varphi \perp \psi$ has AFR (resp. AGR) with respect to Λ and*

$$\Lambda_*(\varphi \perp \psi) \cong \Lambda_*(\varphi) \perp \Lambda_*(\psi).$$

Theorem 4.26. *Let φ be a quadratic form, weakly obedient (resp. obedient) with respect to $\lambda : K \to L \cup \infty$. As before, let $\Lambda = (\lambda, \chi, H) : K \to L \cup \infty$ be a Q-place with first component λ. Let further $E \supset K$ be a field extension and $M : E \to L \cup \infty$ a Q-place with $M|K \supset \Lambda$.*

(a) If φ has AFR (resp. AGR) with respect to Λ, then $\varphi \otimes E$ has AFR (resp. AGR) with respect to M, and $M_*(\varphi \otimes E) = \Lambda_*(\varphi)$.

(b) If actually $M|K = \Lambda$, and if $\varphi \otimes E$ has AFR (resp. AGR) with respect to M, then φ has AFR (resp. AGR) with respect to Λ.

However, can we establish, as in §4.1, whether for two forms φ, ψ over K where φ and $\varphi \perp \psi$ have AGR with respect to $\Lambda : K \to L \cup \infty$, also ψ has AGR with respect to Λ, even when φ and ψ are both strictly regular?

This is a debatable question since (at least for the author) there is no discernible reason why ψ should be obedient with respect to the place λ.

Thus we are forced to make a new turn and follow a different path. The fact that for almost fair reduction, or even almost good reduction, the quasilinear part $QL(\Lambda_*(\varphi))$ could be isotropic, will be an obstacle. We want to exclude this possibility.

Definition 4.27. Let $\Lambda : K \to L \cup \infty$ be a quadratic place and let φ be a form over K. We say that φ has *conservative reduction* (= CR) *with respect to Λ* if φ has AFR with respect to Λ and $QL(\Lambda_*(\varphi))$ is anisotropic.

If φ actually has AGR with respect to Λ, then clearly $QL(\Lambda_*(\varphi)) = \Lambda_*(QL(\varphi))$. Hence φ certainly has CR in this case if φ is in addition regular. If φ only has AFR, then $QL(\Lambda_*(\varphi))$ may be isotropic, even if φ is regular.

As a result of the following lemma we obtain a connection between conservative reduction with respect to Λ and fair reduction with respect to an ordinary place. In the process we will be able to utilize results from §2.3.

Lemma 4.28. *Let $(K \hookrightarrow E, \mu)$ be a realization of Λ, i.e., $E \supset K$ is a field extension and $\mu : E \to L \cup \infty$ is a place with $\hat{\mu}|K \supset \Lambda$. {Note that there exists a realization of Λ with $E \supset K$ even purely transcendental, as demonstrated in §4.3.} If φ has conservative reduction with respect to Λ, then $\varphi \otimes E$ has fair reduction with respect to μ and $\mu_*(\varphi \otimes E) = \Lambda_*(\varphi)$.*

Proof. We use Theorem 4.26(*b*). Let $\varphi = \underset{s \in S}{\perp} \langle s \rangle \otimes \varphi_s$ be a weakly Λ-modular decomposition of φ. For every $s \in S$ we have $\langle s \rangle_E \in Q(\mathfrak{o}_\mu)$. If $s \neq 1$, then $\varphi_s \otimes E$ has FR with respect to μ, and $\mu_*(\varphi_s \otimes E) = \Lambda_*(\varphi_s)$ is hyperbolic. Furthermore, the anisotropy of $QL(\Lambda_*(\varphi)) = QL(\Lambda_*(\varphi_1))$ yields that $\varphi_1 \otimes E$ has fair reduction with respect to μ and $\mu_*(\varphi_1 \otimes E) = \Lambda_*(\varphi_1)$. Hence $\varphi \otimes E$ has FR with respect to μ and

$$\mu_*(\varphi \otimes E) = \Lambda_*(\varphi_1) \perp \frac{\dim \varphi - \dim \varphi_1}{2} \times H = \Lambda_*(\varphi). \qquad \square$$

Theorem 4.29. *Let φ, ψ be quadratic forms over K, both having CR with respect to $\Lambda : K \to L \cup \infty$. Assume that $\varphi < \psi$. Then $\Lambda_*(\varphi) < \Lambda_*(\psi)$.*

Proof. We choose a place $\mu : E \to L \cup \infty$ with $E \supset K$ and $\hat{\mu}|K \supset \Lambda$. By the lemma, $\varphi \otimes E$ and $\psi \otimes E$ both have FR with respect to μ with $\Lambda_*(\varphi) = \mu_*(\varphi \otimes E)$, $\Lambda_*(\psi) = \mu_*(\psi \otimes E)$. The form $\varphi \otimes E$ is a subform of $\psi \otimes E$. By Theorem 2.19 (Corollary 2.26) $\mu_*(\varphi \otimes E)$ is a subform of $\mu_*(\psi \otimes E)$. $\qquad \square$

Now we come to the key definition, which allows us to tame disobedient forms at least a little bit.

Definition 4.30. Let φ be a quadratic form over K and, as before, $\Lambda : K \to L \cup \infty$ a quadratic place.

(a) φ has *stably conservative* reduction (= SCR) with respect to Λ if there exists an $r \in \mathbb{N}_0$ such that $\varphi \perp r \times H$ has conservative reduction with respect to Λ.

(b) In this case $r \times H$ is a subform of $\Lambda_*(\varphi \perp r \times H)$ by Theorem 4.29. We then denote by $\Lambda_*(\varphi)$ the form over L, uniquely determined up to isometry, for which

$$\Lambda_*(\varphi \perp r \times H) = \Lambda_*(\varphi) \perp r \times H,$$

and call $\Lambda_*(\varphi)$ the specialization of φ with respect to Λ.

Remark. $\Lambda_*(\varphi)$ is independent of the choice of r. Namely, if $s \in \mathbb{N}$ is arbitrary then, in the situation of Definition 4.30(a), $\varphi \perp (r + s) \times H = (\varphi \perp r \times H) \perp s \times H$ also has conservative reduction and

$$\Lambda_*(\varphi \perp (r \times s) \times H) = \Lambda_*(\varphi \perp r \times H) \perp s \times H = \Lambda_*(\varphi) \perp (r + s) \times H.$$

Definition 4.31. Similarly we define: a quadratic form φ over K has *stably almost good reduction* (= SAGR) with respect to Λ if there exists an $r \in \mathbb{N}_0$ such that $\varphi \perp r \times H$ has AGR with respect to Λ.

Remark. If φ has SAGR, then $QL(\varphi)$ has good reduction with respect to Λ. Such a form φ has SCR with respect to Λ if and only if $\Lambda_*(QL(\varphi))$ is anisotropic, and then $\Lambda_*(QL(\varphi)) = QL(\Lambda_*(\varphi))$.

Finally we can derive a full analogue of the additive statements in Theorem 4.8 and Theorem 4.9.

Theorem 4.32. *Let φ and ψ be quadratic forms over K. Assume that φ is* strictly regular *and that it has* SAGR *with respect to $\Lambda : K \to L \cup \infty$.*
(a) If ψ has SCR *with respect to Λ, then $\varphi \perp \psi$ has* SCR *with respect to Λ and*

$$\Lambda_*(\varphi \perp \psi) \cong \Lambda_*(\varphi) \perp \Lambda_*(\psi).$$

(b) If $\varphi \perp \psi$ has SCR *with respect to Λ, then ψ has* SCR *with respect to Λ.*

Proof. (a) Choose $r \in \mathbb{N}_0$, $s \in \mathbb{N}_0$ such that $\varphi \perp r \times H$ has AGR and $\psi \perp s \times H$ has CR with respect to Λ. From Theorem 4.25 it follows immediately that $\varphi \perp \psi \perp (r+s) \times H$ has CR with respect to Λ and that

$$\Lambda_*(\varphi) \perp r \times H \perp \Lambda_*(\psi) \perp s \times H = \Lambda_*(\varphi \perp \psi \perp (r + s) \times H).$$

Therefore,

$$\Lambda_*(\varphi) \perp \Lambda_*(\psi) = \Lambda_*(\varphi \perp \psi).$$

(b) We choose integers $r, t \in \mathbb{N}_0$ with $t \geq r$ and such that $\rho := \varphi \perp r \times H$ has AGR and that $\tau := (\varphi \perp \psi) \perp t \times H$ has AFR. We have

$$\rho \perp \psi \perp (t - r) \times H \cong \tau.$$

Therefore,

$$\psi \perp (t - r + \dim \rho) \times H \cong (-\rho) \perp \tau.$$

By part (a) of the Theorem, $(-\rho) \perp \tau$ has SCR with respect to Λ. Thus ψ has SCR with respect to Λ. $\qquad\square$

Furthermore we can extend Theorem 4.29 and part of Theorem 4.26 to stably conservative reduction.

Theorem 4.33. *Let φ, ψ be quadratic forms over K which have* SCR *with respect to Λ. Assume that $\varphi < \psi$. Then $\Lambda_*(\varphi) < \Lambda_*(\psi)$.*

Proof. We choose an $r \in \mathbb{N}$ such that $\varphi \perp r \times H$ and $\psi \perp r \times H$ have CR. By Theorem 4.29 we have

$$\Lambda_*(\varphi) \perp r \times H = \Lambda_*(\varphi \perp r \times H) < \Lambda_*(\psi \perp r \times H) = \Lambda_*(\psi) \perp r \times H.$$

Hence, $\Lambda_*(\varphi) < \Lambda_*(\psi)$. $\qquad\square$

Theorem 4.34. *Let $K \subset E$ be a field extension and let $\Lambda : K \to L \cup \infty$, $M : E \to L \cup \infty$ be quadratic places with $M|K \supset \Lambda$. Let φ be a quadratic form over K which has* SCR *with respect to Λ. Then $\varphi \otimes E$ has* SCR *with respect to M and $M_*(\varphi \otimes E) = \Lambda_*(\varphi)$.*

Proof. We choose an $r \in \mathbb{N}_0$ such that $\varphi \perp r \times H$ has CR with respect to Λ. By Theorem 4.26(a), $\varphi_E \perp r \times H$ has AFR with respect to Λ and

$$M_*(\varphi_E \perp r \times H) \cong \Lambda_*(\varphi \perp r \times H) = \Lambda_*(\varphi) \perp r \times H.$$

In particular, $QL(M_*(\varphi_E \perp r \times H))$ is anisotropic. Thus $\varphi_E \perp r \times H$ has CR with respect to M. Therefore φ_E has SCR with respect to Λ and

$$M_*(\varphi_E) \perp r \times H \cong \Lambda_*(\varphi) \perp r \times H.$$

It follows that $M_*(\varphi_E) \cong \Lambda_*(\varphi)$. $\qquad\qquad\qquad\qquad\qquad\qquad\qquad\square$

4.5 Generic Splitting of Stably Conservative Specialized Quadratic Forms

For a quadratic form φ over a field k, which has SCR with respect to a quadratic place $\Gamma : k \to L \cup \infty$, we will show how the splitting of $\Gamma_*(\varphi)$ with respect to extensions of the field L, conservative for $\Gamma_*(\varphi)$, is regulated by an arbitrarily given generic splitting tower $(K_r \mid 0 \le r \le h)$ and quadratic places coming from the fields K_r (see Theorem 4.39 below).

For this purpose we need the concept of composition of two quadratic places.

Definition 4.35. Let $\Lambda = (\lambda, H, \chi) : K \to L \cup \infty$ and $M = (\mu, D, \psi) : L \to F \cup \infty$ be two quadratic places. Let

$$H_0 := \{x \in H, \; \chi(\alpha) \in D\}.$$

Clearly

$$M \circ \Lambda := (\mu \circ \lambda, \; H_0, \; \psi \circ (\chi | H_0))$$

is a quadratic place from K to F, called the *compositum* of Λ and M.

The theory of these composita is fraught with danger. We will not look at its details, as we would like to advance as quickly as possible towards the core of a generic splitting theory of stably conservative specialized forms $\Lambda_*(\varphi)$. This is possible just with Definition 4.35 and §4.4, although many things would become clearer conceptually after a detailed study of the formal properties of composita of quadratic places.

We only note the following trivial fact.

Remark. If $\Lambda' : K \to L \cup \infty$, $M' : L \to F \cup \infty$ are quadratic places with $\Lambda' \supset \Lambda$, $M' \supset M$, then $M' \circ \Lambda' \supset M \circ \Lambda$.

Theorem 4.36. *Let $\Gamma : k \to L \cup \infty$ be a quadratic place and φ a quadratic form over k which has SCR with respect to Γ. Let $r \in \mathbb{N}_0$ and let $K \supset k$ be a field extension which is φ-conservative and generic for the splitting off of r hyperbolic planes. Then the following statements are equivalent:*

(1) $\text{ind}(\Gamma_*(\varphi)) \geq r$.

(2) *There exists a quadratic place* $\Lambda : K \to L \cup \infty$ *with* $\Lambda_*|k \supset \Gamma$.

Proof. (2) $\Rightarrow$ (1): By Theorem 4.34, $\varphi \otimes K$ has SCR with respect to Λ and $\Lambda_*(\varphi \otimes K) \cong \Gamma_*(\varphi)$. Now $r \times H < \varphi \otimes K$ and thus, by Theorem 4.33, also $r \times H < \Lambda_*(\varphi \otimes K)$. Hence $\Gamma_*(\varphi)$ has Witt index $\geq r$.

(1) $\Rightarrow$ (2): We carry out the proof in two steps.

(a) We choose a purely transcendental extension $E \supset k$ and a place $\mu : E \to L \cup \infty$ with $\widehat{\mu}|k \supset \Lambda$. This is possible by Theorem 4.17. Then we choose a generic splitting field $F \supset E$ of $\varphi \otimes E$ for the splitting off of r hyperbolic planes. Finally we choose an $s \in \mathbb{N}_0$ such that $\varphi \perp s \times H$ has CR with respect to Γ. By Lemma 4.28, $(\varphi \perp s \times H)_E = \varphi_E \perp s \times H$ has FR with respect to μ, and

$$\mu_*(\varphi_E \perp s \times H) = \Gamma_*(\varphi \perp s \times H) = \Gamma_*(\varphi) \perp s \times H.$$

E is a generic splitting field of $\varphi_E \perp s \times H$ for the splitting off of $r + s$ hyperbolic planes. The form $\Gamma_*(\varphi) \perp s \times H$ over L is nondegenerate since by assumption (1) its quasilinear part is anisotropic. Moreover, this form has Witt index $\geq r + s$ by assumption (1). Thus, by the generic splitting theory of §2.4, there exists a place $\sigma : F \to L \cup \infty$ which extends μ. For the associated quadratic place $\widehat{\sigma} : F \to L \cup \infty$ we have $\widehat{\sigma}|E \supset \widehat{\mu}$ and thus $\widehat{\sigma}|k \supset \widehat{\mu}|k \supset \Gamma$. One should observe now that F is also a generic splitting field for the splitting off of r hyperbolic planes from φ (instead of φ_E). So for *this* field F instead of the given field K we have a proof of statement (2).

(b) The fields K and F are specialization equivalent over k. We choose a place $\rho : K \to F \cup \infty$ over k and form the compositum $\Lambda = \widehat{\sigma} \circ \widehat{\rho} : K \to L \cup \infty$. We verify that $\Lambda|k \supset \Gamma$ and are done.

We have $\widehat{\rho} = (\rho, Q(\mathfrak{o}_\rho), \breve{\rho})$, $\widehat{\sigma} = (\sigma, Q(\mathfrak{o}_\sigma), \breve{\sigma})$, where $\breve{\rho}, \breve{\sigma}$ are the characters $Q(\mathfrak{o}_\rho) \to Q(L)$, $Q(\mathfrak{o}_\sigma) \to Q(L)$, induced by ρ, σ, i.e., $\breve{\rho}(\langle a \rangle) = \langle \rho(a) \rangle$ for $a \in \mathfrak{o}_\rho^*$, and $\breve{\sigma}(\langle b \rangle) = \langle \sigma(b) \rangle$ for $b \in \mathfrak{o}_\sigma^*$. Hence $\Lambda = (\sigma \circ \rho, H_0, \chi)$ with

$$H_0 = \{\langle a \rangle \in Q(K) \mid a \in \mathfrak{o}_\rho^*, \ \langle \rho(a) \rangle \in Q(\mathfrak{o}_\sigma)\}$$

and $\chi = \breve{\sigma} \circ (\breve{\rho}|H_0)$.

Let $\Gamma = (\gamma, D, \eta)$. For $a \in k^*$ with $\langle a \rangle \in D$ we have $\langle a \rangle_K \in Q(\mathfrak{o}_\rho)$ and $\breve{\rho}(\langle a \rangle_K) = \langle a \rangle_F$, since ρ is a place over k. Since $\widehat{\sigma}|k \supset \Lambda$ we have $\langle a \rangle_F \in Q(\mathfrak{o}_\sigma)$, hence $\langle a \rangle_K \in H_0$ and furthermore

$$\chi(\langle a \rangle_K) = \breve{\sigma}(\langle a \rangle_F) = \eta(\langle a \rangle).$$

Thus we have verified that $\Lambda|k \supset \Gamma$. $\square$

As a result of §4.4 (Theorems 4.32 and 4.34) the following corollary of Theorem 4.36 is obvious.

Scholium 4.37. *Assume that statements* (1) *and* (2) *of the theorem hold. Let* $\varphi \otimes K \cong \psi \perp r \times H$. *Then* ψ *has SCR with respect to* Λ *and*

$$\Gamma_*(\varphi) \cong \Lambda_*(\psi) \perp r \times H.$$

Now nothing stands in the way of a generic splitting theory for $\Gamma_*(\varphi)$. We note an easy lemma.

Lemma 4.38. *Let* $\Lambda : K \to L \cup \infty$ *be a quadratic place and* φ *a quadratic form over* K *which has* CR *(resp.* SCR*) with respect to* Λ. *Further, let* $j : L \hookrightarrow L'$ *be a field extension, i.e., a trivial place, which is* $\Lambda_*(\varphi)$-*conservative. Then* φ *also has* CR *(resp.* SCR*) with respect to* $\widehat{j} \circ \Lambda$ *and*

$$(\widehat{j} \circ \Lambda)_*(\varphi) = \Lambda_*(\varphi) \otimes_L L'.$$

Proof. Let $\Lambda = (\lambda, H, \chi)$. We have $\widehat{j} = (j, Q(L), Q(j))$, where $Q(j)$ denotes the homomorphism from $Q(L)$ to $Q(L')$, induced by j. We obtain

$$\widehat{j} \circ \Lambda = (j \circ \lambda, H, Q(j) \circ \chi),$$

and of course $\mathfrak{o}_{j \circ \lambda} = \mathfrak{o}_\lambda$. This yields the assertion for conservative reduction. The extension to the case of SCR is obvious. □

Theorem 4.39. *Let* φ *be a quadratic form over* k *and* $\Gamma : k \to L \cup \infty$ *a quadratic place with respect to which* φ *has* SCR. *Let* $(K_r \mid 0 \leq r \leq h)$ *be a generic splitting tower of* φ *with higher indices* i_r *and higher kernel forms* φ_r. *Finally, let* $L' \supset L$ *be a* $\Gamma_*(\varphi)$-*conservative field extension of* L. *We choose a quadratic place* $\Lambda : K_m \to L' \cup \infty$ *which extends* $\widehat{j} \circ \Gamma$ *(i.e.,* $\Lambda \vert k \supset \widehat{j} \circ \Gamma$*), and which in the case* $m < h$ *cannot be extended to a quadratic place from* K_{m+1} *to* L'. *Then*

$$\mathrm{ind}(\Gamma_*(\varphi) \otimes L') = i_0 + \cdots + i_m.$$

The form φ_m *has* SCR *with respect to* Λ *and*

$$\ker(\Gamma_*(\varphi) \otimes L') = \Lambda_*(\varphi_m).$$

Proof. By Lemma 4.38 we may replace Γ by the quadratic place $\widehat{j} \circ \Gamma : K \to L' \cup \infty$ and may thus assume without loss of generality that $L' = L$. We finish by traversing the proof of Theorem 2.39 once more and replacing γ by Γ, λ by Λ and FR by SCR everywhere. □

If char $L \neq 2$ then Theorem 4.39 is essentially the same theorem as Theorem 4.20. We have thus found a new proof for that theorem, which is clearly different from the proof in §4.3. There, a theorem about bad reduction (Theorem 4.19) was an important tool, for which we have not developed a counterpart which is valid when char $L = 2$.

We have reached a point where we can envisage applications in the theory of quadratic forms, and in particular in the generic splitting theory, which necessitate the employment of quadratic places, similar to what we did in Chapter 3 for ordinary places. This would go beyond the scope of this book, however. From a scientific point of view we would also be treading on virgin soil. To my knowledge there exist only two publications, [36] and [37], in which quadratic places are used. Both works assume characteristic $\neq 2$.

References

1. Arason, J.K., Pfister, A.: Beweis des Krullschen Durchschnittsatzes für den Wittring. Invent. Math. **12**, 173–176 (1971)
2. Aravire, R., Baeza, R.: The behavior of quadratic and differential forms under function field extensions in characteristic two. J. Algebra **259**(2), 361–414 (2003)
3. Arf, C.: Untersuchungen über quadratische Formen in Körpern der Charakteristik 2. (Tl. 1.). J. Reine Angew. Math. **183**, 148–167 (1941)
4. Baeza, R.: Ein Teilformensatz für quadratische Formen in Charakteristik 2. Math. Z. **135**, 175–184 (1974)
5. Baeza, R.: Eine Bemerkung über Pfisterformen. Arch. Math. **25**, 254–259 (1974)
6. Baeza, R.: Quadratic forms over semilocal rings. Lecture Notes in Mathematics. 655. Berlin-Heidelberg-New York: Springer-Verlag. VI, 199 p. (1978)
7. Baeza, R.: The norm theorem for quadratic forms over a field of characteristic 2. Commun. Algebra **18**(5), 1337–1348 (1990)
8. Bass, H.: Lectures on topics in algebraic K-theory. Notes by Amit Roy. Tata Institute of Fundamental Research Lectures on Mathematics, No. 41. Tata Institute of Fundamental Research, Bombay (1967)
9. Bass, H.: Clifford algebras and spinor norms over a commutative ring. Amer. J. Math. **96**, 156–206 (1974)
10. Bourbaki, N.: Eléments de Mathématique, Algèbre, Chap.9: Formes sesquilinéaires et formes quadratiques. Hermann, Paris (1959)
11. Bourbaki, N.: Eléments de Mathématique, Algèbre commutative, Chap.6: Valuations. Hermann, Paris (1964)
12. Bourbaki, N.: Eléments de Mathématique, Algèbre, Chap.5: Corps commutatifs. Hermann, Paris (1967)
13. Cartan, H., Eilenberg, S.: Homological algebra. Princeton University Press, Princeton, N. J. (1956)
14. Elman, R., Karpenko, N., Merkurjev, A.: The algebraic and geometric theory of quadratic forms, *American Mathematical Society Colloquium Publications*, vol. 56. American Mathematical Society, Providence, RI (2008)
15. Elman, R., Lam, T.Y.: Pfister forms and K-theory of fields. J. Algebra **23**, 181–213 (1972)
16. Endler, O.: Valuation theory. Springer-Verlag, New York (1972)
17. Engler, A.J., Prestel, A.: Valued fields. Springer Monographs in Mathematics. Springer-Verlag, Berlin (2005)
18. Fitzgerald, R.W.: Quadratic forms of height two. Trans. Amer. Math. Soc. **283**(1), 339–351 (1984)
19. Hoffmann, D.W.: Isotropy of quadratic forms over the function field of a quadric. Math. Z. **220**(3), 461–476 (1995)

M. Knebusch, *Specialization of Quadratic and Symmetric Bilinear Forms*,
Algebra and Applications 11, DOI 10.1007/978-1-84882-242-9,
© Springer-Verlag London Limited 2010

20. Hoffmann, D.W.: On quadratic forms of height two and a theorem of Wadsworth. Trans. Amer. Math. Soc. **348**(8), 3267–3281 (1996)
21. Hoffmann, D.W.: Sur les dimensions des formes quadratiques de hauteur 2. C. R. Acad. Sci. Paris Sér. I Math. **324**(1), 11–14 (1997)
22. Hoffmann, D.W.: Splitting patterns and invariants of quadratic forms. Math. Nachr. **190**, 149–168 (1998)
23. Hoffmann, D.W., Laghribi, A.: Isotropy of quadratic forms over the function field of a quadric in characteristic 2. J. Algebra **295**(2), 362–386 (2006)
24. Hurrelbrink, J., Rehmann, U.: Splitting patterns of excellent quadratic forms. J. Reine Angew. Math. **444**, 183–192 (1993)
25. Hurrelbrink, J., Rehmann, U.: Splitting patterns of quadratic forms. Math. Nachr. **176**, 111–127 (1995)
26. Izhboldin, O.T., Kahn, B., Karpenko, N.A., Vishik, A.: Geometric methods in the algebraic theory of quadratic forms, *Lecture Notes in Mathematics*, vol. 1835. Springer-Verlag, Berlin (2004). Edited by Jean-Pierre Tignol
27. Jacobson, N.: Lectures in abstract algebra. Vol III: Theory of fields and Galois theory. D. Van Nostrand Co., Inc., Princeton, N.J.-Toronto, Ont.-London-New York (1964)
28. Kahn, B.: Formes quadratiques de hauteur et de degré 2. Indag. Math. (N.S.) **7**(1), 47–66 (1996)
29. Kersten, I., Rehmann, U.: Generic splitting of reductive groups. Tohoku Math. J. (2) **46**(1), 35–70 (1994)
30. Knebusch, M.: Isometrien über semilokalen Ringen. Math. Z. **108**, 255–268 (1969)
31. Knebusch, M.: Grothendieck- und Wittringe von nichtausgearteten symmetrischen Bilinearformen. S.-B. Heidelberger Akad. Wiss. Math.-Natur. Kl. **1969/70**, 93–157 (1969/1970)
32. Knebusch, M.: Specialization of quadratic and symmetric bilinear forms, and a norm theorem. Acta Arith. **24**, 279–299 (1973)
33. Knebusch, M.: Generic splitting of quadratic forms. I. Proc. London Math. Soc. (3) **33**(1), 65–93 (1976)
34. Knebusch, M.: Generic splitting of quadratic forms. II. Proc. London Math. Soc. (3) **34**(1), 1–31 (1977)
35. Knebusch, M.: Symmetric bilinear forms over algebraic varieties. In: G. Orzech (Ed.) Proc. Conf. Quadratic Forms, Kingston 1976, Queen's Pap. pure appl. Math. 46, 103–283 (1977)
36. Knebusch, M.: Generic splitting of quadratic forms in the presence of bad reduction. J. Reine Angew. Math. **517**, 117–130 (1999)
37. Knebusch, M., Rehmann, U.: Generic splitting towers and generic splitting preparation of quadratic forms. In: Quadratic forms and their applications (Dublin, 1999), *Contemp. Math.*, vol. 272, pp. 173–199. Amer. Math. Soc., Providence, RI (2000)
38. Knebusch, M., Rosenberg, A., Ware, R.: Structure of Witt rings and quotients of Abelian group rings. Amer. J. Math. **94**, 119–155 (1972)
39. Knebusch, M., Scharlau, W.: Algebraic theory of quadratic forms, Generic methods and Pfister forms, *DMV Seminar*, vol. 1. Birkhäuser Boston, Mass. (1980). (Notes taken by Heisook Lee.)
40. Kneser, M.: Witts Satz über quadratische Formen und die Erzeugung orthogonaler Gruppen durch Spiegelungen. Math.-Phys. Semesterber. **17**, 33–45 (1970)
41. Kneser, M.: Witts Satz für quadratische Formen über lokalen Ringen. Nachr. Akad. Wiss. Göttingen Math.-Phys. Kl. II pp. 195–203 (1972)
42. Kneser, M.: Quadratische Formen. Springer (2002)
43. Lam, T.Y.: The algebraic theory of quadratic forms. W. A. Benjamin, Inc., Reading, Mass. (1973)
44. Lam, T.Y.: Introduction to quadratic forms over fields, *Graduate Studies in Mathematics*, vol. 67. American Mathematical Society, Providence, RI (2005)
45. Lang, S.: Introduction to algebraic geometry. Interscience Publishers, Inc., New York-London (1958)
46. Lang, S.: Algebra. Addison-Wesley Publishing Co., Inc., Reading, Mass. (1965)
47. Lorenz, F.: Quadratische Formen über Körpern. Lecture Notes in Mathematics, Vol. 130. Springer-Verlag, Berlin (1970)

48. Milnor, J.: Algebraic K-theory and quadratic forms. Invent. Math. **9**, 318–344 (1969/1970)
49. Milnor, J.: Symmetric inner products in characteristic 2. In: Prospects in mathematics (Proc. Sympos., Princeton Univ., Princeton, N.J., 1970), pp. 59–75. Ann. of Math. Studies, No. 70. Princeton Univ. Press, Princeton, N.J. (1971)
50. Milnor, J., Husemoller, D.: Symmetric bilinear forms, *Ergebnisse der Mathematik und ihrer Grenzgebiete*, vol. 73. Springer-Verlag, New York (1973)
51. Orlov, D., Vishik, A., Voevodsky, V.: An exact sequence for $K_*^M/2$ with applications to quadratic forms. Ann. of Math. (2) **165**(1), 1–13 (2007)
52. Pfister, A.: Quadratic forms with applications to algebraic geometry and topology, *London Mathematical Society Lecture Note Series*, vol. 217. Cambridge University Press, Cambridge (1995)
53. Ribenboim, P.: Théorie des valuations, vol. 1964. Les Presses de l'Université de Montréal, Montreal, Que. (1965)
54. Roy, A.: Cancellation of quadratic forms over commutative rings. J. Algebra **10**, 286–298 (1968)
55. Scharlau, W.: Quadratic and Hermitian forms, *Grundlehren der Mathematischen Wissenschaften*, vol. 270. Springer-Verlag, Berlin (1985)
56. Springer, T.A.: Quadratic forms over fields with a discrete valuation. I. Equivalence classes of definite forms. Nederl. Akad. Wetensch. Proc. Ser. A. **58** = Indag. Math. **17**, 352–362 (1955)
57. Vishik, A.S.: On the dimensions of quadratic forms. Dokl. Akad. Nauk **373**(4), 445–447 (2000)
58. Witt, E.: Theorie der quadratischen Formen in beliebigen Körpern. J. reine angew. Math. **176**, 31–44 (1936)

Index

Printed in the United States
by Baker & Taylor Publisher Services